Molecular Exercise Physiology

Molecular Exercise Physiology: An introduction is the first student-friendly textbook to be published on this key topic in contemporary sport and exercise science. It introduces sport and exercise genetics and the molecular mechanisms by which exercise causes adaptation. The text is linked throughout to real life sport and exercise science situations such as 'What makes people good at distance running?', 'What DNA sequence variations code for a high muscle mass?' or 'By what mechanisms does exercise improve type 2 diabetes?'

The book includes a full range of useful features, such as summaries, definitions of key terms, guides to further reading, review questions, personal comments by molecular exercise pioneers (Booth, Bouchard) and leading research in the field, as well as descriptions of research methods. A companion website offers interactive and downloadable resources for both student and lecturers.

Structured around central themes in sport and exercise science, such as nutrition, endurance training, resistance training, exercise, and chronic disease and ageing, this book is the perfect foundation around which to build a complete upper-level undergraduate or postgraduate course on molecular exercise physiology.

Henning Wackerhage, PhD is a Senior Lecturer in Molecular Exercise Physiology at the University of Aberdeen. His research interest is molecular exercise physiology in general and specifically the function of the Hippo pathway in skeletal muscle.

Molecular Exercise Physiology
An introduction

Edited by
Henning Wackerhage

Routledge
Taylor & Francis Group

LONDON AND NEW YORK

First published 2014
by Routledge
2 Park Square, Milton Park, Abingdon, Oxon OX14 4RN

and by Routledge
711 Third Avenue, New York, NY 10017

Routledge is an imprint of the Taylor & Francis Group, an informa business

British Library Cataloguing in Publication Data
A catalogue record for this book is available from the British Library

Library of Congress Cataloging in Publication Data

ISBN: 978-0-415-60787-2 (hbk)
ISBN: 978-0-415-60788-9 (pbk)
ISBN: 978-0-203-13214-2 (ebk)

Typeset in Adobe AGaramond Pro
by Swales & Willis Ltd, Exeter, Devon, UK

Printed and bound by CPI Group (UK) Ltd, Croydon, CR0 4YY

Contents

Figures

Tables

Foreword

I decided to study the subject of sport and exercise science, or kinesiology, for various reasons. One of the major ones was that the urge to understand how exercise training works in order to better train others and myself. However, while I learned much about the practicalities of sport and exercise, I gained little insight into the actual mechanisms that mediate the adaptation exercise, mainly because this knowledge was lacking at the time.

Over the years I then read much about molecular and cellular biology and gained an idea about the mechanisms of adaptation and about the role of genetics in relation to sport and exercise. This is especially due to several influential teachers, in particular Alois Mader in Cologne and Mike Rennie and Graeme Hardie in Dundee, the latter being the discoverer of AMPK. Also while I was at Dundee, Neil Spurway suggested writing a book entitled *Genetics and Molecular Biology of Muscle Adaptation*, which was published by Elsevier. Neil really taught me how to write a book, and without the experience of the Elsevier book I would have struggled to write this one. Together with my past PhD students James Higginson, Philip Atherton, Kevin Watt and Rob Judson, and present PhD students Vanessa De Mello and Abdalla Diaai plus postdoc Roby Urcia we also learned to apply cellular and molecular biology methods when addressing exercise physiology questions. With our research we contributed our little bit during a period where molecular exercise physiology moved from the fringes into the mainstream of sport and exercise science and kinesiology.

When the idea for this book was suggested by Routledge publishers I was immediately interested, as a comprehensive textbook that covered many key aspects of molecular exercise physiology and that was suitable for undergraduate students towards the end of their degree was lacking. I was also fortunate that Keith Baar, Jatin Burniston, Stuart Galloway, Stuart Gray, Kian-Peng Goh, Lee Hamilton, Angela Koh, Arimantas Lionikas, Steven Rot, Mhairi Towler and Peer Wulff were all willing to co-write chapters and many students and colleagues proofread our texts. John Greenhorn and Jessica Wettstein took electron microscope images specifically for this book, Ourania Varsou and Christian Schwarzbauer produced MRI images of the nervous system, Abraham Acevedo-Arozena provided a microscopic image of a neuromuscular junction and Elaina Duguid-Collie provided microarray images. I am also indebted to the University of Aberdeen which allowed me to dedicate time to this project and similarly to my partner, parents and friends, who were understanding and supportive. And last and most, rather than least, I am indebted to Frank W Booth, Claude Bouchard, Marius Sudol,

Grahame Hardie, Michael N Hall and Terje Lømo, who have made seminal discoveries and wrote personal accounts and comments for this book. I am very grateful for this and hope that readers will enjoy reading their texts as much as I do.

We, the author team, realize that this book cannot be perfect, as this field is moving fast and we will have overlooked some papers that we should have included. For this we apologize to the colleagues affected. We have tried hard to write this book in plain English so that undergraduate students, postgraduate students as well as lecturers and researchers are able to gain an overview of molecular exercise physiology topics without having to learn too much jargon. The book has been split into 12 chapters with the hope that some colleagues will adapt it as a textbook for a module in molecular exercise physiology, which is a key aim that we wanted to achieve.

Finally, writing this book has made me aware once again that sport and exercise are key to our physiological and psychological wellbeing and for this reason I am looking forward to a period of particularly physically active living, now that we have finished this project.

Henning Wackerhage

Contributors

Keith Baar, PhD   Keith is an Associate Professor at the University of California in Davis. Keith investigates how resistance and endurance exercise are sensed by skeletal muscle and how this leads to an increase in muscle mass and fatigue resistance. He also works on muscle and ligament engineering. In his free time he puts the theory into practice by cycling, running and fitness training in Davis.

Jatin G Burniston, PhD   Jatin is a Reader of Molecular Physiology at the Research Institute for Sport and Exercise Sciences, Liverpool John Moores University, UK. He is a pioneer in the application of proteomic techniques in exercise physiology and has a keen interest in the use of -omic approaches. Jatin's research principally focuses on muscle proteome responses to aerobic exercise and characterization of post-translational modifications using mass spectrometry. Jatin has trained in Muay Thai for approximately 20 years and also enjoys a broad range of other sporting activities, ranging from weightlifting to hill walking.

Stuart Galloway, PhD   Stuart is a Reader in Exercise Physiology/Metabolism within the Health and Exercise Sciences Research Group at the University of Stirling. His research interests focus on factors influencing whole-body and skeletal muscle carbohydrate and fat metabolism. He uses exercise training and nutritional interventions to examine metabolic responses to exercise in healthy adults as well as in clinical populations. Stuart is an alpine skier, hill runner and mountain biker and he is also a bagpipe player in a competing band.

Kian-Peng Goh, PhD   Kian-Peng is a Consultant Endocrinologist at the Division of Endocrinology at Alexandra Health, Singapore. A firm believer in the importance of exercise, he went on to pursue his interest in exercise physiology after completing his specialty training in endocrinology. While his sole research interest is the role of exercise in the prevention and treatment of chronic diseases such as diabetes, his sporting interest is varied and ranges from bagging munros in Scotland to scuba-diving in the South China Sea.

Stuart Gray, PhD   Stuart is a Senior Lecturer at the University of Aberdeen. His two research interests are the loss of muscle mass during normal ageing, which is termed sarcopenia, and the effects of high-intensity exercise on post-prandial lipidaemia and the activation of the immune system. Stuart enjoys racquet sports and regularly runs and cycles to keep fit.

D Lee Hamilton, PhD   Lee is a Lecturer at the University of Stirling in the Health and Exercise Sciences research group. His research interest is primarily the contribution that kinase phosphorylation cascades play in exercise and nutrition adaptations in skeletal muscle. Having been a boxer he has first-hand experience of controlling diet to make weight and understands how important appetite control is to diet success. He now attempts to keep in shape with resistance training and a range of boxing training methods.

Angela Koh, PhD   Angela is a Senior Consultant Endocrinologist at the Division of Endocrinology at Alexandra Health, Singapore. She has worked at the Joslin Diabetes Center's Islet Transplantation and Cell Biology section and the University of Alberta's Clinical Islet Transplant Program. Her research interest is in difficult diabetes, especially type 1 diabetes. Her latest exercise craze is in kickboxing and high-intensity interval training.

Arimantas Lionikas, PhD   Arimantas is a Senior Lecturer at the School of Medical Sciences of the University of Aberdeen. His research focus is the genetic mechanisms underlying muscle mass and the properties of muscle fibres, such as their number, size and proportion of different types. Arimantas got involved in Greco wrestling at secondary school and still continues his exploration of various wrestling styles.

Stephen M Roth, PhD   Stephen is an Associate Professor in Kinesiology at the University of Maryland. His research interests are in the genetic aspects of exercise responses and adaptation. Dr Roth has three children and is desperately trying to keep up with them in his free time.

Mhairi Towler, PhD   Mhairi is Director of award winning animation production company 'Vivomotion'. Based in Dundee, Vivomotion (www.vivomotion.co.uk) make 2D and 3D animations to help scientists communicate their work. Prior to this, Mhairi carried out postdoctoral research on the protein AMPK in the laboratories of Prof. Grahame Hardie and Dr Henning Wackerhage at the College of Life Sciences, University of Dundee, with a focus on the cell signalling response to exercise. Mhairi enjoys flamenco dancing and swimming.

Henning Wackerhage, PhD   Henning is a Senior Lecturer in Molecular Exercise Physiology at the University of Aberdeen. His research interest is generally Molecular Exercise Physiology and specifically the function of the Hippo pathway in skeletal muscle. He is an ex triathlete and now walks, climbs, winter climbs and sea kayaks the hills of, and seas around Scotland.

Oliver Witard, PhD   Oliver is a Lecturer in the Health and Exercise Sciences research group at the University of Stirling. His research focuses on the adaptive response of human skeletal muscle to exercise and nutrition with a special interest in protein recommendations for maximizing the gain of muscle mass. In his youth and early 20s, he played football to a semi-professional level. Now he enjoys distance running.

Peer Wulff, MD   Peer is a Professor of Physiology and heads the Neurobiology research group at the Christian-Albrechts University of Kiel. He is interested in neuronal circuits underlying memory formation in the brain. Peer enjoys water sports, including surfing and kayaking.

At the end of the chapter you should be able to:

- Define molecular exercise physiology and describe its origins.

- Understand basic wet laboratory research calculations, methods, equipment and terms.

1 Introduction to molecular exercise physiology

Henning Wackerhage

INTRODUCTION

Molecular exercise physiology is an emerging sub-discipline within exercise physiology. The term is a shortened version of the term 'molecular and cellular exercise physiology' which was used, among others, by Frank W. Booth (Booth, 1988), a pioneer in this area. In the first part of this chapter we define molecular exercise physiology, distinguish it from exercise biochemistry and trace its roots in molecular biology and exercise physiology.

The second part of this chapter is focused on the practicalities of wet laboratory research and is written for those who have no previous experience in wet lab research; this includes many sport and exercise science graduates.

DEFINITION AND ORIGINS OF MOLECULAR EXERCISE PHYSIOLOGY

Throughout the history of science, scientific disciplines have diversified into scientific sub-disciplines. In exercise physiology one emerging sub-discipline is molecular exercise physiology. The term is a shortened version of 'molecular and cellular exercise physiology', which was first used by Frank W. Booth to describe research in this area (Booth, 1988). The definition we use is:

> Molecular exercise physiology is the study of exercise physiology using molecular biology methods.

Molecular exercise physiologists especially study the signal transduction mechanisms that cause adaptation to exercise and the genetics of sport and exercise-related traits (Spurway and Wackerhage, 2006).

How did molecular exercise physiology evolve? In exercise physiology the historical trend was, as in other life sciences in the twentieth century, to move the frontier of research from the whole-body level via organ systems towards cells and molecules (Baldwin, 2000). This does not mean that molecular studies now replace organ systems research. Instead the molecular research complements the research done at the organ and whole-body level. This enables researchers to answer questions that previous generations of exercise physiologists were unable to answer due to the lack of suitable methods. Thus molecular exercise physiology is an extension of and complement to classical exercise physiology.

The era that preceded and prepared the ground for molecular exercise physiology was the era of biochemistry of exercise. While the whole-body and organ-level exercise physiologists frequently used non-invasive methods that involved little wet lab research, the exercise biochemists made wet lab research their main method of study. As a consequence, researchers needed to learn wet lab methods and had to equip their laboratories with the necessary consumables and equipment. This included chemicals, pipettes, pH meters, centrifuges, spectrophotometers and microscopes. Human exercise physiologists with a skeletal muscle focus also needed to learn the muscle biopsy technique which was introduced to the field by Jonas Bergström or had to use animal models. This was necessary because tissue samples, especially from skeletal muscle, were needed for subsequent biochemical or histochemical analysis. Key researchers active during the exercise biochemistry era of the 1970s and 1980s include the team of the late Philip D Gollnick, Bengt Saltin, Frank W Booth, John O Holloszy, David L Costill and Stefano Schiaffino (Hamilton and Booth, 2000). In addition, researchers such as Robert T Wolfe and Mike J Rennie used stable isotope methods to measure the incorporation of labelled amino acids into human muscle protein in the early 1980s. The above list of leading exercise biochemists is subjective and there are undoubtedly others who deserve a place in the biochemistry of exercise 'hall of fame'.

Exercise biochemists have used especially spectrophotometric and fluorometric metabolite assays, enzyme assays and histochemistry to quantify or visualize metabolites and enzyme activities. Their research has shown that muscle fibre properties, mitochondria, enzymes, metabolites and protein synthesis all adapt to exercise and that elite athletes differ in their type I and II muscle fibre proportions. This knowledge complemented the evidence on the function of the respiratory and cardiovascular systems during exercise which was mainly gained using organ systems physiology. However, the biochemical methods used did not allow the exercise biochemists to answer more fundamental, mechanistic questions such as: What mechanisms cause adaptation to exercise? How do variations in the DNA sequence of an individual link to her or his exercise capacity? To answer these questions exercise physiologists needed to add molecular biology methods to their tool box.

Molecular biology is the part of biology where explanations are looked for at the level of molecules (Morange, 2009). It is an important discipline in biology because organs and whole organisms are made from molecules. Thus life itself can arguably be reduced to the function of molecules. This is perhaps best demonstrated by the fact that researchers are now well on their way to being able to engineer molecules to produce simple, living organisms (Gibson et al., 2008). This raises ethical concerns, but if these researchers manage to create synthetic life then this would prove the hypothesis that life itself is a function of molecules.

The key drivers of the molecular biology era were new methods that allowed biologists to isolate, visualize, quantify and manipulate the molecules that form organisms. The introduction of new molecular biology methods paved the way for landmark discoveries. Some of the most important examples of links between molecular methods and often Nobel Prize-worthy discoveries include the following:

- The analysis of the structure of many proteins and above all of the DNA double helix by James D Watson and Francis Crick in 1953 (Nobel Prize in Physiology or Medicine in 1962) was only possible because Max von Laue (Nobel Prize in Physics in 1914) and others had introduced and developed X-ray crystallography in the early twentieth century.
- Linking the DNA sequence of individuals to their phenotype was made much easier by the introduction of the polymerase chain reaction (PCR) by Kary B Mullis in the 1980s (Nobel Prize in Chemistry in 1993). The PCR method allowed biologists to amplify small DNA fragments. These could then be further analysed by improved DNA sequencing, which was introduced by Frederick Sanger in the late 1970s (Nobel Prize in Chemistry in 1980).
- The study of gene function was made possible because biologists learned to manipulate or engineer DNA in the 1970s. This allowed researchers to knock out genes or to overactivate (knock in) genes, especially in mice, and to study the effect on phenotypes such as exercise capacity. Mario R Capecchi, Sir Martin J Evans and Oliver Smithies won the Nobel Prize in Physiology and Medicine for their methods of introducing genetic mutations in mice in 2007.
- The study of signal transduction events that link signals to adaptive responses was made possible because of the ability to detect individual proteins. The method used to do this is termed Western bloting and involves separating proteins in an electrical field followed by detection with specific antibodies. The term 'Western blot' is a pun on Southern blot, which is a method used to detect DNA fragments introduced by Edwin Southern in 1975. Another key development was the development and use of antibodies to label proteins on membranes as part of a Western blot or in cells by immunocytochemistry.
- The study of gene expression depended critically on the ability to extract RNA and to detect it via Northern blotting, again a pun on Edwin Southern's DNA blotting method. Today Northern blots have been largely been superseded by reverse transcriptase polymerase chain reaction (RT-PCR), which is a variation of the PCR method to amplify and or measure RNA (cDNA) instead of DNA.

To summarize, especially in the 1970s and 1980s many key methods in molecular biology were introduced. Researchers in all life sciences soon started to use these emerging methods to make fundamental discoveries in their respective fields.

Exercise physiologists joined the molecular biology club relatively late, roughly in the mid 1980s. The leading pioneer was Frank W Booth (Box 1.1) who trained in the laboratories of Charles Tipton and John O Holloszy and progressed from exercise biochemistry to molecular exercise physiology. Probably the first time a *bona fide* molecular biology method was used by an exercise physiologist to study an exercise physiology problem was the experiment where Peter Watson, a graduate student, Joseph Stein, a molecular biologist, and Booth used Northern blotting to measure α-actin mRNA in rat skeletal muscle during hindlimb immobilization of rats (Watson et al., 1984). Another breakthrough achievement by the Booth lab was the direct gene transfer into the skeletal muscle of live rodents (Wolff et al., 1990; Thomason and Booth, 1990). This was achieved in parallel with another team, who did a similar experiment at the same time, and enabled researchers subsequently to modulate the activity of genes in order to test their effect on exercise-related variables. In the first method-focused papers only marker genes were expressed, but the method was later used to elegantly demonstrate, for example, that the overexpression of a hyperactive form of the Akt kinase, also known as PKB, resulted in a pronounced skeletal muscle fibre hypertrophy (Pallafacchina et al., 2002). Another major achievement by the Booth lab was to publish the first report on the use of gene expression microarrays for an exercise-related research question, which was to compare the gene expression of a 'slow' soleus with a 'fast/white' vastus muscle (Campbell et al., 2001). In this study they collaborated with Eric Hoffman's microarray facility at the Children's Hospital in Washington DC, a leading microarray laboratory.

BOX 1.1 FRANK W BOOTH: THE ADVENT OF MOLECULAR BIOLOGY TECHNIQUES IN EXERCISE BIOLOGY

A brief history of one early scientist in physical inactivity and exercise

My philosophy of science has evolved under a number of teachings. A conundrum I often ponder is, do men make history or does history make men? From my viewpoint, this thought-provoking phrase is a truncated tribute to Karl Marx who wrote, 'Men make their own history, but they do not make it as they please, they do not make it on their self-selected circumstances, but on their circumstances existing already, given and transmitted from the past.' Certainly circumstances throughout my training and career have guided my philosophical evolution, but I can't deny that genes play an important role as well.

I started liberal arts at Denison University with intent to pursue law. Circumstances led me elsewhere. My biology classes gripped my attention. Serendipity had my University begin a swim team in my sophomore year (the only way I could have made the squad), and the assistant swim coach, Robert Haubrich, was also my academic biology advisor. My senior paper required for biology majors in 1964 was on adaptations to exercise. Two events occurred together: (1) My applications to medical school were not successful so I had to look at an alternative, and (2) Professor Haubrich gave me a flyer about a new exercise physiology PhD program under Charles Tipton at University of Iowa. I got on a train from Columbus, Ohio and went to Iowa City to view the exercise physiology program, whose described course of study had intrigued my interest so much, that I decided to visit and eventually enter in 1965, to replace my interest in medical school. Other students in the Tipton program were James Barnard, Ken Baldwin, and Ron Terjung.

"Tip," a name we respectfully called Charles Tipton, taught me to go after mechanisms and supported my desire to challenge existing dogmas and policies, as this was the period of student discontent over the Vietnam War. University life in the late 1960s was so tremulous with continual protest rallies and the killing of student protesters at Kent State University, that University administers had to accept student challenges as acceptable on their campus (as opposed to today where minimal protest occurs over the loss of academic freedoms and civil liberties on University campus in response to forced compliance to regulations from the non-democratic, non-elected bureaucrats). Also during the Cold War period (February 1945–August 1991), the Russians, on October 4, 1957, launched Sputnik I, an orbiting satellite that greatly accentuated the continual threat that the U.S. feared from the Soviet Union. The same rocket that launched Sputnik could send a nuclear warhead anywhere in the world in a matter of minutes (as I saw in the fictional classic 1964 movie *Dr. Strangelove or: How I Learned to Stop Worrying and Love the Bomb*, which showed a nuclear holocaust from an insane general, in an insanely funny comedy). In 1958, the U.S. Congress passed the National Defense Education Act, whose outgrowth funded Tip's exercise physiology program, and my graduate student salary. The Russians and U.S. were in a space race. The first human walk on the moon in 1969 was from the U.S., which further fueled my interest to the physiological effects of lack of gravity (physical inactivity), along with Jere Mitchell's and Bengt Saltin's classic Dallas bedrest study published in *Circulation* in 1968, fostered my decision to take a first post-doc at the School of Aerospace Medicine. John Holloszy's 1967 paper on the biochemical adaptations of exercise led me to my second post-doc experience to join Baldwin and Terjung at John's lab, a time during which John taught me critical thinking.

After finishing my work with Holloszy, I had a choice of two faculty positions in medical schools (Wayne State in Detroit, or the new University of Texas Medical School at Houston). I selected the latter because it was the home of NASA. Soon after arriving in Houston in 1975, molecular biology was coming to the forefront. I saw the potential of explaining inactivity mechanisms in terms of genes. In the early 1980s, I was fortunate to be present when molecular biology began to appear as a tool in biology at the University of Texas Medical School in the Texas Medical Center, which was located in Houston, Texas. Baylor College of Medicine was one block from my medical school and I listened to many seminars at Baylor during which I began to connect molecular techniques with my research in exercise and inactivity. My first grad student, Peter Watson, collaborated with Joe Stein in biochemistry and endocrinology, whose lab was just a few doors down from mine, to measure mRNAs. We used dot-blot hybridization of P^{32}-labeled plasmids containing the skeletal muscle α-actin cDNA in isolated RNA on a nitrocellulose membrane and published the work in the *American Journal of Physiology* in 1984. Concurrently, in 1980, Kary Mullis invented the polymerase chain reaction (PCR), a method for multiplying DNA sequences *in vitro*, and I began using this technique when it became commercialized.

I will end with where I started: Do humans make history or does history make humans perform some event? I think it is both, like gene–environmental interaction determining phenotype. History and human curiosity interact to determine how well scientific understanding can explain why physical inactivity is an actual contributor to chronic disease and longevity. As Robert Frost, Pulitzer Prize for Poetry, United States Poet Laureate, ended his poem "Stopping by Woods on a Snowy Evening":

The woods are lovely, dark and deep.

But I have promises to keep,

And miles to go before I sleep,

And miles to go before I sleep."

So my journey of life continues, as I have miles and miles to go before I sleep.

Frank Booth, University of Missouri, August, 2013

The next step in genetic modification was to generate transgenic mice (i.e. mice in which a gene is either knocked out or overactivated ('knocked in')). In the context of exercise physiology, pioneering papers reported that the knockout of myostatin resulted in a pronounced muscle hypertrophy, termed double muscling (McPherron et al., 1997) and that the knockout of the myoglobin gene had little or no effect on exercise capacity (Garry et al., 1998). The myostatin paper by See-Jin Lee's team has currently been cited nearly 2000 times, highlighting the impact that such papers can achieve.

Many of the fundamental exercise physiology questions had been answered by the organ systems exercise physiologists, but major questions remained. One of the most important, unanswered questions was: how does exercise training work? Or more specifically, what are the mechanisms that mediate the adaptation to exercise? Up to the 1980s all the answers to this central question were at best semi-scientific and frequently vague concepts involving overload or a misinterpretation of the glycogen 'supercompensation' or 'overcompensation' time course. From the 1980s onwards it became increasingly clear to some pioneers that adaptation to exercise involves gene expression changes and that signal transduction pathways link the signals associated with exercise to the adaptive response. In this context protein phosphorylation, a discovery for which Edmond H Fischer and Edwin G Krebs won the Nobel Prize in Chemistry in 1992, turned out to be the key mechanism. To study this, exercise physiologists needed to learn methods such as Western blotting to detect proteins; the subsequent introduction of phospho-specific antibodies allowed researchers to also measure the phosphorylation of signalling proteins with relative ease as no radioactive tracers were required. One of the first papers to characterize a protein in the context of exercise was published by a Japanese team (Yamaguchi et al., 1985); it has remained obscure despite being technically well advanced for the time.

Breakthrough discoveries were made later. With respect to the muscle growth response to exercise, Keith Baar and Karyn A Esser analysed the translational or protein synthesis activity in muscle after high-intensity electrical stimulation and measured the phosphorylation of the growth-regulating kinase p70S6K (Baar and Esser, 1999). They found indirect evidence for an involvement of p70S6K, which is a member of the growth-regulating mTOR pathway, in the regulation of muscle growth. With respect to the adaptation to endurance exercise, the discovery of AMP kinase by D Grahame Hardie, a former PhD student of Sir Philip Cohen at Dundee in Scotland, was a key discovery. Hardie collaborated with William W Winder to demonstrate for the first time the activation of AMP kinase in response to endurance exercise (Winder and Hardie, 1996). Hardie and others then elucidated the many acute and chronic adaptations that are trigged by AMP kinase activation and thereby identified AMP kinase as a major regulator of adaptive responses to changed energy turnover. Later, a team led by Bruce Spiegelman identified the transcriptional co-factor PGC-1, which was later linked to AMP kinase, as a key regulator of mitochondrial biogenesis (Puigserver et al., 1998; Lin et al., 2002). This added a major player to the transcriptional regulators that were known to regulate the expression of mitochondrial genes (Hood et al., 1994; Scarpulla, 1997). Another important discovery was the link between the release of calcium from the sarcoplasmic reticulum of a muscle to muscle fibre phenotype regulation by the calcineurin pathway. This was work performed by Eva R Chin and others in the lab of R Sanders Williams and gave important insight into how 'slow' and 'fast' muscle fibre genes are regulated by calcium (Chin et al., 1998). In Italy, Stefano Schiaffino, continuing the analysis of muscle fibres with a more molecular approach, linked the ERK pathway to fibre phenotype regulation (Murgia et al., 2000) and the kinase Akt (also known as PKB) to muscle hypertrophy (Pallafacchina et al., 2002).

The other important major strand of molecular exercise physiology is **sport and exercise genetics**, also termed kinesiogenetics. In this field the groups led by Claude W Bouchard in Canada and the United States and the one led by Vassilis Klissouras in Greece used twin and family studies to estimate the heritability of many exercise-related traits. The Bouchard lab then also, almost single handed, moved sport and exercise genetics into the molecular biology era by pioneering the molecular analysis of the human DNA sequence. Bouchard gives a commentary about the development of sport and exercise genetics in Box 1.2.

BOX 1.2 CLAUDE BOUCHARD: FROM DESCRIPTIVE TO PREDICTIVE EXERCISE GENOMICS

The first attempts at identifying genetic markers associated with human exercise-related phenotypes or traits were quite naïve by today standards. They started in the late 1960s with studies based on polymorphisms in red blood cell proteins and enzymes as well as variations at the HLA loci (genes important for immune function). Later, skeletal muscle enzyme variants were screened in the search for significant associations. Few of these studies were reported and none yielded significant genetic predictors of performance or fitness traits.

As DNA screening technologies became available, a few family-based genome-wide linkage scans and a large number of candidate gene studies were published. One can argue that little progress was truly made during that period of time. Only a few papers went beyond a descriptive association stage to begin exploring the functional implications of particular alleles.

More recently, high-throughput transcriptomic and genotyping technologies have impacted the domain of human exercise genomics. Transcript (mRNA) abundance in skeletal muscle was used to define gene expression profiles best correlated with performance or fitness traits. Transcriptomics (measuring the expression of all genes) provided new and more potent candidate genes that are being investigated for sequence variants and their associations with relevant traits. New candidates are also being derived from genome-wide association studies based on large number of measured and imputed single-nucleotide polymorphisms. The first studies were observational in nature and concluded generally based on cross-sectional association patterns with alleles or genotypes. However, we are beginning to see exercise genomic research designed to address the ability to predict the responsiveness to acute or regular exercise. In other words, we are currently seeing a major paradigm shift from single gene, correlative studies to unbiased, genome-wide mapping approaches based on experimental data with a focus on prediction.

Interestingly, the genotyping arrays commonly used today for genome-wide screens of allelic variants are designed to capture common polymorphisms, typically with an allele frequency of 5% and more in the population of interest. There are about 10 million of these polymorphisms in the human genome and each person carries about 3–4 million of these common variants.

(Continued)

BOX 1.2 (CONTINUED)

These common alleles have been exposed to selection pressures for thousands of generations since the advent of early human *Australopithecus* populations. Alleles with extreme effects, either positive or negative, have become fixed or eliminated over this evolutionary journey such that common polymorphisms are unlikely to exert dramatic effect size taken individually. Yet they often associate or predict with low effect sizes exercise-related traits.

The next generation of exercise genomic study designs will have to include not only these common variants but also the estimated 200 000 to 500 000 DNA variants unique to an individual or his or her family lineage. The genomic screens should also include the hundreds of copy number variants of variable nucleotide sizes as they have the potential to impact biology and behavior as well. Other genomic features that should be incorporated in future exercise biology studies include hundreds of insertion/deletions, multiple splice site disruptions, DNA sequence variants impacting micro RNA, other small RNA, as well as enhancer binding sequences.

Finally, the main aim of exercise genomics should always be kept in sight: to explain in greater depth human variation in exercise-related traits and adaptation to acute and chronic exercise. This represents a highly complex undertaking. It requires that exercise genomic studies be properly powered from a statistical point of view. Findings need to be supported by adequately powered replication studies. Functional studies using cell-based models, appropriate mouse strains and transgenics, and human experimental investigations are a necessity if alleles and genotypes are to be causally related to phenotypes. The future of exercise genomics looks exciting but the path forward will not be linear and will be replete with challenges and obstacles.

Claude Bouchard, Pennington Biomedical Research Center, Baton Rouge, Louisiana, February 2012

Claude Bouchard first utilized electrophoretic methods to genotype individuals in the context of sport and exercise (Bouchard et al., 1989), first used methods to analyse mitochondrial DNA sequence variants (Dionne et al., 1991) and first studied DNA sequence variations of nuclear-encoded genes (Deriaz et al., 1994). Together with Robert M Malina and Louis Pérusse, Bouchard also wrote a major textbook in sport and exercise genetics entitled *Genetics of Fitness and Physical Performance* (Bouchard et al., 1997), which covered both classical and molecular genetics studies in relation to exercise.

In the 1990s other exercise physiologists learned to extract the genomic DNA of individuals and to perform mainly PCR-based genotyping assays to study candidate genes. In 2000 Alun G Williams, in a study led by Hugh E Montgomery, reported in the journal *Nature* links between the ACE gene insertion/deletion (I/D) genotype and muscle performance (Williams et al., 2000). This study was much discussed at the time and brought candidate gene association studies to the attention of many exercise physiologists. However, common DNA sequence variations, known as polymorphisms, of candidate genes generally only influence a small percentage of the variation of sport and exercise-related traits. As a consequence, hundreds if not thousands of individuals are needed to statistically power such studies. Because high subject numbers are not achievable for most exercise physiology laboratories, many candidate gene association studies

Figure 1.1 Prof. Dr Frank W Booth and Prof. Dr Claude Bouchard: Two leading molecular exercise physiology pioneers. Frank W Booth, University of Missouri (left) introduced many molecular biology methods to exercise physiology and thus pioneered mechanistic adaptation research. Claude Bouchard, Pennington Biomedical Research Center (right) has not only made major contributions to classical sport and exercise genetics but his team has also pioneered the usage of DNA sequence analysis methods to study sport and exercise genetics.

Frank W Booth   Claude Bouchard

were statistically underpowered and the results were thus often unreliable, as evidenced by the fact that repeat studies frequently reported different or no associations.

Other researchers identified rare candidate gene mutations that, in contrast to the majority of candidate gene variations, had a major effect on athletic performance. Here two findings stand out: An activity-enhancing mutation of the EPO receptor was identified in a Finnish family and was shown to have a large effect on the haematocrit of heterozygous carriers (de la Chapelle et al., 1993). One member of that family, Eero Mäntyranta, won three Olympic gold medals for cross-country skiing and it seems likely that the aforementioned EPO receptor mutation was a major factor. In a second study, a knockout mutation in the myostatin gene was correlated with a greatly increased muscle mass and strength of a toddler (Schuelke et al., 2004). In general, however, genetic variations that have a large effect on performance are rare and it seems that most sport and exercise-related traits depend on the variations in the DNA of many genes which all contribute with a small percentage to the variation of the trait.

Candidate gene association studies gave a limited insight into the genetic causes of the variability of exercise-related traits between individuals. Over time it became clear that exercise physiologists needed to scan most or all of the DNA sequence in order to identify the variations in the DNA sequences that caused variability of exercise-related traits between individuals. The first to pioneer genome-wide DNA sequence analysis in a sport and exercise context was again Claude Bouchard in collaboration with

Jamie A Timmons (Timmons et al., 2010; Bouchard et al., 2011). In these studies DNA sequence variation microarrays were used to search for DNA sequence variations that can explain variations in endurance trainability. However, this is surely only an intermediate step as the sequencing of whole human genomes is now achievable and increasingly affordable (Wheeler et al., 2008). Soon, exercise physiologists will be able to sequence the entire genomes of individuals to trace common and unique genetic variations that can explain the variation of sport and exercise-related traits. At that point the challenge will no longer be the analysis of the DNA sequence but instead the computational analysis of the DNA sequence in order to link variations in the DNA. At this point ethical questions must be considered as a whole-genome analysis may uncover DNA sequence variations that are associated with an increased disease risk (Wackerhage et al., 2009).

To summarize, molecular exercise physiology is a sub-discipline of exercise physiology in which molecular biology methods are used to try to answer exercise physiology questions. The two key foci of research are molecular genetics (i.e. the attempt to link variations in the DNA sequence to sport and exercise-related traits) and the analysis of signal transduction mechanisms that can explain adaptation to exercise. Frank W Booth (Figure 1.1) pioneered molecular adaptation research while Claude Bouchard (Figure 1.1) first used molecular genetics methods in the context of sport and exercise.

INTRODUCTION TO WET LABORATORY RESEARCH

After the introduction to molecular exercise physiology we will now shift the focus of this chapter to wet laboratory research, which is the basis for all molecular exercise physiology. This part of the book may be skipped by those who already have wet laboratory training. However, many of those who study or work in sport and exercise science or kinesiology do not have these skills and thus we feel that it will be helpful to introduce basic wet laboratory research in this book. A good additional source for practical wet laboratory information is the book entitled *At the Bench: A Laboratory Navigator*, written by Kathy Barker (Barker, 2005). In this section we cover the following topics, as they are particularly relevant for molecular exercise physiologists:

- Samples (human, animals, cultured cells)
- Basic wet laboratory skills
- Microscopic methods.

In Chapters 2 and 3 we will also cover DNA, RNA and protein methods.

Samples

Molecular exercise physiologists need to generate samples in order to study the molecules in these samples. Such samples can be of human origin (especially blood and skeletal muscle) or can be any tissue from animals, primary cultured cells or cell lines. In Figure 1.2 we explain the equipment used to obtain skeletal muscle samples and show examples of different muscle samples.

Figure 1.2 Skeletal muscle methods and models. (a) Bergström biopsy needle set up for taking a human skeletal muscle biopsy with suction. The arrow shows the window in the hollow needle into which the muscle protrudes. At the top is the inner biopsy punch with a sharp cutting end which is inserted into the hollow needle to cut off the biopsy sample which protrudes into the window of the hollow needle. (b) Mouse tibialis anterior cross-section stained using an ATPase stain with alkaline preincubation. Histochemical stains or immunohistochemical stains are used to visualize the different types of muscle fibre (see Chapter 4 for further information on muscle fibre types). (c) Isolated mouse extensor digitorum fibre cultured in suspension. The arrows highlight clusters of activated satellite cells that are visible after several days in proliferation medium. (d) Activated mouse satellite cells or myoblasts cultured in a plastic dish. These cells readily differentiate and fuse to form (e) multinucleated myotubes. The differentiation of activated satellite cells or myoblasts into myotubes and fully differentiated muscle fibres is termed myogenesis and can be replicated in cell culture.

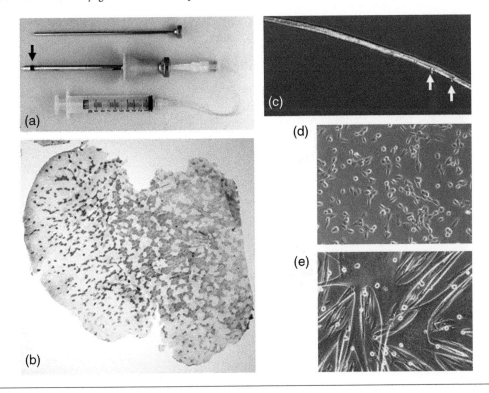

We will now briefly describe how the different types of samples are generated.

Muscle biopsy as an example of a human tissues sample

In humans, samples can realistically only be obtained from a few organs, with blood and to a lesser extent skeletal muscle being the only samples that are routinely obtained. Here we discuss the skeletal muscle biopsy method as an example of how human tissue samples are obtained. The biopsy method was re-introduced by Jonas Bergström in the 1960s and allowed exercise physiologists to obtain skeletal muscle samples for biochemical and molecular biology analysis. Practically, the researcher first disinfects and anaesthetizes the skin, makes a small cut with a scalpel blade and then inserts a biopsy needle into the muscle. In many cases the **Bergström needle** is used (Figure 1.2a). In order to obtain a larger piece of muscle, suction can be applied to the Bergström needle or alternatively a **conchotome trochar** can be used. A less invasive method involves using a **microbiopsy needle** which allows researchers to obtain muscle samples weighing up to ≈50 mg; this would be enough for RNA extraction or Western blots (Hayot et al., 2005).

Mouse hindlimb muscles as an example for an animal tissue sample

Exercise physiologists are mainly interested in the exercising human, but human interventions are limited because samples from tissues such as the brain or heart cannot be obtained from living humans. Thus exercise physiologists use animals, mostly rodents, to study the effect of interventions such as synergist ablation or transgenic modification which are not possible in humans. Moreover, in animals, samples can be obtained from organs that cannot be obtained from living humans. Here, we briefly give an introduction to the dissection of the commonly used mouse hindlimb muscles:

- tibialis anterior (TA, a stained cross-section is shown in Figure 1.2b)
- extensor digitorum longus (EDL)
- gastrocnemius (Gas)
- soleus (Sol).

For the dissection of these muscles a dissection microscope and a dissection kit, which includes fine scissors and tweezers, are required. First, the skin above the gastrocnemius is pulled upwards until it tears and then the lower hindlimb muscles are exposed. After that the tibialis and extensor digitorum longus muscles on the front of the leg are dissected under a dissection microscope followed by the dissection of the gastrocnemius and soleus muscles at the back. Skeletal muscles are then weighed and frozen either directly in liquid nitrogen (unsuitable for microscopy as freezing artefacts occur but fine for subsequent RNA and protein extraction) or in isopentane cooled in liquid nitrogen to near freezing (preferred for microscopy).

Cell culture

Cultured cells are frequently used in biochemical and molecular biology experiments as they allow many manipulations without ethical concerns. Two major types of cell culture can be distinguished:

- **Primary** cell cultures involve the culture of cells that are obtained, for example, from a human muscle biopsy or from a rodent. These cells can only divide a finite number of times and thus need to be generated periodically from fresh tissue. Examples of cultured primary cells are shown in Figure 1.2c–f.
- **Cell lines** are cells with genetic mutations that make them immortal. These cells can be used indefinitely and are generally well characterized.

Cultured cells allow a maximum of manipulation with few ethical concerns, especially if established cell lines are used as no animals need to be sacrificed and no biopsies need to be obtained. Although established cell lines and primary cell cultures can be obtained from every organ, the downside is that the research may be criticized as 'reductionist' because, for example, cultured myoblasts are very different from living human skeletal muscles. In addition, extensive training and equipment is needed to establish and perform cell culture and it is expensive.

An example of a primary cell culture is the culture of whole extensor digitorum longus or soleus muscle fibres, together with satellite cells in their niche (Collins and Zammit, 2009). This technique is difficult to perform but the satellite cells on the muscle fibres survive for several days and during that time progress from quiescent to activation; they then make the decision to either differentiate, as happens during repair and hypertrophy, or self-renew to replenish the pool of satellite cells (Zammit et al., 2004). Muscle fibres with satellite cells can even be transplanted.

The C2C12 mouse myoblasts cell line is an example of a commonly used cell line. The original C2 cells were originally obtained from a mouse hindlimb almost 40 years ago by David Yaffe, then around 1980 Helen Blau subcultured the C2C12 cell line. C2C12 and other cell lines can be purchased from the European Collection of Cell Cultures (ECACC) or the American Type Culture Collection (ATCC) or in many cases they are passed on by colleagues. Uniquely, the mononucleated C2C12 myoblasts can be grown to a high density (confluence) and then a reduction of serum leads to their terminal differentiation into multinucleated myotubes, which partially resemble muscle fibres.

All long-term cell cultures need to be grown under sterile conditions using aseptic technique. For this, surfaces and hands are frequently sprayed with 70% ethanol as a disinfectant, a tissue culture hood and sterile plasticware (i.e. pipettes, dishes etc.) are used and some researchers add antibiotics such as penicillin and streptomycin (commonly abbreviated as Pen-Strep) to their media. It is essential to prevent microorganisms entering the culture media and growing alongside the cultured cells. Tissue culture media can be purchased readymade as solutions of salts and nutrients. In addition, sera such as fetal bovine serum are added to stimulate growth and cell proliferation via the growth factors, mitogens and hormones that are included in such sera.

Unfortunately, the perfect sample which can be obtained non-invasively from any organ of a human after any intervention without ethical concern does not exist. For this reason researchers need to decide on the best sample for their purpose based on ethical concerns, possible interventions, relevance for humans and cost. The different sample types are summarized in Table 1.1.

In practice a combination of samples is commonly used in molecular exercise physiology and related fields to answer a research question. A typical strategy could be to first investigate the activation of a protein or gene by exercise-related signals and its function in cell culture. Rodent models might then be used to study the activation or function of the protein or gene in relation to exercise *in vivo*. Finally, the activation of the protein or gene by exercise would then be verified in humans and human studies could also be conducted to test the benefit of interventions that target the molecule.

Basic wet laboratory skills

Once samples have been generated they need to be analysed. For this, basic wet laboratory skills are required, which are rarely taught in any detail as part of sport and exercise

Table 1.1 Advantages and disadvantages of different sample types for molecular exercise physiology research

	ADVANTAGES	DISADVANTAGES
Human tissues	• Most relevant species • Humans are best for exercise and nutrition interventions	• Ethical concerns • Genetic variability • Invasive methods required to obtain samples • Many tissues (e.g. heart, brain) cannot be obtained • Pharmacological and genetic activation/knockout studies are often not possible
Animal tissues	• Samples from all organs can be obtained • Inbred mice are genetically homogenous • Pharmacological, transgenic, exercise and nutritional interventions are feasible	• Ethical concerns • Animals are not humans • It is expensive to maintain animals
Cultured cells	• Pharmacological and transgenic interventions are comparatively easy to do • Allows to study one cell type in isolation	• Cultured cells differ from human organs • Very limited for exercise interventions • Expensive equipment and running costs

science or kinesiology degrees. For this reason we will now introduce some key wet laboratory topics and skills.

Units

Molecular exercise physiologists frequently measure small volumes, amounts and concentrations. These are given as SI units (International System of units) – litres, kilograms and moles, with prefixes used to determine fractions thereof. These prefixes can be confusing so we have listed the most relevant prefixes in Table 1.2.

Table 1.2 Prefixes used for concentrations, amounts and volumes

SPELLED OUT	ABBREVIATION	NUMERICAL
Millimolar/gram/litre	mM, mg, mL	0.001 or 10^{-3} mol/L or g or L
Micromolar/gram/litre	μM, μg μL	0.000001 or 10^{-6} mol/L or g or L
Nanomolar/gram/litre	nM, ng, nL	10^{-9} mol/L or g or L
Picomolar/gram/litre	pM, pg, pL	10^{-12} mol/L or g or L

Measuring liquids

Molecular exercise physiologists, like other wet laboratory researchers, spend much of their time measuring and transferring liquids. To measure small amounts, (micro-) pipettes are used. They come in three main sizes:

- 1–10 μL (μL is microlitres; white tips),
- 10–100 μL (yellow tips),
- 100 μL–1000 μL (blue tips).

For the polymerase chain reaction (PCR), pipettes that measure 0.1–1 µL are required. The volumes that have been pipetted are then kept in one of the following containers:

- PCR tubes (0.2 or 0.5 mL),
- Eppendorf tubes (1.5 mL),
- small Falcon tubes (15 mL),
- large Falcon tubes (50 mL) and
- glass bottles (50 mL–2 L).

Weighing solids

Solids are weighed using balances. In molecular exercise physiology laboratories a **fine balance** that weighs down to 0.1 mg (milligram; 0.0001 g) is sufficient as mouse muscles, biopsies and small samples weigh roughly from <10 mg to 100 mg. It is useful to have another balance to weigh from 0.1 g to 500 g or 1 kg, as larger amounts are frequently needed for solutions. Solids are weighed using different-sized **spatulas** and weighing boats.

Making up solutions

Once amounts of solids or liquids have been measured, they are frequently combined to make up solutions. The solvent for most solutions is **water**, which comes in different forms. Tap water contains impurities such as ions and bacteria and thus it is frequently purified for use in research. The following types of purified water are commonly used in wet laboratories:

- **Deionized water**: the water is run through columns that remove ions.
- **Double-distilled water** (ddH_2O). The double distillation removes ions and bacteria. This is the standard laboratory water.
- **Ultrapure water** (e.g. MilliQ water): This highly purified water can be obtained by running it through deionizing columns and fine filters. Ultrapure water is used for example for some types of electrophoresis.

Concentrations and solutions

For solutions the right concentration of each constituent needs to produced. Concentrations can be stated as follows:

- **Molar concentration** in mol/L (M);
- **Mass** or **volume concentration** (e.g. mL/L or g/L).
- **Percentage concentration** (%). In addition to the percentage 'v/v' is commonly added to state that it is a volume diluted in a volume and 'w/v' for weight in volume. For example, 0.1 mL in 1 mL is a 10% v/v concentration while 0.1 g in 1 mL is a 10% v/w concentration.
- **Relative concentration**. This is commonly used for antibodies and expressed as the ratio of parts. For example, a 1 in 1000 antibody concentration is equivalent 1 µL of antibody diluted in 999 µL of antibody buffer.

Conversion of a molar concentration into mass per volume

Molar concentrations are the preferred choice as one mole is defined as 6×10^{23} molecules. Thus the molar concentration ratios reflect the ratio of the numbers of molecules the solution. The drawback of molecular concentrations is that there is no instrument to measure molar concentrations directly and thus molar concentrations have to be converted into masses or volumes, which are measurable. The conversion to mass is done using the following equation:

$$m = c_i \times \text{MW} \times v/1$$

where m is mass in g; MW is molecular weight (g/mol); c_i is molar concentration (mol/L); v is desired volume in litres divided by one litre (as the molar concentration is given per litre).

This looks more complicated than it is. For example, assume the aim is to produce 200 mL (0.2 L) of a 1.5 mol/L NaCl (MW = 58 g/mol) solution. Entering these data in the above formula yields:

$$m = 1.5 \text{ mol/L} \times 58 \text{ g/mol} \times 0.2 \text{ L/1} = 17.4 \text{ g}$$

In practice, weigh out 17.4 g of NaCl, place it into a flask and fill up to 100 mL with water. Label the flasks '1.5 mol/L NaCl', and with your initials and the date.

pH measurement, adjustment and buffers

In many cases the pH of the solution is given and needs to be adjusted. The pH is defined as the negative decadic logarithm of the free hydrogen ion concentration. Thus a pH of 1 means a high and a pH of 12 a low free hydrogen ion concentration. The pH of a solution is measured using a calibrated pH meter, adjusted by adding acids and bases, and can be stabilized by buffers. The key component of a pH meter is a pH electrode to determine the pH of a solution. To obtain precise pH readings take your time for each reading as pH electrodes are slow, ensure that the solution is stirred by using magnetic stirrer bars and if in any doubt recalibrate the instrument.

The acids or bases used to adjust pH are usually in the molar range. Hydrochloric acid (HCl) and sodium hydroxide (NaOH) are the most commonly used acid and base to adjust the pH of a solution, respectively. Many acids and bases come as liquid stock solutions or pellets and need to be diluted first. Table 1.3 gives information about how to dilute acids and bases to 1 M solutions. Greater or lower concentrations can be achieved by adding more or less of the acid or base/salt.

The pH of solutions is generally buffered by buffers that bind hydrogen ions at a pH around their acid dissociation constant pK_a. For example, a commonly used buffer is Tris, which is the abbreviation for Tris(hydroxymethyl)-aminomethane. Tris buffer has a pK_a of 8.08. At a pH of 8.08 the addition of acids or bases changes the pH of a Tris buffer least and and Tris buffer is commonly used to stabilize the pH of solutions with desired pH values between roughly 7.1 and 8.9.

ACID OR BASE	FORMULA	MILLILITRES OR GRAMS PER LITRE FOR A 1 MOL/L (M) SOLUTION*
Hydrochloric acid (36%)	HCl	86.2 mL
Acetic acid, glacial (99.5%)	CH_3COOH	57.5 mL
Acetic acid (36%)	CH_3COOH	159.5 mL
Sodium hydroxide	NaOH	40 g (white pellets)
Potassium hydroxide	KOH	56.11 g (pellets)

Table 1.3 Common acids and bases at stock concentration and how to produce a 1 M acid or base

*Use the volume or weight below and fill up to 1 litre using ddH_2O.

From Barker (2005).

Stock solutions

Some frequently used solutions are made up as high highly (e.g. 10-fold) concentration stock solutions. If the stock solution is 10-fold concentrated then take 1 part of the stock solution and add 9 parts of ddH_2O to make up the final solution with the correct concentrations. This is termed a 1 in 10 dilution.

Dilution of stock solutions in variable volumes

To treat cells or organs with nutrients, drugs or inhibitors concentrated stock solutions are prepared which need to be diluted to achieve the desired concentration in a culture dish, organ bath or litre of blood. The equation to calculate the volume of a stock solution that needs to be added to achieve a desired concentration is calculated using the following formula:

$$c_1 v_1 = c_2 v_2$$

where c_1 is the desired concentration, v_1 is the volume the stock solution is added to, c_2 is the concentration of the stock solution, v_2 is the volume of the stock solution that needs to be added to v_1 to achieve c_1.

The formula works for molar and mass concentrations. For example, the task is to dilute a 10 mg/mL myostatin stock solution to a 25 μg/mL myostatin solution in a 3 mL well of a cell culture dish. This is solved as follows:

- Change all concentrations (c) and volumes (v) to one prefix unit, respectively: The stock concentration c_1 is 10 mg/mL, The final concentration c_2 is 0.025 mg/mL (i.e. 25 μg/mL converted to mg/mL).
- Enter the known values into the formula: $v_1 \times$ 10 mg/mL = 3 mL × 0.025 mg/mL.
- Rearrange the formula to obtain the unknown value: v_1 = (3 mL × 0.025 mg/mL)/10 mg/mL = 0.0075 mL or 7.5 μL.
- Thus you will need to add 7.5 μL of a 10 mg/mL myostatin stock solution in order to get a final concentration of 25 μg/mL myostatin solution in a 3 mL well.

Serial dilutions

Serial dilutions are used for preparing concentrations that are too low for reliable pipetting or weighing or for preparing dilutions series to perform, for example, a concentrate–effect experiment. The dilution is given as a 1 in 'number' dilution. For example, a one in ten dilution starting with a 10 mmol/L stock has the following dilutions: 10 mmol/L diluted one in ten is 1 mol/L which diluted one in ten is 100 μmol/L, 10 μmol/L, 1 μmol/L, 100 nmol/L and so on. To achieve this, for example 10 mL of a 10 mmol/L stock concentrations is prepared first. Next 1 mL of the stock is taken and added to 9 mL of water or buffer which results in a 1 mol/L one in ten dilution. After mixing, 1 mL of the 1 mmol/L solution is diluted again in 9 mL of water or buffer to achieve 100 μmol/L, and so on until the desired end dilution or concentration series is achieved.

Microscopic methods

At the end of this chapter we will briefly discuss the preparation of samples for microscopy and the different types of microscopy as this is not covered elsewhere but is commonly used by molecular exercise physiologists. Microscopy allows molecular exercise physiologists, for example, to visualize the localization and expression of individual molecules and to measure changes in cells for example before and after exercise or in response to the knockout of genes that are involved in mediating adaptation to exercise.

Section cutting

Samples for microscopy are, as we discussed earlier, commonly frozen in isopentane cooled to near freezing with liquid nitrogen in order to avoid freezing artefacts. Such samples are then cut, either frozen with a cryostat(-microtome) or embedded in paraffin (wax) and cut with a microtome. Both methods allow sections to be cut at typically 5–10 μm which is a hundredth of a millimetre or less. The sections then adhere to a glass slide. Sections are then usually fixed with chemicals such as **paraformaldehyde** to preserve the tissue and crosslink proteins. In some cases it is possible to perform a stain without fixing samples first.

Stains

There are many methods used to stain samples. The simplest methods are **tinctorial (dye-based) stains**, the most common stain being the **Haematoxylin and Eosin** (commonly abbreviated as H&E) stain, where haematoxylin stains the nucleus in purple and eosin stains proteins in pink. **Histochemical stains** are used by exercise physiologists to distinguish type I/slow/highly oxidative from type II/intermediate, fast/less oxidative muscle fibres. The general principle is that enzymes within the tissue convert a substrate into a dye and the dye intensity after an incubation period indicates the enzyme activity in that location. Histochemical stains such as the ATPase stain with acid or alkaline preincubation and the NADH-TR for mitochondrial enzyme activity are commonly used to distinguish different types of muscle fibre.

Specific molecules can be visualized using **immunohistochemistry** (antibody-based stains for tissues) or **immunocytochemistry** (antibody-based stains of cells). For these stains a **primary antibody** binds to the protein of interest (antigen) and a **secondary antibody** conjugated to a visualization protein binds to and visualizes the location of the primary antibody and hence the protein of interest. In general, there are two types of antibodies:

- **Monoclonal antibodies** are all identical antibodies made by clones of one founder cell. In practice they are made from myeloma (cancer) cells that are fused with spleen cells that have been exposed to the antigen against which antibodies are generated.
- **Polyclonal antibodies** are generated by adding an antigen such as a protein or protein fragment typically into mice, rabbits or goats. The immune system of the host then generates several types of antibodies against the antigen. The antibodies are then purified from the serum of the host.

The quality of the antibody depends on its **affinity**, which is the strength of binding, and the **specificity**, which is whether the antibody only binds to the antigen or also to other molecules. The **epitope** is the part of the antigen that is recognized by the antibody.

After incubation with the primary antibody, the primary antibody solution is washed off and the slide is incubated with the **secondary antibody**, which is raised against antibodies of the species in which the primary antibody is raised. Secondary antibodies are conjugated with either a fluorophore (fluorescent dye) or a dye-generating enzyme such as horseradish peroxidase. Fluorescent secondary antibodies are detected using a fluorescence microscope. Secondary antibodies that are conjugated with a dye-generating enzyme are incubated with a substrate, commonly DAB (3,3′-diaminobenzidine tetrahydrochloride). Horseradish peroxidase then converts DAB to a brown dye which can be seen under a light microscope.

Microscopes

After the stains, the samples are then visualized with a microscope; there are various different types (Figure 1.3). **Light microscopes** are the workhorses of microscopy laboratories. Typical research light microscopes have lenses with magnifications ranging typically from ×5 to ×100. Higher magnification (×40–×100) lenses are frequently **oil immersion** lenses where a drop of immersion oil needs to be placed on the coverslip. This forms an oil bridge between the coverslip and lens, which reduces refraction and improves image quality. For unstained samples, such as living, cultured cells, filters are used to induce light interference which generates contrast. **Phase-contrast** and **Nomarski interference contrast** are two of the methods used to generate contrast in unstained cells. **Inverted (light) microscopes** are used to visualize, for example, living cells on the bottom of culture flasks. Inverted microscopes have the lenses located below the sample while the lighting is above.

Figure 1.3 Microscopy laboratory equipment. (a) A cryostat(-microtome) which is used to cut frozen sections. (1) The frozen sections are 'glued' to a metal block using a viscous solution that freezes quickly. The sample and block are then inserted into the block holder (2). A rotating wheel (3) moves the block holder, block and sample up and down as well as forward by several micrometres during each rotation. During the downward movement a small section is cut off by a knife. This adheres to a glass slide. (b) Inverted light microscope used to visualize cultured cells in plastic dishes. The lenses are located below the sample. (c) Fluorescence and light microscope. (d) Transmission electron microscope.

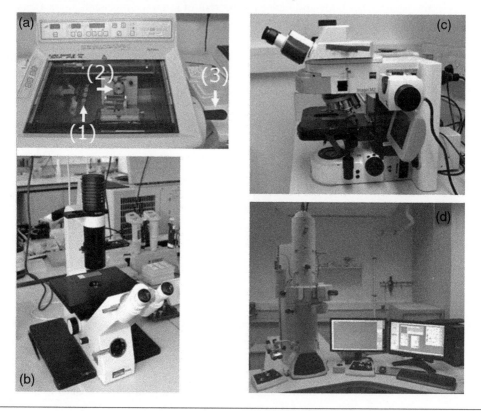

Fluorescence microscopy is a variation of light microscopy. It is used to visualize molecules that are made fluorescent by fluorescent dyes or antibodies. Fluorescence refers to the phenomenon where electromagnetic radiation of one wavelength (usually light of a specific colour) sent into a sample (excitation) causes the sample to emit light at a different wavelength. In order to achieve this, fluorescence microscopes require expensive, additional components, including a suitable light source, an **excitation filter** to select the colour/wavelength of the light going into the sample and an **emission filter** to select the colour/wavelength of the light that makes it from the sample to the eye and/or camera. It is crucial that the filters on a fluorescence microscope match the fluorophores (i.e. fluorescent dyes) used or else the fluorescent stain cannot be detected.

Confocal microscopy is a variation of light and fluorescence microscopy. The drawback of light and fluorescence microscopy is that samples are commonly thicker than the focal plane (i.e. the plane or slice where the sample is perfectly focused). Confocal microscopes use a pinhole to eliminate light that is out of focus and allow very sharp images and a reconstruction of a spatial 3D image.

The resolution limit of light microscopy is ≈0.2 μm because of the long wavelength of visible light. If an electron beam with a much lower wavelength is used instead of visible light then a much higher resolution can be achieved. There are two types of **electron microscopy**:

- **Transmission electron microscopy (TEM)**. Here, a beam of electron goes through a sample and the emitted electrons are captured with a camera. It used to be the standard method to demonstrate satellite cells (Mauro, 1961) and was used in several studies to analyse mitochondria and sarcomeres. Because the images are relatively difficult to quantify, more molecular methods such as immunohisto-chemistry are preferred today for many applications.
- **Scanning electron microscopy**. This type of electron microscopy is used to scan the surface of an object.

SUMMARY

We define molecular exercise physiology as the study of exercise physiology using molecular biology methods. It is a sub-discipline of exercise physiology and can be seen as a complement or extension of classical exercise physiology rather than a replacement. The two major pioneers in this field were Frank W Booth, who intro-duced many gene expression and transgenic methods to exercise physiology, and Claude Bouchard, who started relating an individual's DNA sequence to sport and exercise-related traits.

In molecular exercise physiology, human and animal samples as well as cultured primary cells and cell cultures are used. All samples have advantages and disadvantages in relation to relevance to humans, genetic variability, possibilities for exercise, pharma-cological and transgenic manipulation, ethical concerns and tissues from which samples can be taken. Thus comprehensive research projects are usually carried out using a mix of samples. Key wet laboratory skills are measuring volumes, weights, concentrations and the production of solutions, dilutions and the adjustment of the pH. Microscopy is frequently used in molecular exercise physiology to visualize tissues, cells and molecules which can be stained with dye, during histochemical reactions and, most specifically, with antibodies. Images are then obtained with light microscopes, fluorescence micro-scopes or electron microscopes.

REVIEW QUESTIONS

1   What is the difference between molecular exercise physiology and exercise biochemistry?
2   State three major exercise physiology research questions that can only be answered by using molecular biology methods.
3   How many grams do you need to prepare (a) 100 mL of a 5 mol/L NaCl (MW = 58 g/mol); (b) 0.5 mL of a 20 mmol/L glucose (MW = 180 g/mol); (c) 100 mL of a 7.5% w/v acrylamide solution?

4　You plan to dilute a 10 mmol/L AICAR stock solution to a 5 µmol/mL solution in a 3 mL well. What volume of the 10 mmol/L AICAR stock solution do you need to pipette into the 3 mL well?

5　Compare and contrast the staining of different skeletal muscle fibres with a histochemical method and an immunohistochemical stain.

FURTHER READING

Barker K (2005). *At the bench: a laboratory navigator,* 2nd edn. Cold Spring Harbor Laboratory Press.

REFERENCES

Baar K and Esser K (1999). Phosphorylation of p70(S6k) correlates with increased skeletal muscle mass following resistance exercise. *Am J Physiol* 276, C120–C127.

Baldwin KM (2000). Research in the exercise sciences: where do we go from here? *J Appl Physiol* 88, 332–336.

Barker K (2005). *At the bench: a laboratory navigator,* 2nd edn. Cold Spring Harbor Laboratory Press.

Booth FW (1988). Perspectives on molecular and cellular exercise physiology. *J Appl Physiol* 65, 1461–1471.

Bouchard C, Chagnon M, Thibault MC, Boulay MR, Marcotte M, Cote C and Simoneau JA (1989). Muscle genetic variants and relationship with performance and trainability. *Med Sci Sports Exerc* 21, 71–77.

Bouchard C, Malina RM and Perusse L (1997). *Genetics of fitness and physical performance.* Human Kinetics, Champaign, IL.

Bouchard C, Sarzynski MA, Rice TK, Kraus WE, Church TS, Sung YJ, et al. (2011). Genomic predictors of the maximal O uptake response to standardized exercise training programs. *J Appl Physiol* 110, 1160–1170.

Campbell WG, Gordon SE, Carlson CJ, Pattison JS, Hamilton MT and Booth FW (2001). Differential global gene expression in red and white skeletal muscle. *Am J Physiol Cell Physiol* 280, C763–C768.

Chin ER, Olson EN, Richardson JA, Yang Q, Humphries C, Shelton JM, et al. (1998). A calcineurin-dependent transcriptional pathway controls skeletal muscle fiber type. *Genes Dev* 12, 2499–2509.

Collins CA and Zammit PS (2009). Isolation and grafting of single muscle fibres. *Methods Mol Biol* 482, 319–330.

de la Chapelle A, Traskelin AL and Juvonen E (1993). Truncated erythropoietin receptor causes dominantly inherited benign human erythrocytosis. *Proc Natl Acad Sci U S A* 90, 4495–4499.

Deriaz O, Dionne F, Perusse L, Tremblay A, Vohl MC, Cote G and Bouchard C (1994). DNA variation in the genes of the Na,K-adenosine triphosphatase and its relation with resting metabolic rate, respiratory quotient, and body fat. *J Clin Invest* 93, 838–843.

Dionne FT, Turcotte L, Thibault MC, Boulay MR, Skinner JS and Bouchard C (1991). Mitochondrial DNA sequence polymorphism, $\dot{V}O_{2max}$, and response to endurance training. *Med Sci Sports Exerc* 23, 177–185.

Garry DJ, Ordway GA, Lorenz JN, Radford NB, Chin ER, Grange RW, et al. (1998). Mice without myoglobin. *Nature* 395, 905–908.

Gibson DG, Benders GA, Andrews-Pfannkoch C, Denisova EA, Baden-Tillson H, Zaveri J, et al. (2008). Complete chemical synthesis, assembly, and cloning of a *Mycoplasma genitalium* genome. *Science* 319, 1215–1220.

Hamilton MT and Booth FW (2000). Skeletal muscle adaptation to exercise: a century of progress. *J Appl Physiol* 88, 327–331.

Hayot M, Michaud A, Koechlin C, Caron MA, Leblanc P, Prefaut C, and Maltais F (2005). Skeletal muscle microbiopsy: a validation study of a minimally invasive technique. *Eur Respir J* 25, 431–440.

Hood DA, Balaban A, Connor MK, Craig EE, Nishio ML, Rezvani M and Takahashi M (1994). Mitochondrial biogenesis in striated muscle. *Can J Appl Physiol* 19, 12–48.

Lin J, Wu H, Tarr PT, Zhang CY, Wu Z, Boss O, et al. (2002). Transcriptional co-activator PGC-1 alpha drives the formation of slow-twitch muscle fibres. *Nature* 418, 797–801.

Mauro A (1961). Satellite cell of skeletal muscle fibers. *J Biophys Biochem Cytol* 9, 493–495.

McPherron AC, Lawler AM and Lee SJ (1997). Regulation of skeletal muscle mass in mice by a new TGF-beta superfamily member. *Nature* 387, 83–90.

Morange M (2009). History of molecular biology. In *Encyclopedia of life sciences.* John Wiley & Sons, Chichester.

Murgia M, Serrano AL, Calabria E, Pallafacchina G, Lomo T and Schiaffino S (2000). Ras is involved in nerve-activity-dependent regulation of muscle genes. *Nat Cell Biol* 2, 142–147.

Pallafacchina G, Calabria E, Serrano AL, Kalhovde JM and Schiaffino S (2002). A protein kinase B-dependent and rapamycin-sensitive pathway controls skeletal muscle growth but not fiber type specification. *Proc Natl Acad Sci U S A* 99, 9213–9218.

Puigserver P, Wu Z, Park CW, Graves R, Wright M and Spiegelman BM (1998). A cold-inducible coactivator of nuclear receptors linked to adaptive thermogenesis. *Cell* 92, 829–839.

Scarpulla RC (1997). Nuclear control of respiratory chain expression in mammalian cells. *J Bioenerg Biomembr* 29, 109–119.

Schuelke M, Wagner KR, Stolz LE, Hubner C, Riebel T, Komen W, et al. (2004). Myostatin mutation associated with gross muscle hypertrophy in a child. *N Engl J Med* 350, 2682–2688.

Spurway NC and Wackerhage H (2006). *Genetics and molecular biology of muscle adaptation.* Elsevier/Churchill Livingstone, Edinburgh.

Thomason DB and Booth FW (1990). Stable incorporation of a bacterial gene into adult rat skeletal muscle in vivo. *Am J Physiol* 258, C578–C581.

Timmons JA, Knudsen S, Rankinen T, Koch LG, Sarzynski M, Jensen T, et al. (2010). Using molecular classification to predict gains in maximal aerobic capacity following endurance exercise training in humans. *J Appl Physiol* 108, 1487–1496.

Wackerhage H, Miah A, Harris RC, Montgomery HE and Williams AG (2009). Genetic research and testing in sport and exercise science: A review of the issues. *J Sports Sci* 1–8.

Watson PA, Stein JP and Booth FW (1984). Changes in actin synthesis and alpha-actin-mRNA content in rat muscle during immobilization. *Am J Physiol* 247, C39–C44.

Wheeler DA, Srinivasan M, Egholm M, Shen Y, Chen L, McGuire A, et al. (2008). The complete genome of an individual by massively parallel DNA sequencing. *Nature* 452, 872–876.

Williams AG, Rayson MP, Jubb M, World M, Woods DR, Hayward M, et al. (2000). The ACE gene and muscle performance. *Nature* 403, 614.

Winder WW and Hardie DG (1996). Inactivation of acetyl-CoA carboxylase and activation of AMP-activated protein kinase in muscle during exercise. *Am J Physiol* 270, E299–E304.

Wolff JA, Malone RW, Williams P, Chong W, Acsadi G, Jani A and Felgner PL (1990). Direct gene transfer into mouse muscle in vivo. *Science* 247, 1465–1468.

Yamaguchi M, Nakayama Y and Nishikawa J (1985). Studies on exercise and an elastic protein "connectin" in hindlimb muscle of growing rat. *Jpn J Physiol* 35, 21–32.

Zammit PS, Golding JP, Nagata Y, Hudon V, Partridge TA and Beauchamp JR (2004). Muscle satellite cells adopt divergent fates: a mechanism for self-renewal? *J Cell Biol* 166, 347–357.

2 Genetics, sport and exercise: background and methods

Stephen M Roth and
Henning Wackerhage

INTRODUCTION

In this chapter we will cover **sport and exercise genetics** or kinesiogenomics, which can be defined as the application of genetics to sport and exercise-related traits. Sporting performance depends on nature and nurture or, to use more specific terms, on:

- **genetics** (i.e. inherited and newly occurring variations in the DNA sequence, also referred to as innate talent) and
- **environmental factors** (i.e. factors such as training and nutrition).

The degree to which genetics and environmental factors affect sporting performance or other sport and exercise-related traits varies depending on the trait. For example, the contribution of genetics (the **heritability estimate**) to $\dot{V}O_{2max}$ is $\approx 50\%$ (Bouchard et al., 1999). This means that $\approx 50\%$ of the variation in $\dot{V}O_{2max}$ values can be explained by **genetic or DNA sequence variations** and the other $\approx 50\%$ is due to **environmental factors** such as endurance training or diet. Thus in order to achieve the high $\dot{V}O_{2max}$ values seen in some Olympic endurance athletes it is necessary for an athlete to both carry

DNA sequence variations that code for a high basal $\dot{V}O_{2max}$ and $\dot{V}O_{2max}$ trainability and to train hard. No one has phrased this better than Per-Olof Åstrand, who is reported to have said 'the most important thing an aspiring athlete can do is to choose the right parents'.

Genetics is also important at the intersection of exercise and disease. For example, **sudden death** during sport can frequently be attributed to detrimental mutations or DNA sequence variations in cardiac genes. Such mutations can today be detected using methods such as the polymerase chain reaction (PCR) followed by DNA sequencing and, increasingly, the sequencing of whole genomes. Should the popular, cheap and easy-to-use but not very reliable Physical Activity Readiness Questionnaire (PAR-Q) (Thomas et al., 1992) health pre-exercise screening questionnaire thus be supplemented or even replaced by much more specific genetic tests? Genetic variation also determines the extent by which we can improve our blood glucose or blood pressure with training, This varies a lot and some individuals may even worsen risk factors with exercise training (Bouchard et al., 2012). Should we use genetic tests to determine the likely response to a training programme prior to embarking on such a programme (Roth, 2008)? This is a **personalized medicine** strategy applied to sport and exercise and there is no easy answer to this question. Finally, there are dangers from misusing the emerging genetic knowledge. For example, genetic methods could be used for **gene doping**, and genetic testing could potentially be used, for example, to select embryos that are likely to have special exercise-related traits for implantation into the uterus.

The introduction above highlights the many links between genetics, sport and exercise. In this chapter we will introduce the subject by first giving an introduction to classical and molecular genetics. We will cover twin and family studies that are used to determine the heritability estimates for sport and exercise-related traits. After that we discuss DNA, the information that it contains and how the information encoded in DNA is first transcribed into mRNA and then translated into protein. Mutations of DNA are the source of all genetic variation and we explain how mutations occur and review the different types of mutations. We then explain how to analyse a DNA sequence and describe PCR, which is an important method that allows researchers to amplify a small stretch of DNA. After PCR, either agarose gel electrophoresis or DNA sequencing is used to determine the exact DNA sequence. Common DNA sequence variations can also be detected on a larger scale using 'gene chips' in studies that are termed genome-wide association studies (GWAS). More recently, next-generation sequencing methods have opened up the possibility to sequence whole genomes of individuals, allowing the identification of common and rare DNA sequence variations. Does all of this mean that we open 'Pandora's box' or does our increased understanding of genetics allow us to solve some real problems? We attempt to partially answer this question by discussing the ethics of genetic research in sport and exercise science and the applications that result from it.

CLASSICAL SPORT AND EXERCISE GENETICS

Genetics is defined as the science of **heredity**. Classical geneticists such as Gregor Johann Mendel studied easy-to-measure traits such as eye colour to establish the basic laws of heredity. For example, classical geneticists might determine the eye colour in parents and then measure the frequencies of eye colour in their offspring and formulate laws on the basis of these data. This is descriptive research because it is only the relationships that are described and the mechanisms are not uncovered. In contrast, molecular geneticists aim to uncover how variations in DNA sequences directly cause variations in traits.

Two different kinds of traits can be distinguished:

- **Mendelian** or **monogenic** and **oligogenic** traits are distinguished by clear differences between the variants (e.g. blond, brown, black or red hair) and they depend to a great extent on DNA sequence variations in one (monogenic) or only a few (oligogenic) genes.
- **Polygenic**, **multifactorial** or **continuous traits** show a continuous, often Gaussian variation in the population. Examples are height, weight, strength, muscle mass, aerobic capacity and heart size. As the name suggests, variations in such traits depend usually on the cumulative effect of tens, hundreds or even thousands of DNA sequence variations that affect the expression and/or function of many genes. It is currently hotly debated whether the contributing DNA sequence variations are mostly common or whether rare variations play a decisive role.

Most sport and exercise-related traits, such as endurance running performance, muscle mass or football performance, are polygenic, as these generally follow a continuous Gaussian distribution, rather than just a poor athlete versus elite athlete distribution in the population. This is intuitive because traits such as endurance running performance can be broken down into anthropometric factors, such as muscle fibre distribution, metabolism and cardiovascular capacity. All of these sub-traits also have generally a continuous distribution in the population and many are significantly inherited. Thus it seems likely that performance in many sports depends significantly on hundreds if not thousands of DNA sequence variations and environmental factors, which all add up. As a consequence, the search for the one elite athlete DNA sequence variation with a huge effect size is, for most sports, futile. There are exceptions to this, however; it seems likely that the body height of very tall NBA basketball players is primarily caused by rare DNA sequence variations with a large effect size in the growth hormone system.

One useful strategy is to try to estimate how much of the variability of sport and exercise-related traits such as $\dot{V}O_{2max}$, strength or motor learning depends on genetic or DNA sequence variation and how much depends on environmental variation. The total variability (V_t) of traits depends on three sources:

- V_g or genetic variability (heritability or variability due to DNA sequence variations);
- V_e or environmental variability (training, nutrition);
- e or measurement error (instrument accuracy, precision, etc.), of which genetic variability and environmental variability are the most important. Total variability can thus be expressed as:

$$V_t = V_g + V_e + e.$$

The next question is: How can we estimate **heritability** or in other words the percentage of trait variability that is due to genetic variability (V_g)? Led by Claude Bouchard in the US and Vassilis Klissouras in Europe, researchers have mainly used **twin** but also **family**

studies in order to estimate the heritability of exercise-related traits. This can be done by studying **monozygotic** or 'identical' twins who share almost all of their DNA sequence (Bruder et al., 2008) and **dizygotic** or fraternal twins who share ≈50% of their DNA sequence variations. This implies that all the variation of a trait between monozygous twin pairs is only due to environmental variability (V_e) and error (e), while the variation between dizygotic twins is additionally due to genetic variability (V_g). Variability can be measured as the correlation coefficient r, which can range from 0 (no correlation) to 1 (the values for each twin pair are all identical) and this can be inserted into Falconer's formula (Falconer and MacKay, 2011) to estimate heritability:

$$h^2 = 2\left(r_{MZ} - r_{DZ}\right)$$

As an example, imagine you plan to estimate the heritability of grip strength in a twin study. For this you will need to:

1   recruit pairs of monozygous twins, measure grip strength and calculate the correlation coefficient r_{MZ} and
2   recruit pairs of dizygous twins, measure grip strength and calculate the correlation coefficient r_{DZ}.

If grip strength has a genetic contribution or is heritable then the grip strength of monozygous twin pairs should be better correlated (i.e. r_{MZ} should be larger) than that of dizygous twin pairs (r_{DZ}) because monozygous twins should have virtually no differences in genetic variability due to their identical DNA sequence. For example, assume you have calculated $r_{MZ} = 0.8$ and $r_{DZ} = 0.6$. Thus 2 × (0.8 – 0.6) is 0.4. This means that the heritability, denoted as h^2, for grip strength in this study is estimated to be 0.4 or 40%. In other words, in your study 40% of grip strength is estimated to be due to genetic variability (V_g) while 60% is estimated to be due to environmental variability (V_e) and measurement error (e). An example of data from a real study is shown in Figure 2.1.

Twin studies have been used to estimate the heritability of many sport and exercise-related traits, either with Falconer's formula or other models. We will mention and discuss heritability estimates for endurance and muscle mass and strength-related traits in Chapters 5 and 7.

Twin studies are difficult to perform, however, because twinning rates in most regions are below 20 twins per 1000 live births (Smits and Monden, 2011) and thus it is difficult to recruit sufficiently large cohorts of twins to achieve adequate statistical power for a study. However, because twins are very useful for genetic research, twin registries have been established in several countries, such as the TwinsUK register in London, which currently has ≈12 000 twins registered (www.twinsuk.ac.uk).

Family members other than twins are also genetically related, of course, as is evident from the aforementioned statement by Per-Olof Åstrand that parents determine significantly the sports and exercise-related traits of their offspring. Family studies allow researchers to recruit greater number of subjects but the genetic relationship between family members and thus the formulas to work out heritabilities are more complicated than in twins. The most famous and productive genetic study in sport and exercise genetics is the HERITAGE Family Study (HEalth, RIsk factors, exercise Training And

Figure 2.1 Twin study results. Example of a twin study plot examining heritability of change in strength in response to strength training. Data from monozygous (MZ) twins are shown in black squares while data from dizygous (DZ) twins are shown in white squares. The correlation for MZ twins (r_{MZ}) was 0.49 while the correlation from dizygous twins (r_{DZ}) was 0.22, resulting in an estimated heritability of 0.54 or 54%. Redrawn after Thomis et al. (1998); Smits and Monden (2011).

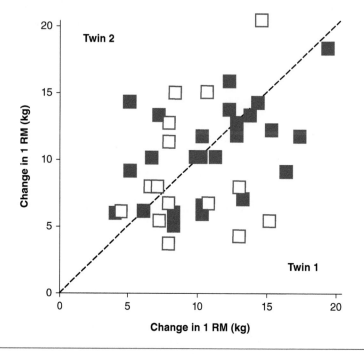

Genetics) which is led by Claude Bouchard (Bouchard et al., 1995), introduced as a pioneer of sport and exercise genetics in Chapter 1. This study has been a unique source of sport and exercise-related genetic information and still continues to be used to generate major output. For this study, parents and adult offspring in 90 Caucasian and 40 African-American families were recruited and endurance exercise trained for 20 weeks. Before and after exercise training many variables were measured, including $\dot{V}O_{2max}$, blood pressure, insulin sensitivity and plasma lipids, and body composition and DNA samples of individuals were taken. This allowed researchers not only to estimate the heritability for measured variables such as $\dot{V}O_{2max}$ and the trainability of $\dot{V}O_{2max}$, but also to subsequently conduct research into the DNA sequence variations that were causing the heritability of the measured sport and exercise-related traits as well as health-related traits. We will cover several key findings of the HERITAGE study in Chapter 5 when we discuss the genetics of endurance performance.

MOLECULAR SPORT AND EXERCISE GENETICS

Articles about genetics are often vague, using terms such as 'genetic variation', 'genetic variability' or 'genetic factors'. So what is behind these terms? In brief, terms such as 'genetic factors' are abstract terms for saying that there are differences in the DNA sequence between humans. For this reason we will mainly use the term 'DNA sequence variations' in this book as it is in our view a more intuitive term. In this section we will first review and extend information on DNA and the human genome and then describe how mutations occur before discussing different types of mutations.

DNA (deoxyribonucleic acid) comprises a coding base, either a purine or pyrimidine, plus a backbone made from a sugar called ribose and a phosphate. There are four different bases or nucleotides in DNA:

- G guanine;
- C cytosine;
- A adenine; and
- T thymine.

In a DNA double-strand helix a G on one strand pairs or hybridizes with a C on the other strand via three hydrogen bonds, while A pairs with T via two hydrogen bonds. Because of that extra hydrogen bond, GC pairs stick together better than AT pairs.

In most cases only the DNA sequence of one strand is given because the other strand has the complementary sequence and thus can be directly inferred from the sequence for the other strand. DNA sequences are written using the single-letter abbreviation for each base (e.g. TAAGCGTAGA). The DNA sequence of a single strand is read and written from the so-called 5′ (five prime, left) end to the 3′ (three prime, right) end of the molecule.

DNA can be present in tightly packed form as **chromosomes** during cell division, but most of the time the DNA is partially unravelled in the nucleus of cells. The DNA in the **human genome** (Ensembl assembly GRCh37.p3, Feb 2009) contains:

- **3 280 481 986 DNA base pairs** (≈3.3 Gb or billion bases for a haploid genome, where only the bases of one chromosome of a chromosome pair are counted). However, the full, diploid genome with all 46 chromosomes contains **≈6.6 Gb**;
- 20 599 known protein-coding genes;
- 895 novel protein-coding genes;
- 8563 RNA genes (i.e. RNAs that are not translated into proteins; for example ribosomes are made from ribosomal RNAs rather than proteins or small regulatory RNAs).

In the human genome less than 2% of the sequence contains genes that encode proteins. Initially it was thought that the remaining DNA had no function; it was referred to as 'junk DNA'. However, a key result of the so-called ENCODE project, which aimed to study the function of DNA, was that we now know that ≈80% of the DNA sequence has a regulatory or other known function (Bernstein et al., 2012). We will cover the functions of regulatory DNA in greater detail in Chapter 3, as they are important for mediating the adaptation to exercise which is the topic of that chapter.

We have used the term 'gene' already several times, but what actually is a gene? Genes are stretches of DNA that encode mostly proteins, the major building blocks of cells, but also RNAs. Cells continually read the information contained in genes and build proteins using a process termed gene expression. For a gene to be expressed as a protein, the DNA of the gene needs to be **transcribed** into **RNA**, which has a slightly different chemical structure from DNA as it is single stranded and comprises C (cytosine), G (guanine), A (adenine) with U (uracil) instead of T (thymine). Genes are read and transcribed into RNA by the enzyme **RNA polymerase** (types I, II, III exist; **RNA polymerase II** is the main one to copy typical genes), which is first directed to specific genes by regulatory mechanisms. RNA polymerase then scans along the chosen gene, reads the DNA and synthesizes an **RNA** copy of the DNA of the gene.

The newly synthesized RNA is then **spliced** by **spliceosomes**, which are enzyme complexes that can cut and join RNA, or by self-splicing. RNA splicing is necessary because genes comprise:

- **exons** – stretches of a gene or newly transcribed RNA where three bases encode one amino acid in the expressed protein and
- **introns** (also known as intervening sequences) – stretches of a gene or newly transcribed RNA that do not encode amino acids.

During splicing, the introns of an RNA molecule are cut out and the exons are joined together. This turns RNA into **mRNA** (messenger RNA) where all-base triplets encode amino acids in the final protein. RNA can also be degraded by various RNA-degrading enzymes termed **RNases**. Thus the concentration of each mRNA at any one time depends on the rates of its transcription, splicing and degradation.

During **translation**, mRNA is translated into a protein. Translation occurs in the cytosol at the **ribosome**. Ribosomes are organelles that bind to mRNA, read the mRNA sequence and add, for each three-base combination, one amino acid to the nascent protein. For this process, amino acids are bound to **tRNAs** (transfer RNAs). This works because the tRNAs pair up with the triplet in the mRNA and the ribosome then attaches the amino acid on the tRNA to the nascent amino acid chain.

Examples of amino acids that are encoded by certain DNA triplets are given in Table 2.1.

The DNA of the human genome is subdivided into **46** chromosomes. Chromosomes only form when cells divide, in order to have a highly packaged form of DNA that can easily be transported to daughter cells (as shown in Figure 2.2). Most of the time DNA is unpackaged and dispersed. Humans have the following chromosomes:

- 44 **autosomes**, numbered 1–22 (these are the 22 pairs of non-sex chromosomes);
- 2 **sex chromosomes, X and Y**, carried as XX in women and XY in men.

For each chromosome or autosome pair, one chromosome originates from the mother and the second from the father. This is an oversimplified description because your mother's and father's chromosomes recombine during meiosis and exchange identical parts in a process termed **homologous recombination**.

Having two copies of each chromosome implies that we also carry two copies of each gene or, using a more technical term, we are **diploid**. The extra gene can be compared to a spare tyre: if a crucial gene on one chromosome is defective due to a loss-of-function mutation, then in many cases there is little effect if the other copy of the gene is intact.

Table 2.1 Examples of amino acids, their one- and three-letter abbreviations and the DNA triplets or codons that encode them

AMINO ACID	ABBREVIATION*	DNA CODONS
Isoleucine	I, Ile	ATT, ATC, ATA
Leucine	L, Leu	CTT, CTC, CTA, CTG, TTA, TTG
Valine	V, Val	GTT, GTC, GTA, GTG
Phenylalanine	F, Phe	TTT, TTC
Methionine	M, Met	ATG
Cysteine	C, Cys	TGT, TGC
Alanine	A, Ala	GCT, GCC, GCA, GCG
Glycine	G, Gly	GGT, GGC, GGA, GGG
Proline	P, Pro	CCT, CCC, CCA, CCG
Threonine	T, Thr	ACT, ACC, ACA, ACG
Serine	S, Ser	TCT, TCC, TCA, TCG, AGT, AGC
Tyrosine	Y, Tyr	TAT, TAC
Tryptophan	W, Trp	TGG
Glutamine	Q, Gln	CAA, CAG
Asparagine	N, Asp	AAT, AAC
Histidine	H, His	CAT, CAC
Glutamic acid	E, Glu	GAA, GAG
Aspartic acid	D, Asp	GAT, GAC
Lysine	K, Lys	AAA, AAG
Arginine	R, Arg	CGT, CGC, CGA, CGG, AGA, AGG

*The one-letter and three-letter amino acid abbreviation is shown.

The term 'allele' is used to denote one form or variant of a gene. Alleles can be abbreviated in different ways. For example for the *ACE* gene the insertion '*I*' allele denotes the insertion of a 287 bp (base pairs) long stretch of DNA while '*D*' marks a deletion allele where these 287 bp are missing. Since each human carries two copies of each gene, the combinations possible for any person are *II*, *ID* or *DD*. These allele combinations are known as the **genotype** of an individual for this genetic variation. Carrying two identical alleles is termed **homozygous** (e.g. *II* or *DD* genotype) while carrying different alleles is termed **heterozygous** (e.g. *ID* genotype).

After discussing DNA and the human genome we now aim to explore the origins of DNA sequence variation between individuals. How do such DNA sequence variations originally occur? The source of genetic variation is DNA **mutations**. Such mutations occur especially during cell division, when the DNA sequence of all chromosomes has to be duplicated so that each daughter set has a full set of chromosomes. The process of doubling up the DNA prior to a cell division is termed **replication**. During replication **DNA polymerases** read and replicate each DNA strand with high fidelity. Nonetheless, in ≈1 per 100 000 bases an error occurs (Pray, 2008). These errors are detected and corrected by proteins in a process termed **proofreading** and repaired by mismatch repair proteins (Pray, 2008). We will not go into detail here, but essentially cells have

Figure 2.2 Set of male chromosomes. Chromosomes represent the highest degree of DNA packaging which only occurs during the metaphase of mitosis/cell division. In other cells the DNA is partially opened up. The image was kindly provided by Gordon Hyslop, Genetics Laboratory Services, Ninewells Hospital, Dundee.

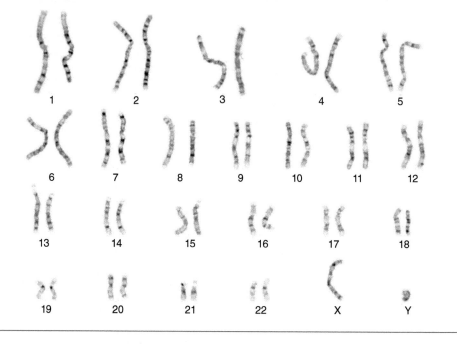

mechanisms to eliminate almost all mutations that occur during replication. In addition, mechanisms such as 'cell cycle checkpoints' prevent the cells with mutated DNA or incomplete chromosome sets from dividing. These checks and balances keep the number of new or *de novo* mutations very low. However, because ≈6.6 billion base pairs need to be replicated during each cell division in millions of dividing cells, mutations and the resultant variations in the DNA sequence are inevitable. The rate of mutation is increased by physical and chemical agents termed **mutagens**, which include radiation, UV light and chemicals such as ethidium bromide.

Many mutations are 'neutral' in that they have no or only a negligible impact, positive or negative, on the fitness and appearance of an organism. Also many mutations occur in the cells of the body but only mutations that occur in germ cells will be passed on to the offspring. Due to positive selection, beneficial mutations or neutral mutations close to beneficial mutations are passed on to future generations and become enriched in a population as carriers will have an advantage. When a mutation becomes common, normally defined as being found in at least 1% of a population, it is defined as a **common allele** or **polymorphism**.

What types of mutations or DNA sequence variations exist? DNA sequence variations can range from a single base substitution, termed **SNP** (pronounced 'snip'; single-nucleotide polymorphism), to gross chromosomal rearrangements, where bits are added or taken from whole chromosomes. The common types of DNA sequence variations are summarized in Table 2.2.

Table 2.2 Simplified overview of the types of DNA sequence variations with examples		
TYPE OF MUTATION	NORMAL ALLELE	MUTATED ALLELE[b]
Substitution (single-nucleotide polymorphism[a]; SNP)	TAG TCA	TAC TCA
Insertion/deletion (also known as indel)	TAG TCA	TAG **CAG CAC** TCA or TAG
Inversion	TAG TCA	TAG **ACT**
Frameshift[c]	TAG TCA	TAG **GTC** A
Copy-number variation	Additions, deletions, or inversions of DNA sequence, often containing entire genes (e.g. number of gene copies differs among individuals)	
Gross chromosomal rearrangements	Additions or deletions of whole chromosome parts	

[a]Polymorphisms are DNA sequence variations found in at least 1% of the population due to repeated inheritance.
[b]Mutated bases are shown in bold.
[c]Three bases or one codon in an exon encode one amino acid in the coded protein. If one or two extra bases are added then a frameshift occurs from this point, which means that all codons and thus the amino acid sequence in the coded protein will change completely after the mutation.

There is a specific terminology to describe mutations or DNA sequence variation (Strachan and Read, 2004). Such DNA sequence variation can be stated as the variation in the DNA sequence or, if it occurs in a gene, as a resultant amino acid sequence variation in the resultant protein. To describe such mutations, a 'g' is frequently used to indicate genomic DNA and a 'p' is used to indicate a protein. The A of the first ATG codon (all genes start with this codon) is +1 and the preceding base is –1 (there is no 0). Thus g.856G→A indicates that the guanine base in position 856 is replaced by an adenine. To give an example for an amino acid sequence variation in a protein, p.M122C indicates that the amino acid methionine in position 122 of the protein is replaced by a cysteine. Be aware that deletions and mutations in introns have special codes which we do not discuss here.

The most abundant sequence variations are SNPs and copy-number variations. In the dbSNP database there are currently ≈38 million validated SNPs that have been unearthed in human genomes (http://www.ncbi.nlm.nih.gov/projects/SNP/). Also it has been estimated that a given individual carries 4 million DNA sequence variations, of which most are SNPs (Bouchard, 2011). Most SNPs are **neutral** as they often lie in regions that have little function. In contrast, functional SNPs in promoters and regulatory regions can either affect gene expression or change the amino acid sequence of a protein or, in an extreme case, replace an amino acid with a premature stop codon so that the resultant protein is incomplete and maybe non-functional. This is, for example, the case in the *ACTN3* p.R577X genotype (X stands for a premature stop codon), which affects speed and power performance (Yang et al., 2003).

The presence and effect of especially large copy-number variations such as insertions, deletions and inversions of large chunks of DNA, often containing genes, has long been underestimated. By comparing the genomes of eight humans, researchers revealed that large copy-number variations are more frequent than was previously thought and may be an important source of genetic variability (Kidd et al., 2008). To summarize, mutations occur mainly during DNA replication and give rise to DNA sequence variations. Such sequence variations include SNPs, insertions, deletions, frameshift and copy-number mutations, and gross chromosomal rearrangements. Many sequence variations are silent, which means that they do not affect the fitness or phenotype of an individual.

MUTATIONS, EVOLUTION AND COMMON VERSUS RARE ALLELES

Mutations can occur in any cell of the body. However, only mutations and the resultant alleles or DNA sequence variations that are present in germ cells will be passed on to the next generation. Evolution will lead to the selection of alleles that are advantageous and eliminate alleles that are detrimental. However, it is important to recognize that 'advantageous' and 'detrimental' refers to the likelihood of reproduction, so if a certain allele causes a disease at old age but does not affect the chance of reproduction then this will be a neutral allele. Some alleles are **dominant**, which means that their presence on just one chromosome will cause a phenotype. Other alleles are **recessive**, which means that they need to be present on both chromosomes for a phenotype to occur.

Currently researchers debate whether inherited diseases, sport and exercise-related and other traits are caused mainly by common alleles, termed polymorphisms, which can be found in at least 1% of the population or by rare alleles which only occur in some families. In medical genetics where deleterious disease alleles are analysed, this is hotly discussed as the **common disease common variant** versus **common disease rare variant** hypotheses (Schork et al., 2009). This discussion has been fuelled by the results of many genome-wide association studies (GWAS) which allow the measurement of hundreds of thousands of common SNPs in a single experiment (see http://www.genome.gov/gwastudies for an online catalogue of GWAS study results). To give an example, in one GWAS study 14 000 patients versus 3000 controls were subjected to the analysis of SNPs. The results confirmed that common disease SNPs exist but these common SNPs only explain a small percentage of the heritability of the diseases investigated (Wellcome Trust Case Control Consortium, 2007). Similarly, a GWAS was performed on 15 000 individuals to look for common SNPs for height, which is a key performance-limiting factor in sports such as basketball (Lettre et al., 2008). The authors identified 12 common loci related to height (note that several SNPs can reside in one locus) but these loci only explained a mere 2% in the variation of height, despite body height being a highly inherited trait. The question is whether the 'missing heritability' in many GWAS studies is mainly because the data analysis underestimates the importance of common alleles, or whether rare alleles play a major role. We have attempted to visualize the rare versus common allele debate for Mendelian or monogenic and polygenic traits in a highly simplified form in Figure 2.3.

In the sport and exercise context there are examples of traits that can be explained mainly by common alleles or where a rare DNA sequence variation has a major effect. An example for a trait whose heritability may depend mainly on SNPs is the trainability of the $\dot{V}O_{2max}$, which is, like $\dot{V}O_{2max}$ itself, ≈50% inherited (Bouchard et al., 1999). In one GWAS led by Bouchard, 473 subjects belonging to 99 families from the HERITAGE Family Study were tested for 324 611 SNPs. The researchers

Figure 2.3 Schematic illustration of the possible contributions of rare and common alleles to Mendelian or monogenic and polygenic traits. (a) Mendelian or monogenic traits such as eye colour are characterized by clear distinctions between the phenotypes such as blue or brown eye colour. (b) In this figure blue circles represent rare alleles and open circles represent common alleles. In Mendelian or monogenic phenotypes one major common or rare allele mainly determines the phenotype but some smaller alleles may contribute to the phenotype. (c) Polygenic traits such as body height, weight or endurance running performance typically have a Gaussian distribution of the phenotype. (d) This is caused by many either rare or common alleles with relatively small effect sizes. This figure is purely hypothetical and it is currently unknown whether common or rare alleles play a more important role for the heritability of such polygenic traits.

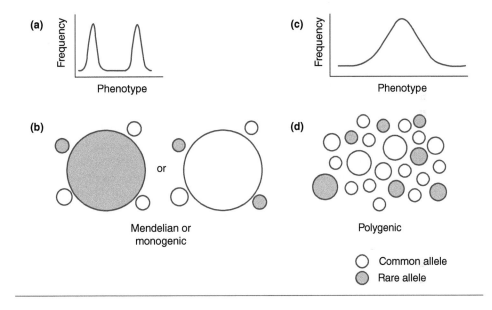

found that 21 SNPs accounted for 49% of the variance in $\dot{V}O_{2max}$ trainability (Bouchard et al., 2011). Given that this value equals the overall heritability of $\dot{V}O_{2max}$ trainability the conclusion is that $\dot{V}O_{2max}$ trainability is entirely due to these 21 common SNPs.

In contrast, a large effect of a rare allele has been demonstrated for the haematocrit which determines the oxygen transport capacity of the blood and hence the $\dot{V}O_{2max}$. The rare allele is a gain-of-function allele of the erythropoietin receptor, which was identified in a Finnish family that includes the treble Olympic gold medalist Eero Antero Mäntyranta. The researchers found that family members who were heterozygous for this mutation had, similar to athletes who have misused erythropoietin, a high haematocrit (de la Chapelle et al., 1993). Interestingly, although being heterozygous for this mutation appears to increase the haematocrit, it appears that being homozygous for this allele is lethal as no homozygous individuals were reported in this study (de la Chapelle et al., 1993). To summarize, both rare and common DNA sequence variations are likely to contribute to sport and exercise-related traits but their relative contributions are largely unknown at the moment.

The issue of the importance of common versus rare DNA sequence variations for sport and exercise-related traits will only be settled once whole genomes of athletes have been sequenced and analysed. Technically, whole genome sequencing has already become a reality (Wheeler et al., 2008) and applying this method to sequence, for example, the genomes of East African endurance runners, West African speed and power athletes and many other outstanding athletes will eventually yield a conclusive answer about the importance of common and rare alleles for sport and exercise-related traits.

In medical genetics, the sequencing of individual genomes is well underway as part of the 1000 Genomes Project (http://www.1000genomes.org). This international collaboration aims to sequence the genomes of 2500 unidentified people from 25 populations around the world using next-generation sequencing and make these data freely available. It will be possible to use this as a reference database to compare the genomes of elite athletes against in order to explore the types of genetic variation that contribute to elite athletic performance.

DNA METHODS: POLYMERASE CHAIN REACTION (PCR)

After introducing sport and exercise genetics in the first part we will now explain the methods and study types that are used to analyse DNA sequences and to determine the relationship between DNA sequences and sport and exercise-related traits.

Analysing DNA sequences is not easy because DNA is a huge molecule and there are only two copies of each DNA molecule (i.e. chromosomes) in most human cells. For example, chromosome 1, the largest human chromosome, has ≈247 000 000 base pairs and there are only two chromosome 1's per cell. Each base weighs roughly 500 g/mol ($\times 2$ for each base pair) and so the molecular weight of chromosome 1 is 500 g/mol $\times$ 2 $\times$ 247 000 000 ≈ 2.47×10^{11} g/mol. Compare this to glucose, which has a molecular weight of 180 g/mol. Thus DNA molecules are too large and their concentration is too low to allow an easy analysis.

The solution to this problem is a method which allows the production of high concentrations of small DNA fragments. The method in question is the **polymerase chain reaction** which is commonly known by its abbreviation as **PCR**. PCR was invented primarily by Kary Mullis, who later won the Nobel Prize in Chemistry in 1993 for his discovery. According to him, the idea for PCR arose as follows:

> Sometimes a good idea comes to you when you are not looking for it. Through an improbable combination of coincidences, naïveté and lucky mistakes, such a revelation came to me one Friday night in April, 1983, as I gripped the steering wheel of my car and snaked along a moonlit mountain road into northern California's redwood country. That was how I stumbled across a process that could make unlimited numbers of copies of genes [not necessarily genes but all sorts of small DNA products], a process now known as the polymerase chain reaction (PCR) (Mullis, 1990).

A PCR reaction requires a so-called **mastermix** which comprises:

- **forward and reverse primers** (≈20 bp short DNA single strands) that bind to the start and end of the DNA fragment selected for amplification,
- **Taq polymerase** which synthesizes a DNA double strand from a single strand starting at the point where the primer binds to a DNA single strand,
- **dNTPs or nucleotides** (i.e. the TTPs, ATPs, GTPs, CTPs that are used to synthesize new DNA),
- **buffer** which includes Mg^{2+} as a co-factor and chemicals that keep the pH stable.

A sample of **genomic DNA** is then added to the mastermix. In practice, genomic DNA is first extracted either from the leukocytes within the blood found in the 'buffy coat' of a not-coagulated, centrifuged blood sample or from so-called buccal or cheek cells that can be obtained by rubbing a cotton-type swab on the inner cheek or by simple mouthwash. Once these cells are obtained, the cell membranes are chemically 'lysed' or broken down, allowing the DNA to be separated from the rest of the cell material. The DNA is then precipitated out of solution and cleaned with ethanol rinses before being rehydrated into solution. DNA is a stable molecule and can be stored in water with inhibitors which prevent it from breaking down at 4°C for many years.

The mastermix and genomic DNA of each individual are then combined in thin-walled (for quick heat transfer), 0.2 mL PCR tubes or plates and inserted into a thermal cycler. The key components of the **thermal cycler** are:

- a **thermal block** with holes that hold the tubes with the mastermix and genomic DNA;
- a **programmable computer** that controls the cycles of rapid heating and cooling within the thermal block.

The PCR reaction is then started and its ingenuity is that the reactions, which lead to the amplification of the DNA between the two primers, are controlled simply by heat rather than repetitive pipetting. The PCR reaction comprises cycles which under ideal conditions double the concentration of the DNA or PCR product with each cycle. The steps of a PCR cycle are (see also Figure 2.4):

1 **Denaturation**: The sample is heated to **90–96°C** which causes the DNA double strand to separate (also known as denaturation or melting). The aim is to obtain single DNA strands because primers can only bind to single strands.
2 **Annealing**: The sample is cooled to **40–65°C** which facilitates the binding of the forward and reverse primers to their target DNA sequence. Primers quickly bind to their complementary DNA sequences because their concentration is high.
3 **Extension**: The sample is warmed to ≈72°C. This temperature is a good working temperature for *Taq* polymerase (an enzyme from the hot-spring organism *Thermus aquaticus* (Saiki et al., 1988)) which will find the sites where primers bind to the DNA single strands. *Taq* then extends the DNA double strand from 5′ to 3′ direction. The PCR cycles are repeated until a sufficient amount of DNA product is produced, commonly 25–35 cycles.

Thus, to summarize, PCR is a method used to amplify a small DNA product of a large DNA molecule (usually chromosome) by cycles of **denaturation**, primer **annealing** and **DNA polymerase extension**.

ANALYSIS OF PCR PRODUCTS

PCR products, which are double-stranded DNA, are invisible to the naked eye and thus need to be visualized or analysed otherwise. This can occur during the PCR reaction or else methods are used to visualize the amplified PCR products or the DNA sequence is determined. PCR products or any other piece of DNA may be analysed or used by one of the following methods:

Figure 2.4 Polymerase chain reaction. (1) Denaturing or melting of a double-stranded DNA into signal strands. (2) Binding of primers (dotted lines) to single DNA strands. (3) *Taq* polymerase recognizes the double strand (primer binding to single strand)–single strand boundary and extends or synthesizes the double strand. (4) *Taq* polymerase continues to synthesize DNA until the end of the single strand or until the temperature is changed. (5) Three double-stranded PCR products after several PCR cycles.

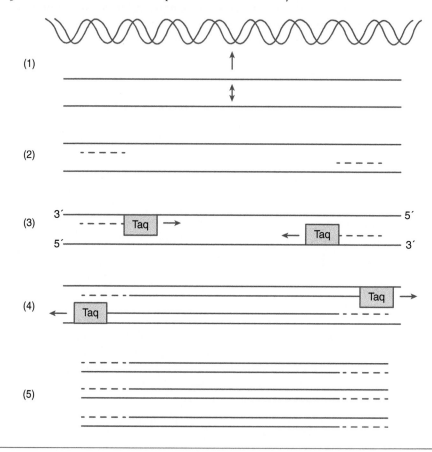

- **DNA fragment separation and visualization**: DNA fragments are separated by size on an **agarose gel** followed by **ethidium bromide–UV light** or other visualization. DNA products will appear as bands in order of their length (Figure 2.5).
- **Restriction enzyme digestion and visualization**: PCR products can also be cut with a **restriction enzyme** and then separated on an agarose gel followed by ethidium bromide–UV light visualization (Figure 2.5). Restriction enzymes that cut specific DNA sequences can be used especially to distinguish alleles in a polymorphism.
- **Fluorescent labelling of alleles**: There are several methods involving fluorophores (fluorescent primers or hybridization probes) in the PCR reaction. At the end of the reaction each DNA variant or allele fluoresces in a different colour (e.g. one allele PCR product in red and the other allele PCR product in green with heterozygotes being yellow (a mixture of red and green)). This is a good strategy (e.g. when many samples are tested in thermal cyclers that can automatically read the fluorescent

colour of each sample). The advantages are that no agarose gels are needed and that the results can be read automatically.

- **DNA sequencing** is performed by the 'Sanger' sequencing method of the DNA product using a DNA sequencer. This will reveal the DNA sequence for the whole DNA/PCR product. It is important for (a) searching for so far undiscovered DNA variants and (b) verifying genotyping methods that rely on the detection of DNA bands of different size.

- **Genetic modification**: Integrate the DNA product into a DNA 'vector' and use it to express it in a cell or an animal. For this a promoter is needed so that the DNA product is transcribed and the vector needs to be **transfected** into cells.

Figure 2.5 Thermal cycler and PCR reaction. (a) Thermal cycler. The arrow points towards the metal block into which the samples are inserted. (b) Electrophoresis of PCR products in an agarose gel. The negatively charged DNA runs towards the positive electrode. The arrows point towards dyes which are mixed with the PCR product. (c) Visualization of an *ACTN3* R577X PCR product which has been digested with DdeI (Yang et al., 2003). The 290 bp PCR product has been cut either once into 205 bp and 85 bp fragments (RR) or twice into 108 bp, 97 bp and 86 bp fragments. Note that it is impossible to distinguish the 85 and 86 bp fragments. *Refers to primer dimers, an artefact that is seen in many gels.

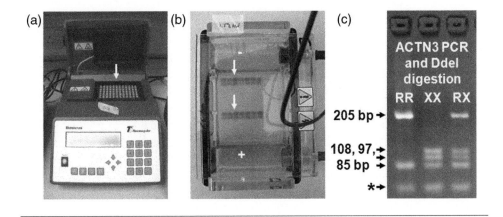

DNA separation and visualization can be used if the size of the PCR product varies between genotypes. For example, in the angiotensin-converting enzyme (ACE) gene there can be either a 287 bp-long insertion (*I*) or this fragment can be missing (the so-called deletion or *D* allele) (Rigat et al., 1990). If primers are designed to amplify the DNA of the insertion plus a bit to either side then the resulting PCR products will vary by 287 bp depending on whether an individual's allele has the insertion or not. The PCR products are then mixed with ethidium bromide or less toxic DNA dye, which allows visualization of the DNA under UV light. The DNA samples are then run in an electrical field through an agarose gel. As DNA is negatively charged, it will migrate towards the positive electrode or anode. The large ACE insertion allele will travel slower through the gel than the smaller ACE deletion allele. Thus individuals homozygous for the insertion allele (*II*) will have a higher band (DNA travels less through the gel), those for the deletion allele a low band and heterozygous individuals will have both a high and a low band. This PCR strategy would work for genotypes such as the *ACE I/D* locus that vary considerably in length, but in reality a different strategy is used.

Restriction enzymes or restriction endonucleases are enzymes that cut DNA at so-called restriction sites, which are short, specific DNA sequences. This can be used, for example, to cut a PCR in fragments if a certain allele is present. For example, the restriction enzyme *Dde*I cuts the following double-stranded DNA sequence:

$$5' \ldots \text{C} {\downarrow} \text{TNAG} \ldots 3'$$

$$3' \ldots \text{GANT} {\uparrow} \text{C} \ldots 5'$$

where 'N' can be any base, 5' and 3' indicates the direction of DNA and the arrows indicate where the cut is made. This can be used to selectively cut specific alleles. If the DNA cuts one allele, then two bands show up on the agarose gel. In contrast, if the DNA does not cut the other allele, then only one band is visible. For example, PCR followed by *Dde*I restriction enzyme digestion is used to identify the *ACTN3* R577X genotype, which is associated with speed and power performance (Yang et al., 2003). Such a genotype is known as a **restriction fragment length polymorphism** or **RFLP**.

The third possibility is to use (Sanger) **DNA sequencing**. Frederick Sanger is a British scientist who has won two Nobel Prizes, one of which was awarded for his method of DNA sequencing. Very briefly, different length DNA fragments are generated from the PCR product that is to be sequenced and fluorescent colour-labelled using dideoxynucleotide triphosphates (ddNTPs). These are then separated on the basis of their length by electrophoresis and the resulting ladder of fluorescent coloured bands represents the DNA sequence of the original DNA product (see Figure 2.6a for an example of the data generated with this method). It is read by a fluorescence detector and stored as DNA sequence on a computer. This is essentially how the human genome and many other genomes have been sequenced. In contrast to the agarose gel methods above, this allows the whole DNA sequence of the PCR product to be read rather than just one specific variant in an allele.

For **genetic modification** PCR-amplified DNA can be inserted into so-called vectors, which are DNA molecules or viruses. Vectors can be introduced into cells by using reagents such as Lipofectamine or by electroporation. If a suitable promoter is present then the DNA product will be transcribed into mRNA and translated into protein. This is used to genetically modify cells. Many variants of transfection exist and there are also many caveats in the experimental strategy. PCR products can also be inserted into mouse stem cells to create transgenic knock-in mice or potentially also into humans for gene therapy or even, the worst case scenario, gene doping.

Next-generation sequencing and personal genomes

The all-important PCR method may soon be superseded by next-generation sequencing methods (Figure 2.6b) that essentially allow the direct sequencing of much larger DNA fragments such as whole chromosomes much quicker. The human genome of

Figure 2.6 DNA sequencing equipment and results. (a) Result of Sanger sequencing. The peaks of the digested and electrophoresed bands are measured and then plotted. This is then automatically plotted as a DNA sequence seen at the top. (b) Next-generation sequencer. Such instruments allow the parallel sequencing of a large number of DNA fragments. In reality the peaks are in one of four colours and one colour indicates one base.

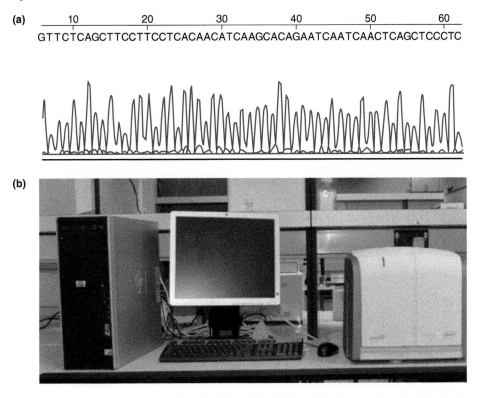

the DNA researcher James D Watson (of Watson and Crick fame) has been sequenced using next-generation sequencing methods at 1/100th the cost of Sanger sequencing methods within two months (Wheeler et al., 2008). As with computers, the speed of DNA sequencing will further increase while the cost will decrease and it will not be long until many of us can afford to have our own genome sequenced (a US$1000 genome has been proposed as a goal). Genetics will then move from being a primarily methodological problem to being a bioinformatics and ethical problem. Next-generation sequencing will also change sport and exercise genetics fundamentally and the key challenge will be to obtain information useful for exercise physiologists from the wealth of information that is contained in sequenced human genomes.

RESEARCH STRATEGIES IN SPORT AND EXERCISE GENETICS

Having explained the most important methods of DNA analysis we will now focus on the design of genetic studies in sport and exercise genetics.

Genetic association studies

In sport and exercise genetics the most common study design is to test whether a known polymorphism in a **candidate gene** is **associated** with a sport or exercise-related trait. Candidate genes are those hypothesized to have some influence on the trait of interest, for example a gene that encodes a protein important in the structure or function of the cells underlying the trait. Polymorphisms are then examined in those candidate genes, the idea being that different alleles could affect the regulation of the gene or the structure of the encoded protein, thus impacting cell function and the trait itself. Many of the associations studied thus far in exercise science have been published in a human gene map for performance and health-related fitness phenotypes that gives an overview over this work (Bray et al., 2009). For example, the *ACTN3* R577X polymorphism was first discovered as part of research into muscular dystrophies, where it was identified as a common polymorphism rather than as a disease-causing mutation (North et al., 1999). Because the ACTN3 protein is expressed in fast type II fibres and is completely absent in individuals with the XX genotype, it was hypothesized that the genotype would be associated with power and strength performance, which was confirmed in an association study (Yang et al., 2003). Subsequent association studies have confirmed that *ACTN3* R577X genotype is associated with power performance in elite-level athletes, but its importance in other contexts is less certain, as detailed in Chapter 7.

There are four different types of genetic association studies:

- **association studies** in the narrow sense of the word;
- **case–control studies**;
- **linkage analysis studies**;
- **genome-wide association studies** (GWAS).

In an **association study** all subjects are genotyped for one genotype and this is then compared to the trait investigated. For example, to test the hypothesis that the angiotensin-converting enzyme (ACE) gene *I/D* (insertion/deletion) genotype is associated with grip strength, the *ACE* genotype needs to be measured with PCR and strength with a grip-strength dynamometer. The data are then statistically analysed to test the hypothesis that grip strength differs between *II*, *ID* or *DD* carriers. Association studies are popular as they are comparatively easy to perform but the problem is that misleading results can easily result from poor study design (Altmuller et al., 2001; Chanock et al., 2007). The most common pitfall in sport and exercise genetics studies is a low number of subjects, as frequently several hundreds if not thousands of subjects are required to yield sufficient statistical power, which is difficult to achieve for sport and exercise science laboratories. For these reasons the early sports and exercise genetics literature suffers from many unreliable, underpowered candidate gene association studies (Bouchard, 2011).

Another factor is the population history of the subjects. While most studies benefit from individuals of different races and ethnicities, this can hinder genetic investigations as different groups have distinct allele and genotype frequencies for many polymorphisms. If diverse subjects are recruited this genetic variation needs to be taken into consideration to minimize misinterpretation of the results. We recommend reading the checklist for genotype–phenotype association reports that was published by Chanock et al. (2007) before embarking on such studies.

Case–control studies are a variation of association study. Unlike gene association studies, in a case–control study the subjects are selected on the presence or absence of a trait and then the researchers measure genotype frequencies of one or several candidate gene alleles within those opposing groups. The aim is to test whether or not a particular genotype is more common in those with a particular trait. The term case–control comes from the medical genetics field, where the cases would be subjects with a disease who are compared to healthy controls. In our field, we rarely focus on genetic disease but instead select cases and controls simply by having different trait values, typically high versus low. For example, we might recruit 1000 subjects for a study of $\dot{V}O_{2max}$ and then select the top and bottom 100 as the cases and controls, respectively. In this situation the assignment of 'case' or 'control' is purely arbitrary.

A third study type aims to identify the genetic variants that are associated with a phenotype when no candidate genes are known. This has commonly been used to narrow down and identify genetic variants that cause an inherited phenotype or disease in families or inbred mouse strains but has also been used for exercise-related traits. The method is **linkage analysis** and the goal is to associate different regions of the genome with the trait of interest by relying on families with shared DNA sequence. The output of a linkage analysis is a figure that shows the so-called 'LOD score', an abbreviation for 'log$_{10}$ of odds,' on the y-axis and the position in the genome, broken down into chromosomes on the x-axis. Without going into the complicated detail of the method, the higher the LOD score the higher the likelihood that a causative genetic variation is located in this genomic region and then genes within these regions can then be identified as possible candidate genes for more detailed analysis. Linkage analysis has been successfully used in identifying causative genes in monogenetic diseases but it is less successful for the identification of phenotypes that depend on many genetic variants or quantitative trait loci (QTLs). Figure 2.7 shows an example where a linkage analysis

Figure 2.7 Quantitative trait locus (QTL) mapping for muscle weight. QTL mapping or genome position on chromosome 6 against LOD score for tibialis anterior muscles as a result of a linkage analysis performed on inbred mouse strains. Redrawn after Lionikas et al. (2010).

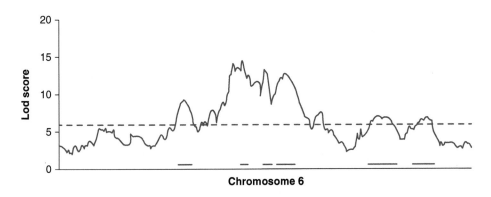

has been performed on chromosome 6 for the muscle size of the tibialis anterior (TA), extensor digitorum longus (EDL), quadriceps femoris (QF), gastronemius (gas) and soleus (sol) muscles of mice. Note that the central region of the chromosome contains the highest associations with muscle mass, thus representing a location to look for candidate genes and polymorphisms.

In **genome-wide association studies** (GWAS) researchers use so-called SNP chips (microarrays) that measure hundreds of thousands of SNPs all over the human genome. Similar to linkage analysis studies the goal is to identify possible genes and polymorphisms that contribute to a trait rather than hypothesize candidate genes in advance. For high-quality GWAS, large sample sizes, reliably measured sport and exercise-related traits and a large, high-quality set of SNPs are required. In the last few years many GWAS have been performed and the results of GWAS are accessible online: http://www.genome.gov/gwastudies. This searchable website stores the results of published GWAS studies and allows, for example, to see whether a gene is linked to a certain trait or vice versa. At the moment there are some exercise and body composition-related results listed and therefore the list is relevant for sport and exercise scientists. GWAS are commonly collaborative studies where tens of thousands of individuals are tested. For example, in one study 2.2 million SNPs were measured in 15 821 individuals to determine, among other traits, the SNPs that are associated with body height (Lettre et al., 2008). These studies quantify the variation of traits that are due to common SNPs. They do not, however, identify the variation that is due to other types of polymorphisms or rare alleles which are not included on the SNP chip. Other information from GWAS is the identification of candidate genes (i.e. genes that carry SNPs associated with the trait). Even if the measured SNPs in a gene only explain a small fraction of the variation of the trait it is likely that these genes contain other genetic variants that explain much more of the variation. Eventually such questions will be resolved by the sequencing of individual genomes (Wheeler et al., 2008), which will reveal common and rare alleles.

Identification of candidate genes

In this section we will discuss two different strategies that aim at identifying the function of a gene. The first strategy is termed the **Mendelian candidate gene** strategy. It is usually applied if there is a striking, inherited phenotype for which a candidate gene is known or clearly anticipated. In sport and exercise genetics an example of a candidate gene approach is a study where the high haematocrit of a Finnish family was linked to the erythropoietin (EPO) pathway, specifically to a mutation in the EPO receptor (de la Chapelle et al., 1993). Only a small number of key genes could be anticipated as causing the remarkably phenotype in these family members and careful analysis identified the mutation in one of these genes. A second example is a child with an extraordinarily high muscle mass which appeared to be inherited as other members of the family were reported to be remarkably strong. The researchers hypothesized and confirmed that a mutation in the myostatin gene was probably responsible (Schuelke et al., 2004). We will discuss these two studies in Chapters 5 and 7, respectively. Successful candidate gene studies are major publications but since few sport and exercise-related traits are Mendelian or monogenetic, such studies are the exception.

The second strategy is **transgenic** research. Transgenic studies are used to determine the function of a gene in an animal or cell model. In the sport and exercise context this could, for example, be useful if a gene was only expressed directly after endurance exercise but its function was unknown. The standard model to identify the function of such a gene is to **knock-in** or **knockout** a gene either in cultured cells *in vitro* or animals *in vivo*. The most common *in vivo* transgenic models are transgenic mice. In 2007 Capecchi, Evans and Smithies won the Nobel Prize in Physiology or Medicine 'for their discoveries of principles for introducing specific gene modifications in mice by the use of embryonic stem cells'. Here is some key terminology in relation to transgenic technology:

- The result of a **knockout** is that the gene is not expressed. This can be achieved by removing an exon or the whole gene from the genome. The resulting defects or changes in the phenotype can be attributed to the knocked out gene. In sport and exercise genetics the myostatin knockout mouse with its remarkable muscle mass doubling is a prominent example of a knockout organism (McPherron et al., 1997).
- **Knock-in** is to overexpress a gene or make constitutively active a mutant of a gene. For example, it has been shown that an overexpression of PGC-1α increases not only mitochondrial biogenesis but also slow muscle fibre formation (Lin et al., 2002).
- **Inducible** means that a gene construct can be induced (i.e. switched on or off), for example by giving a drug such as tamoxifen or doxycycline. The advantage of this method is that the function of the gene can be studied at different time points, such as development, adulthood and old age. This targeted treatment is essential if a knockout or knock-in is developmentally lethal or results in animal suffering.
- **Tissue-specific** gene knock-in or knockout can be achieved by using a tissue-specific promoter, such as the muscle creatine kinase promoter to express a gene in muscle fibres but not satellite cells or other tissues in the body. This avoids lethality because a gene knock-in or knockout might be lethal in organs other than the one studied.

There are now a plethora of transgenic mice with different exercise-related phenotypes and we will discuss these in Chapters 5 and 7.

SPORT AND EXERCISE GENETICS: APPLICATIONS AND ETHICS

At the end of this chapter we discuss the practical applications of genetic research in sport and exercise science and ethical considerations. On this topic the British Association for Sport and Exercise Sciences has published a position stand and related review (Wackerhage et al., 2009). On the topic of personalized exercise prescriptions we have also published our view (Roth, 2008). Here, we will discuss genetic performance tests and genetic tests for anti-doping testing as two applications of genetic research in the sport and exercise science context to introduce some of the ethical aspects in sport and exercise genetics.

Genetic performance tests

Classical performance tests are used to measure sport and exercise-related variables such as $\dot{V}O_{2max}$ or maximal strength to measure current or predict future performance in sport or to make personalized exercise prescriptions (Roth, 2008). Given that virtually all sport and exercise-related variables are partially inherited, similar information can be potentially obtained by conducting genetic performance tests. One of the earliest genetic tests to be offered commercially was the *ACTN3* R577X genotyping test (Genetic Technologies Ltd, 2007). Similarly, other companies offer genetic performance tests based on one or a few polymorphisms which have been linked to sport and exercise-related traits in the literature. Recently, a genetic $\dot{V}O_{2max}$ trainability test has been developed which is presumably based on the results of much more powerful $\dot{V}O_{2max}$ predictor gene studies (Timmons et al., 2010; Bouchard et al., 2011) (http://www.xrgenomics.co.uk/). These tests can be purchased online and customers will need to send a DNA sample to these companies.

Do such genetic performance tests make sense? Commercial tests based on one or a few polymorphisms are frequently oversold because few genetic variants are conclusively linked to sports and exercise-related traits and the contribution of the genetic variation to the trait is generally small. For example, the *ACTN3* R577X test only predicts a minor fraction of the overall variation of sprinting performance and ≈80% of the population carry a functional *R* allele anyway. In other words, this and similar single polymorphism tests yield little useful information when compared to classical performance tests such as a jump and reach, 100 m sprint or Wingate test. However, the information gained from each genetic test for a trait is additive and thus if companies tested for a wide range of sport and exercise-related polymorphisms and calculated a genetic score then this might provide some useful information about a trait (Ahmetov et al., 2009). The tests of $\dot{V}O_{2max}$ predictor genes, discussed in Chapter 5, appear to predict a high proportion of the inherited $\dot{V}O_{2max}$ trainability and should therefore be meaningful if these data are confirmed (Timmons et al., 2010; Bouchard et al., 2011). It seems inevitable that the commercial sequencing of whole human genomes for under US$1000 will soon be available. At that stage genetic performance testing will be a problem for bioinformaticians, who will ultimately be able to analyse the sequenced genomes for DNA variations that are linked to sport and exercise-related traits. Thus, taken together, genetic performance testing has reached an intermediate stage between near-useless single polymorphism testing and the data mining of whole human genomes which potentially contains all the DNA sequence variations that determine sport and exercise-related traits.

What are the ethical concerns associated with genetic performance testing? There are two main differences between genetic and classical performance testing (Wackerhage et al., 2009): First, genetic tests can be performed as soon as genomic DNA can be obtained and the predictive power of the test does not change with advancing age because the DNA sequence of an individual is constant. This implies that meaningful genetic tests could potentially be conducted on embryos or minors and, in contrast to classical performance tests, would predict variables such as muscle strength or $\dot{V}O_{2max}$ just as well as if such genetic tests were performed on an adult. The consequences could be embryo selection based on such tests or that children may be discouraged from starting a sport they like. Second, there is the risk that currently unknown disease associations might be identified at a later stage. An example is the *APOE4* allele, which was initially associated with changed lipid metabolism but later with the much more serious late-onset familial Alzheimer's disease (Strittmatter et al., 1993). Thus a test for a muscle growth gene could also turn into a test for a cancer growth gene if researchers made such a link after the test was performed. For the above reasons we recommend that genetic performance tests should not be performed on embryos or minors but only on adults who have given their consent after counselling, as is standard practice for medical genetic tests. Finally, in some countries such as the United States it is legally forbidden to discriminate on the basis of genetic tests (Genetic Information Nondiscrimination Act, GINA). This law would be applicable, for example, if a professional team requested to genetically test potential players prior to making the decision whether or not to sign them.

Genetic tests for anti-doping testing

Genetic tests termed '**genetic fingerprinting**' could potentially be used to connect biological samples such as contaminated syringes or blood bags to individual athletes accused of drug doping for performance enhancement (Tamaki and Jeffreys, 2005). The method is based on the detection of 'human tandem repeats', which are repeated DNA sequences. The number of repeats varies from 6 to >100 base pairs and can differ enormously between individuals. Thus PCR-based and other methods of measuring the number of several tandem repeats for several loci can give a unique genetic fingerprint for an individual (Tamaki and Jeffreys, 2005). These tests are currently in routine use in forensic science.

A second application of genetic testing for anti-doping is if an individual carries an unusual mutation such as the erythropoietin receptor mutation, which can lead to an unusually high haematocrit (de la Chapelle et al., 1993). Such an individual may be barred from competition because their haematocrit level exceeds the limit at which athletes are allowed to compete. However, because this is due to genotype and not manipulation, such individuals should be allowed to compete if they can demonstrate they carry mutations that are causing the phenotype. After all, talent in sport means to be carrier of several common and unusual genetic variations that are beneficial for a sport.

Gene doping

After a recent modification the World Anti-Doping Agency (WADA) now defines **gene doping** as follows (World Anti-Doping Code, 2009):

The following, with the potential to enhance sport performance, are prohibited:

1   The transfer of nucleic acids or nucleic acid sequences;
2   The use of normal or genetically modified cells;
3   The use of agents that directly or indirectly affect functions known to influence performance by altering gene expression. For example, peroxisome proliferator activated receptor δ (PPARδ) agonists (e.g. GW 1516) and PPARδ–AMP-activated protein kinase (AMPK) axis agonists (e.g. AICAR) are prohibited.

Gene doping has been discussed as a potential new threat to sport because some transgenic animals can display a striking athletic phenotype, as is the case of the PEPCK mouse, which has a 40% higher $\dot{V}O_{2max}$ and greatly enhanced running performance (Hakimi et al., 2007). Some researchers and journalists predict that such manipulations might soon be used in humans as well. However, it is much easier to mutate the DNA of one embryonic stem cell that gives rise to a transgenic mouse than to safely and effectively gene-manipulate somatic cells within the human body with a low chance of detection. Perhaps motivated by the media hype about gene doping, the German coach Thomas Springstein enquired in an email about a gene vector called Repoxygen and a Chinese scientist was reported to offer genetic manipulation to athletes (Friedmann, 2010). Currently it seems possible that ignorant individuals may attempt interventions with DNA, but safe, well-dosed, performance-enhancing and hard-to-detect gene doping is unlikely to become a reality soon. After all, if gene doping for rogue researchers was within reach then most monogenetic diseases would have been cured by now, which is not the case. In other words, gene doping attempts are unlikely to be safe, effective and hard to detect.

SUMMARY

Sport and exercise-related traits such as the $\dot{V}O_{2max}$ or strength are significantly inherited. This means that variations in the DNA sequence partially explain the variation of these traits. Twin and family studies can be used to calculate heritability estimates for sport and exercise-related traits. Most cells in the human bodies carry ≈6.6 Gb (gigabases) of DNA organized as 23 chromosome pairs or 46 chromosomes. Different types of mutations in the DNA sequence can occur, especially during cell division, and if such mutations spread in a population they are known as polymorphisms. There is an intense debate about whether sport and exercise-related traits and other traits are mainly affected by polymorphisms or by rare DNA sequence variations. The polymerase chain reaction (PCR) allows researchers to produce large amounts of a small stretch of DNA. The DNA-duplicating reactions of a PCR cycle are controlled by temperature and involve the denaturation of the DNA double strand, the annealing of primers and the extension by *Taq* polymerase. The amplified DNA fragments, termed PCR products, can then be separated and visualized under UV light by using electrophoresis of agarose gels or by DNA sequencing. In genetic research either association studies (e.g. measuring a genotype in different groups) or mechanistic studies (e.g. transgenic mice) can be carried out. Recent developments include next-generation sequencing, which opens up the possibility to sequence individual human genomes, and genome-wide association studies (GWAS). Examples of genetic tests are genetic performance tests, genetic tests for personalized exercise prescription and genetic tests for anti-doping research. While not all current tests are useful, it is inevitable that these tests will increasingly be used. Users should be aware of ethical issues such as the possibility to perform genetic tests for embryo selection. Another misuse of genetic research might be gene doping. While rogue individuals may attempt to use DNA in sport it is unlikely that safe, well-dosed, performance-enhancing and hard-to-detect gene doping will become a reality in the near future.

REVIEW QUESTIONS

1   Design a twin study to estimate the heritability of grip strength.
2   What are the common types of DNA sequence variation and how do they occur? What is the difference between a mutation and a polymorphism?
3   Compare and contrast a case–control study with a genome-wide association study (GWAS).
4   Briefly describe the polymerase chain reaction. What is it used for?
5   Explain and compare two methods that are used to analyse PCR products (i.e. DNA sequences that are amplified by the PCR reaction).
6   How is genetic testing envisioned to assist with prediction of athletic performance? What are some ethical concerns?

FURTHER READING

Bouchard C, Malina RM, and Perusse L (1997). *Genetics of fitness and physical performance.* Human Kinetics, Champaign, IL.

Bouchard C and Hoffman EP (2011). *Genetic and molecular aspects of sports performance,* Vol. 18 of *Encyclopedia of sports medicine.* John Wiley & Sons, Chichester.

Pescatello LS and Roth SM (2011). *Exercise genomics.* Springer, New York.

Roth SM (2007). *Genetics primer for exercise science and health.* Human Kinetics, Champaign, IL.

Roth SM (2008). Perspective on the future use of genomics in exercise prescription. *J Appl Physiol* 104, 1243–1245.

Wackerhage H, Miah A, Harris RC, Montgomery HE and Williams AG (2009). Genetic research and testing in sport and exercise science: A review of the issues. *J Sports Sci* 1–8.

REFERENCES

Ahmetov II, Williams AG, Popov DV, Lyubaeva EV, Hakimullina AM, Fedotovskaya ON, et al. (2009). The combined impact of metabolic gene polymorphisms on elite endurance athlete status and related phenotypes. *Hum Genet* 126, 751–761.

Altmuller J, Palmer LJ, Fischer G, Scherb H and Wjst M (2001). Genomewide scans of complex human diseases: true linkage is hard to find. *Am J Hum Genet* 69, 936–950.

Bernstein BE, Birney E, Dunham I, Green ED, Gunter C and Snyder M (2012). An integrated encyclopedia of DNA elements in the human genome. *Nature* 489, 57–74.

Bouchard C (2011). Overcoming barriers to progress in exercise genomics. *Exerc Sport Sci Rev* 39, 212–217.

Bouchard C, Leon AS, Rao DC, Skinner JS, Wilmore JH and Gagnon J (1995). The HERITAGE family study. Aims, design, and measurement protocol. *Med Sci Sports Exerc* 27, 721–729.

Bouchard C, An P, Rice T, Skinner JS, Wilmore JH, Gagnon J, et al. (1999). Familial aggregation of $\dot{V}O_{2max}$ response to exercise training: results from the HERITAGE Family Study. *J Appl Physiol* 87, 1003–1008.

Bouchard C, Sarzynski MA, Rice TK, Kraus WE, Church TS, Sung YJ, et al. (2011). Genomic predictors of the maximal O uptake response to standardized exercise training programs. *J Appl Physiol* 110, 1160–1170.

Bouchard C, Blair SN, Church TS, Earnest CP, Hagberg JM, Hakkinen K, et al. (2012). Adverse metabolic response to regular exercise: is it a rare or common occurrence? *PLoS ONE* 7, e37887.

Bray MS, Hagberg JM, Perusse L, Rankinen T, Roth SM, Wolfarth B, et al. (2009). The human gene map for performance and health-related fitness phenotypes: the 2006–2007 update. *Med Sci Sports Exerc* 41, 35–73.

Bruder CE, Piotrowski A, Gijsbers AA, Andersson R, Erickson S, Diaz de Ståhl T, et al. (2008). Phenotypically concordant and discordant monozygotic twins display different DNA copy-number-variation profiles. *Am J Hum Genet* 82, 763–771.

Chanock SJ, Manolio T, Boehnke M, Boerwinkle E, Hunter DJ, Thomas G, et al. (2007). Replicating genotype-phenotype associations. *Nature* 447, 655–660.

de la Chapelle A, Traskelin AL and Juvonen E (1993). Truncated erythropoietin receptor causes dominantly inherited benign human erythrocytosis. *Proc Natl Acad Sci U S A* 90, 4495–4499.

Falconer DS and MacKay TFS (2011). *Introduction to quantitative genetics,* 4 ed. Longmans Green, Harlow, Essex.

Friedmann T (2010). How close are we to gene doping? *Hastings Cent Rep* 40, 20–22.

Genetic Technologies Ltd (2007). ACTN3 sport performance test™. http://www.gtg.com.au/archives/migration/2/010/5/ACTN3%20web%20brochure.pdf.

Hakimi P, Yang J, Casadesus G, Massillon D, Tolentino-Silva F, Nye CK, et al. (2007). Overexpression of the cytosolic form of phosphoenolpyruvate carboxykinase (GTP) in skeletal muscle repatterns energy metabolism in the mouse. *J Biol Chem* 282, 32844–32855.

Kidd JM, Cooper GM, Donahue WF, Hayden HS, Sampas N, Graves T, et al. (2008). Mapping and sequencing of structural variation from eight human genomes. *Nature* 453, 56–64.

Lettre G, Jackson AU, Gieger C, Schumacher FR, Berndt SI, Sanna S, et al. (2008). Identification of ten loci associated with height highlights new biological pathways in human growth. *Nat Genet* 40, 584–591.

Lin J, Wu H, Tarr PT, Zhang CY, Wu Z, Boss O, et al. (2002). Transcriptional co-activator PGC-1 alpha drives the formation of slow-twitch muscle fibres. *Nature* 418, 797–801.

Lionikas A, Cheng R, Lim JE, Palmer AA and Blizard DA (2010). Fine-mapping of muscle weight QTL in LG/J and SM/J intercrosses. *Physiol Genomics* 42A, 33–38.

McPherron AC, Lawler AM and Lee SJ (1997). Regulation of skeletal muscle mass in mice by a new TGF-beta superfamily member. *Nature* 387, 83–90.

Mullis KB (1990). The unusual origin of the polymerase chain reaction. *Sci Am* 262, 64–65.

North KN, Yang N, Wattanasirichaigoon D, Mills M, Easteal S and Beggs AH (1999). A common nonsense mutation results in alpha-actinin-3 deficiency in the general population. *Nat Genet* 21, 353–354.

Pray LA (2008). DNA replication and causes of mutation. *Nature Education* 1.

Rigat B, Hubert C, Alhenc-Gelas F, Cambien F, Corvol P and Soubrier F (1990). An insertion/deletion polymorphism in the angiotensin I-converting enzyme gene accounting for half the variance of serum enzyme levels. *J Clin Invest* 86, 1343–1346.

Roth SM (2008). Perspective on the future use of genomics in exercise prescription. *J Appl Physiol* 104, 1243–1245.

Saiki RK, Gelfand DH, Stoffel S, Scharf SJ, Higuchi R, Horn GT, et al. (1988). Primer-directed enzymatic amplification of DNA with a thermostable DNA polymerase. *Science* 239, 487–491.

Schork NJ, Murray SS, Frazer KA and Topol EJ (2009). Common vs. rare allele hypotheses for complex diseases. *Curr Opin Genet Dev* 19, 212–219.

Schuelke M, Wagner KR, Stolz LE, Hubner C, Riebel T, Komen W, et al. (2004). Myostatin mutation associated with gross muscle hypertrophy in a child. *N Engl J Med* 350, 2682–2688.

Smits J and Monden C (2011). Twinning across the developing world. *PLoS ONE* 6, e25239.

Strachan T and Read AP (2004). *Human molecular genetics,* 3rd edn. Garland Science, New York, Oxford.

Strittmatter WJ, Saunders AM, Schmechel D, Pericak-Vance M, Enghild J, Salvesen GS, et al. (1993). Apolipoprotein E: high-avidity binding to beta-amyloid and increased frequency of type 4 allele in late-onset familial Alzheimer disease. *Proc Natl Acad Sci U S A* 90, 1977–1981.

Tamaki K and Jeffreys AJ (2005). Human tandem repeat sequences in forensic DNA typing. *Leg Med (Tokyo)* 7, 244–250.

Thomas S, Reading J and Shephard RJ (1992). Revision of the Physical Activity Readiness Questionnaire (PAR-Q). *Can J Sport Sci* 17, 338–345.

Thomis MA, Beunen GP, Maes HH, Blimkie CJ, Van LM, Claessens AL, et al. (1998). Strength training: importance of genetic factors. *Med Sci Sports Exerc* 30, 724–731.

Timmons JA, Knudsen S, Rankinen T, Koch LG, Sarzynski M, Jensen T, et al. (2010). Using molecular classification to predict gains in maximal aerobic capacity following endurance exercise training in humans. *J Appl Physiol* 108, 1487–1496.

Wackerhage H, Miah A, Harris RC, Montgomery HE and Williams AG (2009). Genetic research and testing in sport and exercise science: A review of the issues. *J Sports Sci* 1–8.

Wellcome Trust Case Control Consortium (2007). Genome-wide association study of 14,000 cases of seven common diseases and 3,000 shared controls. *Nature* 447, 661–678.

Wheeler DA, Srinivasan M, Egholm M, Shen Y, Chen L, McGuire A, et al. (2008). The complete genome of an individual by massively parallel DNA sequencing. *Nature* 452, 872–876.

World Anti-Doping Code (2009). http://www.wada-ama.org/en/world-anti-doping-program/sports-and-anti-doping-organizations/the-code/.

Yang N, MacArthur DG, Gulbin JP, Hahn AG, Beggs AH, Easteal S, et al. (2003). ACTN3 genotype is associated with human elite athletic performance. *Am J Hum Genet* 73, 627–631.

3 Signal transduction and adaptation to exercise: background and methods

Jatin G Burniston, Mhairi Towler and Henning Wackerhage

INTRODUCTION

In this chapter we aim to answer questions such as 'why and how do we adapt to exercise?' We begin by highlighting limitations to the classical view of supercompensation in exercise training, before introducing the signal transduction hypothesis of adaptation. Related to this we discuss how exercise generates signals that are sensed by sensor proteins, conveyed and computed by signal transduction proteins and how adaption regulators regulate transcription, translation or protein synthesis, protein degradation and other cellular functions such as the cell cycle. All this makes cells and organs adapt to exercise. In the second half of the chapter we introduce RT-qPCR, which is a method used to measure mRNAs, and Western blotting, which is used to measure proteins. The chapter concludes with a discussion of high throughput approaches, including microarrays and proteomics, and their application in exercise physiology. To prepare for this chapter you should read Chapter 1, which introduces molecular exercise physiology and basic wet laboratory research methods, and Chapter 2, which introduces sport and exercise genetics and explains PCR (polymerase chain reaction) which is the basis for RT-qPCR.

WHY DO WE ADAPT TO EXERCISE?

Physiological adaptations are changes that occur within individuals in response to external factors such as exercise and other environmental factors, such as altitude. In the history of adaptation research one early idea is the **overload** concept proposed by Julius Wolff, who linked the loading of bones to their adaptation in an 1892 book entitled *The Law of Bone Remodelling* (Wolff, 1892). This hypothesis is now known as Wolff's law. It can be extended to other organs if the meaning of the term overload is extended beyond mechanical overload. Stating the overload principle for sport is of course correct but the overload principle is just stating the obvious, namely that exercise training is required for adaptation. Moreover it does not explain the mechanisms by which bones or other organs respond and adapt.

A different theory, viewed by some as a mechanistic explanation of adaptation, is the so-called **supercompensation** or **overcompensation** hypothesis (Koutedakis et al., 2006). The supercompensation hypothesis is the transposition of the **general adaptation syndrome** proposed by Hans Selye applied to exercise training. It describes a decline of an often undefined y-axis variable during exercise and its recovery after exercise. The recovery, however, does not just reach pre-exercise levels but overshoots, which is termed supercompensation. Such supercompensation is regarded as the adaptation to exercise (Koutedakis et al., 2006), but this does not shed light on the mechanisms responsible and the whole idea has at least four flaws:

- Supercompensation happens for muscle glycogen after endurance exercise and feeding (Bergstrom and Hultman, 1966). However the time course for most other systems before, during and after exercise is different: for example neither mitochondria, capillaries nor neurons are lost during exercise, as would be expected if the supercompensation hypothesis were true.
- The supercompensation hypothesis is a time course but not a mechanism. Using the glycogen time course before, during and exercise as an example, it is well demonstrated that glycogenolysis depletes glycogen during exercise whereas insulin-stimulated glucose uptake and glycogen synthesis are at least partially responsible for the increase of the glycogen concentration after exercise and carbohydrate intake.
- The supercompensation hypothesis implies that recovery periods are essential for adaptation. In reality the heart adapts to exercise despite continuous contraction. Also skeletal muscles adapt to chronic electrical stimulation applied continuously over several weeks (Henriksson et al., 1989).
- Despite being propagated for decades (Koutedakis et al., 2006) there is little actual evidence that the supercompensation time course is essential for adaptation. In contrast, in this book we quote several hundreds of references that support the alternative hypothesis that signal transduction pathways mediate adaptation to exercise.

Because of the above, the supercompensation hypothesis should no longer be used in an attempt to explain adaptation to exercise. In this chapter we will now introduce the **signal transduction hypothesis of adaptation**. According to this hypothesis, specific sensor proteins detect exercise-related signals, which are then computed by transduction pathways or networks. These early signals regulate downstream events including gene transcription, translation or protein synthesis and protein breakdown. The result is changed organs or organs that have adapted to exercise.

SIGNAL TRANSDUCTION HYPOTHESIS OF ADAPTATION

All organisms need to survive in an environment that is in constant flux. At a whole-organism level inputs from the external environment are sensed and information is relayed via endocrine and nervous systems, which act as specialized adaptation systems for the whole organism. Within these systems there are specific sensor organs (e.g. eye, ear) and computing organs (e.g. nervous system) that allow complex messages to be processed. In order to do so many variables need to be sensed. This input is then computed and cells and the whole organism adapt accordingly. In addition to hormonal and neural signals, individual cells contain the necessary machinery to sense and adapt to changes in their local environment. This intrinsic ability of cells is elegantly illustrated by the specificity of resistance training, where hypertrophy of the muscles recruited during training occurs in the absence of measurable effects in muscles that did not contribute to the exercise. Therefore, exercise-induced muscle hypertrophy cannot be due to systemic factors such as endocrine hormones. Even more convincing evidence for the intrinsic adaptability of cells comes from experiments performed on animal tissues *ex vivo* or cells cultured *in vitro*. Under these circumstances the link between individual cells and the organism or organ is severed completely. Nonetheless, electrical stimulation of skeletal muscle *ex vivo* or myotubes *in vitro* instigates intracellular events that are associated with exercise-induced adaptations such as mitochondrial biogenesis.

In the forthcoming sections we will discuss the intracellular events that link the acute responses to exercise with long-term adaptation. Conceptually, the cellular events in this process can be broken down into three major steps:

1   **Sensing of exercise-related signals**: Sensor proteins detect changes in Ca^{2+}, AMP, glycogen, pO_2, amino acids, force, neurotransmitters and hormones.
2   **Signal transduction**: Proteins form pathways and networks that convey and compute the sensed input.
3   **Effector processes**: Effector proteins regulate transcription, translation or protein synthesis, protein degradation and cellular functions such as the cell cycle. This is the actual adaptation to exercise.

The signal transduction hypothesis of adaptation is illustrated in Figure 3.1 and then explained in more detail in the following text.

Figure 3.1 Signal transduction hypothesis of adaptation. (1) In the first step exercise leads to changes in physical (e.g. tension) and chemical (change of the Ca^{2+} concentration), which are exercise signals. These signals regulate the activity of sensor proteins (SE). In the second step sensor proteins then regulate the activity of signalling proteins (SP) by phosphorylation and other modifications. Signalling proteins are part of signal transduction pathways or networks which convey and compute this information. The third step is the regulation of adaptation regulators (AR) by signal transduction pathways. This can be (2) the activation or inhibition of transcription factors and changes in gene expression, (3) the regulation of translation regulators which up- or downregulate translation or protein synthesis and (4) the regulation of protein breakdown by various mechanisms. Finally, exercise can regulate processes such as cell division or cell death by the aforementioned mechanisms. All these effects together are the adaptation to exercise.

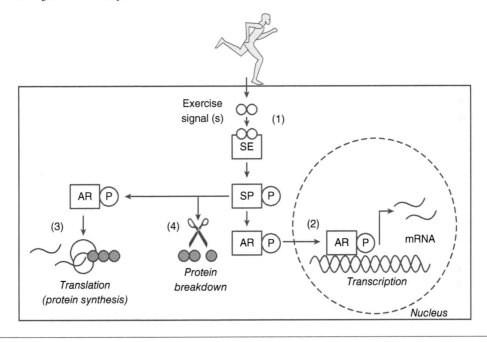

STEP 1: SENSING OF EXERCISE-RELATED SIGNALS

Sensor proteins are found on the cell membrane (e.g. hormone receptors) and also within all cells. They can be further divided into **small molecule sensors, hormone and neurotransmitter receptors** and **other sensors**. Small molecule sensors are proteins that contain a domain capable of binding molecules such as calcium ions (Ca^{2+}) and this enables the cell to sense changes in the concentration of this molecule. We use two examples to explain the function of small molecule sensors: the AMP and glycogen sensor AMPK and the Ca^{2+} sensor calmodulin.

AMPK (AMP-activated kinase) is a protein complex that comprises three different proteins; in other words it is a heterotrimer. This AMPK complex is activated by ADP and AMP binding, which leads to the activation of AMPK via phosphorylation of Thr172 by upstream kinases (see Chapter 4). In contrast, glycogen, a much larger molecule, binds to the glycogen-binding domains and inhibits AMPK (Hardie et al., 2012).

Calmodulin senses Ca^{2+} via four so-called EF-hand motifs (i.e. a series of amino acids to which Ca^{2+} can bind). Increases in Ca^{2+} concentration thus not only trigger muscle contractions but also 'inform' skeletal and cardiac muscle of this event as Ca^{2+} will bind more to calmodulin in contracting muscles. Calmodulin then co-activates proteins such as calcineurin, which promotes the expression of slow muscle fibre genes as an adaptation (Chin et al., 1998).

Cells also sense signals from the superior endocrine and nervous systems via hormone and neurotransmitter receptors. An example of a hormone receptor is the β-adrenergic receptor, which responds to the catecholamines adrenaline and noradrenaline. The receptor protein is interwoven with the cell membrane. Binding of adrenaline to the extracellular domain changes the intracellular region, which activates so-called G-proteins. This instigates a cascade of events including opening of membrane Ca^{2+} channels and stimulation of adenylate cyclase, which is the enzyme that produces cyclic AMP (cAMP). This has immediate effects on muscle metabolism and also triggers metabolic adaptations (Pearen et al., 2008).

Finally, there are other cellular receptors that sense signals other than small molecules or the input from the nervous and endocrine systems. Such sensors measure stretch, electrical potentials, proteins and molecules on the surface of other cells and/or length or tension/force. We will not discuss these sensors in detail. However, it is worth noting that the poorly understood **mechanoreceptor** that links muscle 'loading' to protein synthesis after resistance exercise (Hornberger, 2011) is one of the most important receptors from the point of view of exercise physiology.

STEP 2: TRANSDUCTION OF EXERCISE-RELATED SIGNALS

The second step of signal transduction involves the computation of the signals detected by **sensor proteins. Signal transduction pathways** convey and compute the sensed information using several mechanisms, including:

- **protein–protein interactions**: binding between proteins and other molecules allows information to be conveyed from one protein to the next;
- **protein modifications**: covalent modifications, including phosphorylation, acetylation, glycosylation etc., change the activity of proteins by causing changes in the shape of the protein;
- **translocation**: movement of signal transduction proteins between the nucleus, cytoplasm, membrane, organelles and from inside to outside the cell and vice versa allow signals to be transported through space;
- **synthesis and degradation**: changes in the concentration of signalling proteins can amplify or terminate intracellular signals.

The above mechanisms are linked such that several of the individual mechanisms contribute to each step of signal transduction. For example, protein–protein interaction is necessary for one protein to bind and modify the next. The change in protein shape due to modification may then expose a localization signal that causes the protein to translocate. In some cases modification causes translocation to the nucleus and affects gene transcription. In the case of ubiquitination, modification causes translocation to the proteasome where the protein is degraded and this terminates the signalling event. For the sake of clarity we will discuss the transduction processes individually in more detail below.

Protein–protein and other molecule interactions

Proteins need to be in contact with each other so that they can modify one another and convey information. Furthermore, many cellular functions rely on multimeric protein complexes. An example of this is mTORC1, which consists of several proteins and is a key regulator of ribosomal translation implicated in nutrition and exercise-induced protein synthesis (see Chapter 6). Protein–protein interactions are based on relatively weak chemical bonds and generally involve protein-binding domains, which function like the interfaces of 'Lego' bricks. For example, **WW domains** ('W' is the single-letter code for tryptophan) on some proteins interact with **proline-rich motifs** (e.g. PPxY) on other proteins (Ilsley et al., 2002). The knowledge of such domains and interactions is crucial for understanding signal transduction pathways and thus a personal comment by Marius Sudol, who was one of the discoverers of the WW domain, is included here to give an example how such research is done (Box 3.1). In several cases **transcription factors**, which are proteins that bind to DNA to enhance gene transcription, need to join together in order to recognize and bind DNA sequences. For example, NF-κB proteins are known to need to form a dimer to bind to DNA.

BOX 3.1 MARIUS SUDOL. A BRIEF HISTORICAL ACCOUNT OF THE WW DOMAIN DISCOVERY

The WW domain as a functional module was discovered serendipitously, and the discovery was possible because of the supportive and intensely research-oriented academic environment of my Alma Mater, The Rockefeller University.

In 1978 I was accepted to The Rockefeller University as a graduate student, one of 20 students selected per year from numerous applicants around the world. The day I received the telegram from Dean James Hirsch about my acceptance was one of the happiest days of my life. I was already married and my wife Anna, who was a biologist, joined me two months later and was accepted to the graduate program as well. We made history at the University as the first married couple accepted to the graduate program. I joined the laboratory of Ed Reich, who worked on a gene induced by Rous sarcoma transforming virus, a plasminogen activator gene. Anna joined the laboratory of Zanvil Cohn, who was a prominent immunologist, an august scientist and the most liked mentor at the University. Thanks to his magnetic personality, he was able to assemble a large "family of researchers" who literally transformed the field of cellular immunology into molecular science.

After graduation in 1983, I decided to stay at "The Rock" (as we warmly refer to the University among alumni) and joined the laboratory of Hidesaburo Hanafusa, who worked on the Src oncogene. When I secured my first NIH grant, Hidesaburo generously gave me the freedom to pursue my own research. In the early 1980s, after learning that the Src gene product was a protein tyrosine kinase, almost everybody was hunting for Src kinase substrates. It was an intense and competitive quest among many laboratories worldwide. Over coffee break discussions with Anna and members of the Zanvil Cohn lab, the idea of using Niels Jerne's anti-idiotypic approach

(Continued)

BOX 3.1 (CONTINUED)

for the identification of Src substrates surfaced several times. I was not convinced about this approach because the biochemical and biophysical understanding of the anti-idiotype was only intuitive and quite nebulous. Nevertheless, when two rabbits, which were ordered by one of my colleagues in the Hanafusa lab, became available for free because of a sudden change in his project, I accepted the bunnies that otherwise would be terminated. I injected them with milligram quantities of highly purified IgG raised against the SH3 domain of Yes kinase, one of the closely related Src family kinases. Several months later, the sera from both rabbits were immunoreactive and allowed the functional cloning of a protein that we called Yes kinase-associated protein, YAP. At first we were frustrated because, in our hands, YAP was not phosphorylated on tyrosine by Yes kinase, as we had hypothesized, and it did not modulate Yes kinase activity. Moreover, the complex between YAP and Yes was not robust and was actually difficult to detect. However, when we cloned cDNAs of chicken, mouse and human YAPs, we noticed an insert of 38 amino acids in one of the longer cDNAs. My technician David Lehman was the first to notice this insert and called the unusual cDNA a partially spliced message, which he pronounced in one of the lab meetings as a cDNA of YAP that should be disregarded. Seeing that the insert did not contain any stop codons and did not have splice sites at the junctions, plus it encoded sequence that was repeated as a semi-conserved block of amino acids within the long isoform of YAP, I smiled. The YAP cDNA cloning paper published in the *Oncogene* journal was dedicated to Anna because of her advice on Jerne's approach. The paper with the comparison of YAP cDNA sequences was published in *JBC* with David Lehman as a co-author. The detailed delineation of the WW domain by various computer programs, which were state-of-the-art at that time, was a collaborative effort with my good colleague, Peer Bork from EMBL in Heidelberg, and it was reported in *TiBS* before the *JBC* publication.

Soon after reports of WW domains were published, the PPxY ligand of the WW domain was identified by my student Henry Chen and communicated in *PNAS* by Hidesaburo Hanafusa. I know that Hidesaburo was proud of seeing our innovative study submitted for publication, which he nourished from a polite and self-imposed distance, by thoughtful advice and constant encouragement. Being on the same floor in Dr. Hanafusa's laboratory where Bruce Mayer, a graduate student, had just identified the SH3 domain in Crk oncogene, and witnessing the identification of the PxxP ligand of the SH3 domain by the laboratory of David Baltimore, located two floors below, helped our work tremendously. We simply were at the right place, at the right time and yes, we were well prepared to unveil one of the smallest modular protein domains known today—the WW domain.

Marius Sudol, May 2013

Protein modifications

Signalling proteins are modified by the addition or removal of side groups to specific amino acid residues. This usually occurs via strong covalent bonds and changes the shape (or conformation) of the protein. Several small chemical groups can be attached to different amino acids of a protein. Often, a particular modification may be prerequisite to the next, and several residues may need to be modified in a sequential manner to change the protein's function.

There are probably more than 200 different types of protein modification, although not all of them have well-described roles in signal transduction. We have listed the most commonly studied protein modifications, the enzymes that catalyse these modifications and the amino acids that are modified in Table 3.1.

Table 3.1 Protein modifications, enzymes that catalyse such reactions and the amino acids (residues) within a protein that are modified

MODIFICATION (CHEMICAL GROUP THAT IS ADDED OR REMOVED) AND KEY REVIEW PAPER	MODIFYING ENZYME(S)	AMINO ACIDS (RESIDUES) MODIFIED
Phosphorylation, dephosphorylation (PO_4^{3-} or P_i)	Kinases, phosphatases	Ser, Thr, Tyr (all have an OH group)
Ubiquitination: the addition of a 8.5 kDa peptide termed 'ubiquitin'	Ubiqutin ligases and deubiquitinases	Lys
Sumoylation: the addition of a 12 kDa peptide termed 'Sumo'	Sumo ligases and Sumo-specific proteases	Lys
Acetylation and deacetylation (CH_3CO)	Acetylases and deacetylases	Lys
Methylation and demethylation (CH_3)	Methyltransferases and demethylases	Lys, Arg
Glycosylation: the addition of a glycan such as O-GlcNAc to a protein	Glycosyltransferases	Ser, Thr, Asn
Fatty acylation and prenylation (myristate, palmitate, farnesyl and geranylgeranyl)	Fat-adding and -removing enzymes (e.g. prenylases)	Small amino acids such as Gly or Cys

Cohen (2002); Glickman and Ciechanover (2002); Johnson (2004); Mischerikow and Heck (2011); Erce et al. (2012); Ohtsubo and Marth (2006); Resh (2006).

The most studied protein modification is phosphorylation. Edmond Fischer and Edwin Krebs (Fischer and Krebs, 1955) were the first to make the link between the phosphorylation of a protein and stimulation of its enzymatic activity. They studied the effects of phosphorylation on muscle glycogen phosphorylase activity and were awarded the Nobel Prize in Physiology or Medicine in 1992 for showing that reversible protein phosphorylation is a key regulatory mechanism for this enzyme. Proteins are phosphorylated on **Ser** or **Thr** residues by serine/threonine kinases and on **Tyr** residues by tyrosine kinases (see Table 3.1). Ser, Thr and Tyr each contain a hydroxyl (OH) moiety to which an inorganic phosphate taken from ATP is bound: ATP + protein $\leftrightarrow$ ADP + protein-P. Note that the P stands for inorganic phosphate or PO_4^{3-}, which is not just a phosphorus atom. At physiological pH, inorganic phosphate is a negatively charged group and this changes the conformation (shape) of the protein by pushing away other negative charges and attracting positive charges on amino acids. Such conformational changes explain why phosphorylation affects the function of a protein.

It has been estimated that around one-third of all proteins are phosphorylated (Cohen, 2002), which gives an indication of the importance of this modification as a regulatory mechanism. Indeed, approximately 500 human genes encode protein kinases, which represents about 1.7% of the entire human genome (Manning et al., 2002). Phosphorylation is a reversible modification and phosphorylated proteins can be dephosphorylated by serine/threonine or tyrosine phosphatases. The number of tyrosine kinases is roughly the same as the number of tyrosine phosphatases (Alonso et al., 2004). However, the number of serine/threonine phosphatases ($\approx$40) is much less than the number of known serine/threonine kinases (Cohen, 2002). This difference in the number of phosphatases to kinases is likely due to their different mechanisms of action. Kinases typically are highly selective, which is based on their ability to recognize a particular amino acid motif on the phosphorylated protein. In contrast, phosphatases can usually dephosphorylate many proteins but they target individual proteins for dephosphorylation by interacting with regulator subunits. For example, protein phosphatase-1 (PP1) interacts with the PP1 regulatory subunit 3A to specifically dephosphorylate muscle glycogen phosphorylase (Toole and Cohen, 2007).

Translocation of signal transduction proteins

Signal transduction also depends on the controlled movement of signal transduction proteins within the cell. One of the most important transport events is the transport of a signal transduction protein between the nucleus and the cytoplasm. In many cases such transport depends on the activation of a **nuclear localization signal** (NLS) on a protein. NLS are recognized by proteins that transport protein cargo through **nuclear pores** from the cytosol into the nucleus of a cell. Usually, the activation of NLS involves protein modification or a change in protein–protein interaction, which exposes the NLS. For example, NF-κB is bound to its inhibitor, IκB, when it is in the cytosol. This is because IκB masks the NLS of NF-κB, which prevents it from transiting into the nucleus (Roff et al., 1996).

Synthesis and degradation of signalling proteins

Finally, signals can be transduced by changing the **synthesis** or **degradation** rates of signal transduction proteins. Examples are the hypoxia sensor HIF-1 and the myogenic regulatory factors MyoD, Myf5, myogenin and MRF4, which regulate muscle development or myogenesis. These signals are then terminated by selective degradation of the transcription factors. For example, the MyoD signal is stopped when the ubiquitin ligase MAFbx ligates ubiquitin to MyoD (Tintignac et al., 2005). This modification marks MyoD for degradation by the 26S proteasome. Thus changes in the transcription, translation or degradation of signal transduction proteins frequently occur and cause changes in signal transduction.

STEP 3: EFFECTOR PROTEINS REGULATE EXERCISE-INDUCED ADAPTATION

Exercise-activated signal transduction pathways regulate the concentration or activity of adaptation regulators which in turn regulate transcription, translation, protein breakdown or other cellular processes such as cell division. This results in changes in the concentrations of metabolic enzymes or contractile proteins and thus the function of cells and organs. In other words, this is the final step by which exercise changes our bodies via adaptation.

Regulation of gene expression

Exercise results in the changed transcription of many genes. In general, the transcription of a gene can be broken down into eight steps (Fuda et al., 2009). Without going into all the detail these eight steps control:

- the unwinding of the tightly packed DNA, which is termed **chromatin remodelling**, which makes genes accessible;
- the recruitment of RNA polymerase II to the start of a gene;
- the rate at which RNA polymerase II transcribes the DNA of the gene into RNA; and
- the termination of transcription and recycling of RNA polymerase II.

We will now discuss some key events during transcriptional regulation by exercise or other signals.

Chromatin remodelling or epigenetic regulation: the opening up or packaging of DNA

The DNA in the human genome has been estimated to be almost 2 m long. In order to fit into a tiny nucleus it must be tightly packaged. This DNA packaging is achieved by wrapping DNA around complexes which are built from eight **histone** proteins. DNA together with histone complexes is termed **chromatin** and the DNA wrapped around one histone complex is called a **nucleosome**.

The tight packaging of DNA in chromatin must be unravelled before a gene can be transcribed. The packaging and unpackaging of DNA is known as **chromatin remodelling** and histone modifications are key mechanisms in the regulation of this process. Mapping of packaged and unpackaged DNA on a genome-wide scale (Dunham et al., 2012) has revealed that histone tail modifications are cell specific and mark genes, transcription start sites and stretches of regulatory DNA, via which gene expression is regulated. Indeed, signal transduction pathways modulate chromatin remodelling by methylating (CH_3, methyl group), acetylating (CH_3CO, acetyl group) and phosphorylating histone proteins, especially in the tail regions of histones H3 and H4. The enzymes that catalyse these modifications include histone methyltransferases and histone demethylases as well as histone acetyltransferases and deacetylases. The resultant modifications are abbreviated stating first the histone number, second the amino acid which is modified and finally the type of modification (ac stands for acetylation, me1, me2, m3 for methylation, dimethylation and trimethylation, respectively). For example, H3K27me3 refers to the trimethylation of lysine 27 (K is the one-letter abbreviation for lysine) of histone 3. Histone modifications are one of the many regulatory events that contribute to the adaptation to exercise (McGee and Hargreaves, 2011) and in Chapter 4 we will discuss some histone modifications in skeletal muscle that are important for the development of different muscle fibre types.

Transcription factors, proximal promoters and enhancers

After a stretch of DNA has been opened up by chromatin remodelling it can be accessed by transcription factors to affect the expression of genes. We will now describe this process. A stretch of DNA called the **core promoter** is found close to the start of genes and serves as a docking site for RNA polymerase II as well as proteins known as basic transcription factors that together form the **transcription pre-initiation complex**. The assembly of the transcription pre-initiation complex is sufficient for low levels of gene transcription.

Exercise and otherwise activated signal transduction pathways modulate transcription via sequence-specific transcription factors that bind proximal promoter and enhancer regions and greatly enhance the rate of transcription.

The **proximal promoter** is a region of DNA up to a few hundred base pairs upstream from the core promoter, whereas **enhancer regions** can often be hundreds of thousands of base pairs away from the transcription start site (Sanyal et al., 2012). Loops in the DNA bring distal enhancer regions close to the proximal promoter. Indeed, it is very rare for an enhancer to affect genes that are adjacent to it in the DNA sequence. It has been predicted that there are ≈70 000 regions with promoter-like features and ≈400 000 regions with enhancer-like features in the human genome (Dunham et al., 2012), which gives an indication of the complexity of this process. Indeed, the myogenic regulatory factor MyoD binds to >20 000 DNA sites in muscle cells (Cao et al., 2010).

Estimates suggest there are ≈1800 DNA-binding transcription factors in the human genome (Vaquerizas et al., 2009). Of these, the regulatory transcription factors have a **DNA-binding domain** that recognize a specific **DNA motif**. Some common types of DNA-binding domain include the C_2H_2 zinc-finger, homeodomain, and helix-loop-helix (Vaquerizas et al., 2009). Often it is necessary for transcription factors to form homo- or heterodimers in order to create a correct DNA-binding motif. These transcription factors are further regulated by the binding of co-factors. A much researched co-factor relating to exercise adaptations is PGC-1α (peroxisome proliferator activated receptor co-factor 1α) (Scarpulla, 2002).

Post-transcriptional regulation of mRNA

The primary transcript produced by RNA polymerase II undergoes post-transcriptional processing to become messenger RNA (mRNA) that can be translated into protein. The majority of human genes consist of multiple **exons**, which is the part of the gene that encodes the protein interspersed by **introns** or intervening sequences. Introns are removed by spliceosomes and, depending on which exons are retained, different splice variants of the gene can be created. A well-known example relevant to exercise physiology is the alternative splicing of the insulin-like growth factor I (IGF-I) gene to create mechano growth factor (MGF), which was discovered by Geoffrey Goldspink's group (Yang et al., 1996). The activity of spliceosomes is in part regulated by proteins that recognize and mark different splice sites, but as yet it is unknown how exercise regulates alternative splicing.

Post-transcriptional processing can also be modulated by small RNA species, typically ≈21–26 nucleotides long, known as siRNA (small interfering RNA) and miRNA (micro RNA) (Valencia-Sanchez et al., 2006). miRNAs and siRNAs are produced in different ways but have similar functions on selective mRNA degradation. After their synthesis, miRNA and siRNA form part of the RNA-inducing silencer complex (RISC). The miRNA and siRNA confer specificity to the RISC complex by binding complementary sequences in their target mRNA, which results in either the degradation of the target mRNA or inhibition of mRNA translation. It is estimated that up to 30% of human gene transcripts may be affected by this type of regulation. Changes in miRNA occur in response to both endurance (Safdar et al., 2009) and resistance (McCarthy and Esser, 2007) exercise and so may be involved in modulating the adaptive response to training. Indeed, miRNA expression also differs between high and low responders to resistance training (Davidsen et al., 2011).

Regulation of mRNA translation to synthesize proteins

Mature mRNA is then translated into protein by ribosomes. Ribosomes consist of a small 40S and a large 60S subunit of rRNA (ribosomal RNA that join together to create an 80S ribosome). Ribosomes are controlled by ribosomal proteins, which are regulated especially by the mTOR signalling network. In the late 1990s it was noted that translational regulators were modulated in hypertrophying muscles (Baar and Esser, 1999) and subsequent research has shown that activation of translation via the mTOR pathway is a key mechanism by which resistance exercise increases translation, which eventually causes muscle hypertrophy (see Chapter 6).

Translation occurs in three stages:

1   **initiation**, where the ribosome and mRNA are assembled;
2   **elongation**, where the mRNA is read by the ribosome and translated into an amino acid chain and
3   **termination**.

We will now discuss the regulation of these steps. In particular translation initiation, which is the most tightly regulated by exercise and other signals and is the major method by which resistance exercise causes muscle hypertrophy.

Translation initiation is a multi-step process regulated by **eukaryotic initiation factors (eIF)** and culminates in formation of the 80S initiation complex. Prior to translation the subunits of the ribosome are separate and the 40S subunit must be 'primed' by binding of a transfer RNA (tRNA). The tRNA contains the complementary sequence for the AUG start codon of mRNA and is also attached to the amino acid methionine. Proteins are synthesized beginning from the amine N-terminal and ending with a carboxyl C-terminal, therefore all proteins begin with a methionine residue although this is often removed after translation. The priming of the 40S ribosome is regulated in part by eIF3 and creates a 43S pre-initiation complex. Next eIF4 guides the 5'-end (named 5'-cap) of mRNA into the 43S complex and the ribosome proceeds along the mRNA until the start codon is found. Other eIFs then recruit the 60S subunit to create the 80S initiation complex, and synthesis of the polypeptide chain begins.

When inactive, the initiation factors eIF3 and eIF4 are bound to the signalling proteins p70 S6k and 4E-BP1, respectively. Resistance exercise, nutrients and hormones such as insulin activate the mTOR pathway, which in turn phosphorylates p70 S6k and 4E-BP1. As a consequence, eIF3 and eIF4 detach from p70 S6k and 4E-BP1 and contribute to the assembly of ribosomes. More on this topic in Chapter 4.

After assembly of the 80S initiation complex, elongation follows which is controlled by eukaryotic elongation factors (eEF) and uses the codons of the mRNA as a template to recruit the correct sequence of tRNA. Elongation proceeds in a cycle involving (i) binding of activated tRNA (i.e. tRNA bound with its respective amino acid), (ii) peptide bond formation and (iii) release of the inactive tRNA. The ribosome moves along the mRNA in this manner until a stop codon is reached, at which point the process is terminated by **eukaryotic release factors (eRF)**.

Regulation of protein breakdown

Signal transduction pathways also regulate the lifespan of proteins by controlling their degradation rate. Protein degradation can occur via at least three processes:

- ubiquitin–proteasome pathway;
- autophagy–lysosome pathway; or
- cytosolic proteolytic systems.

The cytosolic proteolytic systems include the caspase and calpain proteases, which are associated with apoptotic cell death and Ca^{2+}-activated proteolysis, respectively. These systems are not strongly implicated in protein turnover, so, for the remainder of this section we will focus on the roles of the ubiquitin–proteasome and the autophagy–lysosome pathways in exercise-induced adaptation.

The ubiquitin–proteasome pathway

The majority of intracellular proteins are degraded by the ubiquitin–proteasome pathway, which involves ubiquitination of target proteins followed by their degradation in the 26S proteasome. Proteins are selected for degradation by attaching ubiquitin to the ε-amino group of a lysine residue or in some cases to the N-terminus of the protein. Ubiquitin itself is a small protein with 76 amino acid residues and it is so named because it is ubiquitous in all tissues. Ubiquitination is an energy-requiring process carried out by a series of three types of enzymes, named E1, E2 and E3. Ubiquitin is first of all 'primed' by an E1-activating enzyme and then transferred to one of several E2-conjugating enzymes, which in turn interact with numerous E3 ligases. The actual ubiquitination of target proteins is performed by E2–E3 pairs, but it is the E3 ligase that confers specificity to the process. Two E3 ligases in skeletal muscle are MuRF1 (muscle ring finger 1) and MAFbx (muscle atrophy F-box), which were first discovered in models of muscle atrophy (Bodine et al., 2001). However, many other E3 ligases also regulate muscle protein degradation, and there are more than 600 different E3 ligases in the human genome (Li et al., 2008). Ubiquitinated proteins are then digested via the 26S proteasome, which is barrel-shaped complex built from more than 30 proteins.

The autophagy–lysosome pathway

Autophagy occurs through various mechanisms that differ in the way they capture proteins or organelles and deliver them to the lysosome for degradation. Macro-autophagy involves entire regions of the cytosol or specific organelles and protein complexes being engulfed by a vacuole known as an autophagosome, which then fuses with the lysosome. Micro-autophagy involves the direct uptake of cytosolic components into lysosomes. In addition, more selective types of autophagy, known as chaperone-mediated autophagy and chaperone-assisted selective autophagy, are able to degrade specific proteins. Although autophagy may seem highly destructive, autophagic processes are involved in the adaptation of muscle to exercise training (He et al., 2012) and mechanical stress (Ulbricht and Hohfeld, 2013).

Summary: the signal transduction hypothesis of adaptation

In the text above, we have described in detail the mechanisms by which exercise causes adaptation. We acknowledge that it is easy to get lost because of all the detail and for this reason we will now summarize the most important points illustrated in Figure 3.1. When we exercise our muscles, tension and the concentrations of Ca^{2+}, AMP and ADP all increase intracellularly, while adrenaline and noradrenaline increase in the circulation and glycogen and oxygen decrease within the cells. All of these exercise signals are sensed by sensor proteins which can be described as the eyes and ears of cells. The plethora of signals is then conveyed and computed by signal transduction pathways and/or networks, which resemble the nervous system of the cells. Such signalling works by protein–protein binding, phosphorylation and other modifications, and signalling molecules move, for example, in-between the nucleus and cytosol. Finally, the exercise-activated downstream signalling molecules regulate adaptation regulators, which include transcription factors, regulators of protein synthesis and protein breakdown and regulators of processes such as cell division and cell death. The combined action of these adaptation regulators changes our organs and this is the adaptation to exercise.

METHODS USED TO INVESTIGATE SIGNAL TRANSDUCTION

In the second part of this chapter we will introduce practical wet laboratory methods that are used to study the signal transduction hypothesis of adaptation and related questions. Specifically, we will introduce RT-qPCR and Western blotting as they are routinely used by molecular exercise physiologists to measure mRNA and protein concentrations. Before reading this text you should be familiar with basic wet laboratory methods which are described in Chapter 1 and with DNA methods, especially the polymerase chain reaction (PCR) which we have covered in Chapter 2.

Real-time quantitative PCR (RT-qPCR)

In the section above we have explained that adaptation to exercise frequently involves the increased or decreased expression of genes or, in other words, changes in mRNA concentrations. Quantitative RT-PCR is the key method to measure mRNA. It involves the extraction of RNA, the reverse transcription into complementary DNA, termed cDNA and then real-time quantitative RT-PCR (RT-qPCR).

The extraction of RNA presents some specific challenges. Chief among which are aggressive RNAse (ribonuclease) enzymes that, when active, quickly degrade RNA into useless fragments. To avoid this it is important to use RNAse-free plasticware and prevent contamination by wearing fresh gloves at all times, and by cleaning surfaces and instruments with an RNAse-inhibiting solution. In addition, buffers and reagents used during the assay should be made with RNAse-free water. This is water treated with diethylpyrocarbonate (DEPC), a chemical that deactivates RNAses.

RNA is extracted with reagents such as Trizol, which are based on the one-step protocol developed by Chomczynski and Sacchi (1987). This method exploits the specific chemical properties of RNA to isolate it from the other macromolecules. The reagent contains phenol, chloroform and the guanidinium thiocyanate, which denatures proteins including RNases. Muscle is minced in this buffer using scissors and then homogenized using an electric homogenizer. Centrifugation of this mixture creates a pellet at the bottom of the tube that contains protein, DNA and organic matter such as lipids. This enables the solution containing RNA to be collected. Salt is then removed from this solution by washing the RNA with alcohol and the RNA precipitates and is collected at bottom of the tube after centrifugation. The RNA concentration is then measure by spectrophotometry. Based on the extinction coefficient of RNA at a wavelength of 260 nm, 1 absorbance unit corresponds to a RNA concentration of 40 μg/mL. RNA absorbance is maximal at 260 nm, whereas absorbance by amino acids is greatest at 280 nm (Figure 3.2) and thus the A_{260nm}/A_{280nm} ratio gives an estimate of the purity of RNA extracted. Pure RNA has an A_{260}/A_{280} ratio of 1.9–2.1, typically a value greater than 1.8 is required for gene expression analysis. Note, however, this measurement does not show whether the RNA is intact or has been degraded by RNAses.

To determine whether RNAses have degraded the RNA, the sample can be run on a denaturing agarose gel. This separates the RNA molecules according to size and allows them to be visualized. The protocol described above extracts total RNA, approximately 80–90% of which is ribosomal RNA (rRNA). Messenger RNA (mRNA) represents approximately 2.5–5% of total RNA, and the remainder is transfer RNA (tRNA). Therefore, when viewed on an agarose gel (Figure 3.2) the major bands seen are rRNA, which in eukaryotic cells are of 28S and 18S size. Individual mRNA are much smaller and less abundant than rRNA and cannot be distinguished on the agarose gel. However, the quality of the 28S and 18S rRNA bands can be used to infer overall sample quality. Generally, these bands should be sharply defined and the intensity of the 28S band should be twice that of the 18S band. If there is streaking or laddering of these bands the quality of the sample has been compromised.

In the next step the isolated RNA is converted into cDNA via reverse transcription using a **reverse transcriptase** enzyme which is found in retroviruses and transcribes RNA into DNA. A common method of reverse transcription involves oligo dT (thymine) primers that recognize the poly-A (poly-adenine) tail of mRNA. Reverse

Figure 3.2 RT-qPCR equipment and results. (a) Thermal cycler for RT-qPCR. The instrument contains a thermal cycler and a fluorescent analyser that measures the fluorescence of each sample once during each cycle. The fluorescence depends on a dye that becomes fluorescent when a double-stranded PCR product is formed. (b) Typical RT-qPCR output showing the increase of fluorescence during each cycle. In the samples that have the highest cDNA concentration for the gene investigated the fluorescence becomes detectable from ≈15 cycles onwards. In contrast, in samples with the lowest cDNA concentration the fluorescence only becomes measurable from ≈30 cycles onwards.

(a)

(b)

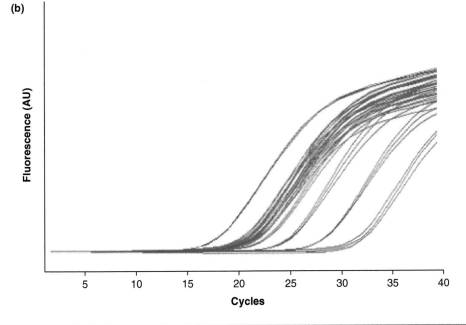

transcriptase then creates cDNA in the 3′ to 5′ direction starting at the poly-A tail. Theoretically, this produces cDNA for all mRNA present in the tissue and the sample can be used to measure the expression of numerous different genes.

The specificity of the RT-qPCR process is due to the use of primers that are complementary to the sequence of the target gene. Two primers are required that flank a region of typically 100–300 bp. If oligo dT primers are used it is best to design primers that isolate a section near the 3′-end, but many other factors also influence primer design. Indeed, the complexity of successful primer design has given rise to numerous software tools to help design suitable primers. In addition, it is important to consider splice variations, potential single-nucleotide polymorphisms (SNPs) and other variations to ensure that different variants of mRNA are all amplified.

Real-time PCR is conducted on a special thermal cycler, which includes a fluorimeter that enables the concentration of the PCR product to be quantified during each cycle. For this, fluorescent reporter molecules are needed, which may be either sequence-specific probes such as Taqman probes or non-specific DNA-binding dyes such as SYBR green. Here, we will describe the use of non-specific DNA-binding dyes. A common reagent is SYBR, which emits fluorescent green signal. The fluorescence of SYBR in solution is very low but its signal multiplies more than 1000-fold when it binds to double-stranded DNA. Thus, as the target sequence is amplified by PCR the concentration of double-stranded DNA multiplies and the fluorescence signal from SYBR increases. The instrument quantifies the fluorescence signal after each synthesis step and the data are plotted against the cycle number to create an amplification profile (Figure 3.2).

After the qPCR programme has finished, the fluorescence signal is monitored while the temperature of the sample is gradually increased. The temperature at which double-stranded DNA dissociates depends upon the length of the DNA sequence. Therefore, if the fluorescence signal is primarily due to numerous copies of an identical sequence there will be a single point/temperature at which the fluorescence signal diminishes. In contrast, if the sample is contaminated with non-specific PCR products of different lengths, there will be numerous thermal transitions in the fluorescence signal.

The common use of RT-qPCR is a semi-quantitative technique because quantitation of a transcript is performed relative to a reference gene. Typical reference genes include GAPDH, tubulin, β_2-microglobulin or 18S ribosomal RNA. The choice of an appropriate reference gene should be based on empirical knowledge that this gene is stably expressed throughout the experiment states. Alternatively, the average between two reference genes is sometimes used. Absolute quantitation of transcripts can also be performed, but is less common in the exercise physiology literature.

Western blotting

The most important component of our cells is arguably protein and changes in, for example, mitochondrial or sarcomeric protein determine our endurance and strength, respectively. Moreover, most signal transduction is done by proteins and for these reasons molecular exercise physiologists measure the concentrations of individual proteins and also protein modifications such as covalent phosphorylation. The method used to measure individual proteins is termed Western blotting. It was developed in the late 1970s, almost simultaneously across the laboratories of George Stark (Renart et al., 1979), Julian Gordon (Towbin et al., 1979) and W Neal Burnette (Burnette, 1981). To this day, it remains one of the most commonly used techniques for measuring the abundance of a protein. The name is a word play coined by W Neal Burnette on 'Southern blot', which is a similar method for DNA developed by Ed Southern. Western blots are typically used to measure the abundance of a protein relative to a reference protein, and are therefore a semi-quantitative assay.

The measurement of proteins with Western blotting, which is also termed immuno-blotting because of the use of antibodies, involves four key steps:

- protein extraction and denaturation;
- separation of proteins on the basis of their size (in kDa) by denaturing gel electrophoresis;
- electrotransfer of proteins from inside the gel onto the surface of nitrocellulose or PVDF membranes;
- detection and visualization of the protein of interest using a primary antibody which binds to the protein and a secondary antibody which binds to the primary antibody and allows visualization.

The primary objective of protein extraction for Western blotting is to completely extract the protein of interest. The highly ordered structure of myofibrils and connective tissue mean skeletal muscle is relatively difficult to prepare for protein assays. The greatest protein yields can be achieved by grinding muscle in liquid nitrogen using a mortar and pestle, and then processing the powder in a Potter–Elvehjem homogenizer. This homogenizer shears cells by forcing them between a Teflon pestle and the wall of the glass tube of the homogenizer. Alternatively, muscles can be cut with scissors and minced with an electrical homogenizer. A typical homogenization buffer for extracting soluble proteins from skeletal muscle is a pH-buffered solution containing a surfactant such as Triton x-100 to disrupt membranes as well as protein interactions. Protease inhibitors and EDTA and EGTA are included to inhibit protein-degrading proteases. If the aim is to measure phosphorylated proteins, then phosphatase inhibitors are included to avoid dephosphorylation during extraction.

After homogenization, the homogenized tissue in buffer is centrifuged and the supernatant is collected, which contains the soluble proteins. The protein concentration of the supernatant is then measured using a protein assay such as the Bradford or BCA assay. The final stage is to denature and negatively charge proteins in preparation for electrophoresis. This is achieved by mixing the sample with Laemmli sample buffer (Laemmli, 1970), which contains a strong negatively charged or anionic detergent called SDS (sodium dodecyl sulfate). The samples are heat denatured by incubation in this solution at $\approx 95°C$ for 3–5 min. Under these conditions, proteins unfold completely and the SDS binds with the proteins and prevents them from refolding when the sample is cooled.

Gel electrophoresis is then used to separate the denatured and negatively charged proteins within each sample. Gels are cast between two glass plates and are constructed from acrylamide polymers, which cross-link to form a molecular sieve. Samples containing the same protein concentrations are loaded into wells at the top of the gel and an electric current is passed through the gel. This causes the negatively charged proteins to migrate toward the positive electrode at the bottom of the gel. Small proteins pass through the gel faster than large proteins and thus the procedure effectively orders proteins according to their size (Figure 3.3).

Figure 3.3 Protein electrophoresis equipment and results. (a) Protein electrophoresis set-up. (1) A Power pack is connected to (2) an electrophoresis tank with protein-loaded gels included. (b) Schematic drawing showing how proteins, which are denatured and negatively charged as a result of SDS treatment, migrate over time towards the positive electrode at the bottom of the gel. This allows the separation of denatured protein on the basis of their molecular weight, given in kDa. (c) Result of an electrophoresis experiment where skeletal muscle proteins have been separated as described. For Western blotting these proteins are transferred onto a membrane and incubated with antibodies against a specific protein which is then visualized.

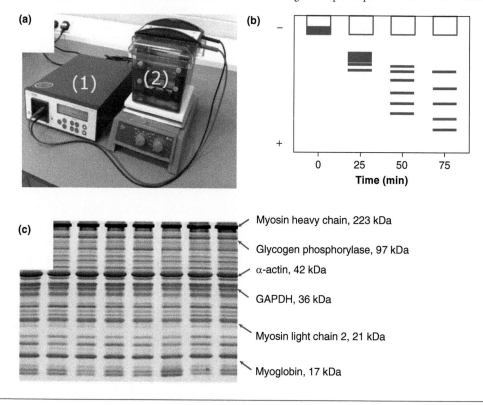

At this stage, the proteins are interwoven within the gel and are inaccessible to antibodies. To overcome this problem, the gel is sandwiched against a sheet of nitrocellulose or PVDF membrane and an electric current is used to elute the proteins from the gel onto the surface of the membrane. This process makes the proteins accessible to antibodies.

Nitrocellulose and PVDF membranes are designed to capture proteins. So, prior to detecting the target protein with an antibody, the membrane must be 'blocked' by incubating it with a protein mixture such as non-fat dried milk. The presence of abundant milk proteins ensures that antibodies incubated with the membrane bind selectively to their target proteins rather than unspecifically to random protein-binding sites on the membrane. Typically, two different antibodies are used in series. The **primary antibody** binds the protein of interest, whereas the **secondary antibody** binds and detects the primary antibody. The secondary antibody is conjugated to a molecule or enzyme that enables it to be visualized. For example, the enzyme horseradish peroxidase (HRP) is commonly used to catalyse a light-emitting reaction which is known as enhanced chemiluminescence (ECL). The light emitted from this process is too weak to be seen

Figure 3.4 Western blot results. Original Western blots showing changes in the inhibitory phosphorylation of the transcriptional co-factor YAP (65 kDa) during the differentiation of mononucleated myoblasts into multinucleated myotubes, which is a key process during skeletal muscle development. While there appear to be some changes in the concentration of total YAP, the greatest change is in the concentration of phospho-YAP Ser127, which increases during differentiation. α-Actin was used as a loading control.

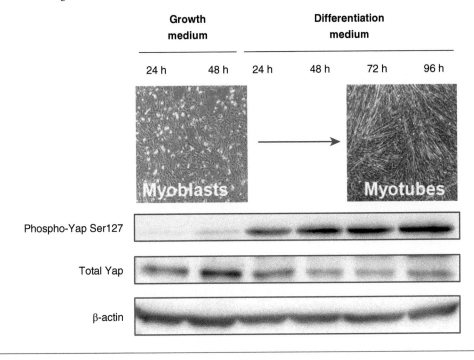

by the naked eye but it can be captured either on photographic film or by a camera system that captures light from the membrane over several minutes to create a digital image (Figure 3.4).

Although care is taken to load equivalent amounts of each sample on to the gel, it is necessary to normalize Western blot data by expressing the quantity of the protein of interest relative to either a reference protein or the total amount of protein on the transfer membrane. Commonly, a highly abundant protein such as the glycolytic enzyme glyceraldehyde 3-phosphate dehydrogenase (GAPDH) or an actin is used as a reference protein. For this, the membrane can be either cut in two if the protein of interest and the reference protein have different molecular weights, or two rounds of antibody staining are performed. Alternatively, total protein staining can be performed directly on the membrane. It is worth noting that a respectable coefficient of variation for Western blotting is about 50% if normalized to a high-abundance reference protein, but can be reduced to about 30% if normalized to a total protein stain (Aldridge et al., 2008).

One of the primary uses of Western blotting is the detection of post-translational modifications. This requires antibodies that detect a particular type of modification at a specific residue in the protein of interest, such as the phosphorylation of threonine 172 of AMPK. Numerous commercial vendors have extensive catalogues of antibodies, which encompass a broad spectrum of proteins and their different modifications.

PERSPECTIVE ON 'OMIC' APPROACHES

The limitations of methods such as RT-qPCR or Western blotting are that only one or a few mRNA or proteins are measured at a time and that the choice of these targets is subjective. An alternative approach is to measure many if not all genetic variations, mRNAs, proteins and molecules in an unbiased approach. Methods such as **next-generation DNA sequencing**, **genome-wide association studies** (GWAS see Chapter 2), **gene expression profiling** or **transcriptomics**, **proteomics** and **metabolomics** allow such an unbiased approach for the DNA sequence, all SNPs, the expression of most genes, changes in many proteins and metabolites, respectively. All of these methods generally result in long lists of data which are then analysed and presented by **bioinformatics**. The above methods can be abbreviated as 'omic' approaches. Here, we discuss gene expression profiling using expression microarrays and proteomics as two key omics methods that are used by molecular exercise physiologists to study the adaptation to exercise and related problems in a non-biased way.

Gene expression profiling using microarrays or gene chips allows researchers to perform the equivalent of thousands of RT-qPCR reactions in one experiment, albeit with lower precision (Figure 3.5). Each gene chip contains tens of thousands of different oligonucleotide probes, to which cDNAs made from mRNAs can bind if present. Each probe is printed in a spot ≈10 μm in diameter, creating a microarray. After being incubated with fluorescent dye-labelled cDNA, the fluorescent signals from each spot are scanned and quantified. Microarrays are specific to the organism and are often able to detect all known and predicted mRNAs or transcripts. An early example relevant to exercise physiology was the comparison of mouse slow- and fast-twitch muscle by Frank W Booth's group (Campbell et al., 2001), which profiled ≈6000 transcripts and discovered transcription factors that were specific to either slow- or fast-twitch muscle. Subsequent developments in our knowledge of the mouse genome as well as improvements in microarray manufacturing processes mean that it is now possible to assay ≈45 000 transcripts (Burniston et al., 2013). Similar microarrays are available for humans. For example, microarrays that screen ≈38 500 human transcripts were recently used to investigate muscle responses to resistance (Gordon et al., 2012) or endurance (Keller et al., 2011) training, respectively.

Bioinformatic and statistical analysis tools are fundamental to the interpretation of genome-wide expression profiling data. Commercial and open access packages are available that link experimental data with information from gene ontology databases and associate each gene with three key characteristics:

- **cellular component**, such as cytosolic, membrane, mitochondrial matrix;
- **biological processes**, such as glycolysis and oxidative phosphorylation; and
- **molecular function**, such as oxidoreductase activity and transcription factor binding.

Such analysis allows researchers to see whether, for example, the expression of genes for oxidative phosphorylation or membrane proteins have changed their expression.

In addition, lists of differentially expressed genes can be mapped against transcription factor binding site and other databases to identify transcription factor-binding sites, signal transduction pathways and other links between the co-regulated genes (e.g. Burniston et al., 2013). This combination of bioinformatics and genome-wide profiling allows researchers to see the bigger picture behind long lists of up- or downregulated

Figure 3.5 Gene expression or transcription microarray analysis. (1) Skeletal muscle or other mRNA is extracted and cDNA is synthesized and labelled with a fluorescent probe. (2) The fluorescence-labelled cDNAs are then added to a microarray chip and hybridized to the probes on the chip in a hybridization oven. (3) The samples are automatically washed and (4) scanned by a microarray scanner. (5) A magnified output of a microarray. Each dot is one probe and the intensity of the probe corresponds to the concentration of mRNA in the original sample.

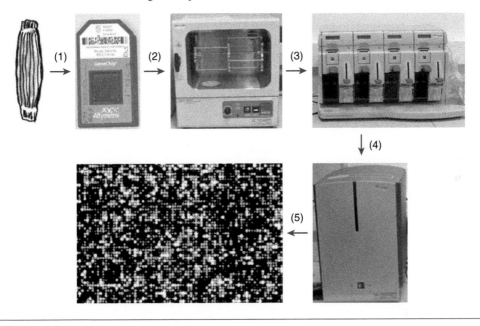

genes. For example, Keller et al. (2011) used genome-wide profiling to study skeletal muscle adaptation to endurance exercise and found that this is associated with enrichment of a gene network directed by a cluster of the transcription factors Runx1, Pax3 and Sox9. In fact, this work goes further and also shows that regulatory miRNA post-transcriptionally regulates these transcription factors. It is probably fair to state that such mechanisms would not have been hypothesized and tested using traditional one-gene-at-a-time approaches. However, future research will show whether these mechanisms are confirmed as important for the adaptation to exercise.

One disadvantage of transcription microarrays is that the probes rely on prior knowledge of the mRNA sequence. A more recent development is to use next-generation sequencing in order to sequence all RNAs. This has been used, for example, to measure the overall RNA response to exercise in thoroughbred horses (Park et al., 2012). In addition to reporting changes in the expression of known genes, this approach discovered over 50 novel transcription factors that currently have unknown functions but were up- or downregulated in response to exercise.

Proteomics is an omics method used to measure, for example, changes in large numbers of proteins in response to exercise. It complements gene expression arrays as the relationship between mRNA and protein is not always linear. The difficulty with proteins, in contrast to RNA extraction and gene expression profiling, is that it is practically impossible to isolate all proteins using a single extraction and also proteins cannot be amplified and the proteome is much more diverse than the transcriptome. The strategy used for protein expression profiling is **2D gel electrophoresis**, which separates proteins based on their isoelectric point and molecular weight. This creates a unique pattern of protein spots and the gel images can be compared to look for differences in the abundance of individual proteins. Protein spots of interest can then be cut from the gel and digested into peptides using the protease trypsin. The peptide mixture is then analysed using a mass spectrometer, which creates a 'peptide mass fingerprint' that can be used to identify the protein (Figure 3.6).

There are many different configurations of mass spectrometer. Conceptually, the simplest mass spectrometer consists of 'source' that transfers ionized peptides into the gas phase and delivers them to a time-of-flight (ToF) mass analyser. As the name suggests, this instrument measures the time it takes for ions to transverse a specified distance. The system is calibrated by recording the time-of-flight of a range of standards of

Figure 3.6 Proteomics strategy. Schematic diagram demonstrating a typical proteomics approach. (a) A skeletal muscle is first homogenized and then the proteins are separated by 2D electrophoresis. (b) Individual proteins are then turned into smaller fragments by tryptic digestion. (c) The protein fragments are then identified by mass spectroscopy. A result showing a pattern for specific protein fragments is shown.

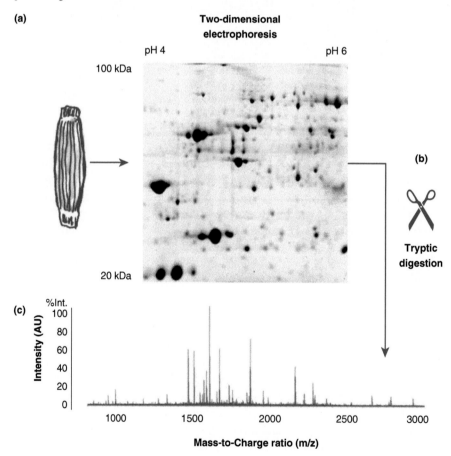

known masses, and unknown peptides can then be analysed. The mass spectrometer creates a spectrum of masses and each peak in the mass spectrum represents the mass of an individual peptide. The list of masses is then compared against a list of predicted masses contained in gene and protein databases and allows the identification of the protein.

We published the first proteomic studies reporting training-induced adaptations in rat (Burniston, 2008) and human (Holloway et al., 2009) skeletal muscle. Samples were separated using 2D electrophoresis, which resolved just over 250 protein spots in human muscle. We observed that proteins were separated into their various species, which may include different splice variants and post-translational states. This is primarily because small changes in amino acid sequence or modifications such as phosphorylation substantially change the molecular weight or isoelectric point of the protein. For example, phosphorylation typically causes the protein to migrate to a more acidic position of the pH scale so this becomes a separate spot on the gel. The major effect of exercise training was to alter the expression of distinct protein species rather than the total abundance of a protein. For example, in human muscle, high-intensity interval training changed both the splice variation and phosphorylation of troponin T, and this occurred in the absence of changes in myosin heavy chain profile. This illustrates how proteomics can be used to investigate muscle phenotype at a finer grain of detail than traditional fibre typing techniques.

Other work relevant to exercise physiology has used proteome mining techniques to catalogue all proteins that can be detected by mass spectrometry. For example, Parker et al. identified over 2000 proteins from biopsy samples of human skeletal muscle. In addition to abundant myofibrillar proteins and metabolic enzymes, the proteins identified included signalling proteins that are mentioned elsewhere in this book, including sensing proteins such as AMPK, signal transduction proteins such as Ca^{2+}-dependent kinases and transcription factors such as Stat3, as well as numerous eukaryotic initiation factors, proteasome subunits and E3 ligases (Parker et al., 2009). These techniques cannot yet be used to compare different samples but it gives a preview of what may soon be possible.

In summary, omics methods allow the subjective or non-biased, large-scale identification of mRNAs or proteins that change their concentration or become modified for example in response to exercise. This allows researchers to develop new hypotheses and gain new insight which is much faster than with one-gene-at-a-time approaches. While DNA and RNA omics approaches now allow investigation of the whole genome or transcriptome, current proteomics technology is capable of seeing perhaps only 1% of the entire proteome, which is nonetheless far greater than what could be achieved using Western blotting and has added important knowledge.

SUMMARY

Our understanding of adaptation to exercise has changed fundamentally, particularly during the last 30 years. The signal transduction hypothesis has replaced earlier hypotheses and is now well supported by experimental data which are discussed throughout this book. Conceptually, the adaptation to exercise can be broken down into three steps. First sensor proteins sense changes in molecules such as Ca^{2+}, AMP, ADP, glycogen and oxygen as well as signals such as hormones and mechanical stress. Second, activated sensor proteins instigate signal transduction events, which convey and compute the sensed information via processes such as protein–protein binding, translocation and protein modifications such as covalent phosphorylation. At the bottom of such signal transduction pathways and networks lie adaptation regulators which regulate transcription, translation or protein synthesis, protein degradation and other functions such as the cell cycle, and thereby cause the adaptation of our cells and organs to exercise. RT-qPCR is a method that allows molecular exercise physiologists to quantify RNA by first reverse transcribing RNA into cDNA. The cDNA is then amplified in a thermal cycler. A fluorescent dye is used that allows measurement of the concentration of the DNA product during each cycle with a fluorometer that is part of the thermal cyclers used for RT-qPCR. Western blotting involves the separation of denatured proteins within a gel followed by detection that involves antibodies raised against the protein of interest. Omic approaches allow the non-biased measurement of DNA, RNA and proteins, which allows the development of new hypotheses and gives an idea about the big picture.

REVIEW QUESTIONS

1   Discuss the shortcomings of earlier hypotheses that attempt to explain the adaptation to exercise.
2   Summarize the signal transduction hypothesis of adaptation to exercise by drawing and explaining a schematic diagram.
3   How does RT-qPCR work?
4   How does a Western blot work?
5   Compare and contrast a Western blot with proteomics.

FURTHER READING

Burniston JG and Hoffman EP (2011). Proteomic responses of skeletal and cardiac muscle to exercise. *Exp Rev Proteomics* 8, 361–77.

REFERENCES

Aldridge GM, Podrebarac DM, Greenough WT and Weiler IJ (2008). The use of total protein stains as loading controls: an alternative to high-abundance single-protein controls in semi-quantitative immunoblotting. *J Neurosci Methods* 172, 250–254.

Alonso A, Sasin J, Bottini N, Friedberg I, Friedberg I, Osterman A, et al. (2004). Protein tyrosine phosphatases in the human genome. *Cell* 117, 699–711.

Baar K and Esser K (1999). Phosphorylation of p70(S6k) correlates with increased skeletal muscle mass following resistance exercise. *Am J Physiol* 276, C120–C127.

Bergstrom J and Hultman E (1966). Muscle glycogen synthesis after exercise: an enhancing factor localized to the muscle cells in man. *Nature* 210, 309–310.

Bodine SC, Latres E, Baumhueter S, Lai VK, Nunez L, Clarke BA, et al. (2001). Identification of ubiquitin ligases required for skeletal muscle atrophy. *Science* 294, 1704–1708.

Burnette WN (1981). "Western blotting": electrophoretic transfer of proteins from sodium dodecyl sulfate—polyacrylamide gels to unmodified nitrocellulose and radiographic detection with antibody and radioiodinated protein A. *Anal Biochem* 112, 195–203.

Burniston JG (2008). Changes in the rat skeletal muscle proteome induced by moderate-intensity endurance exercise. *Biochim Biophys Acta* 1784, 1077–1086.

Burniston JG, Meek TH, Pandey SN, Broitman-Maduro G, Maduro MF, Bronikowski AM, et al. (2013). Gene expression profiling of gastrocnemius of "minimuscle" mice. *Physiol Genomics* 45, 228–236.

Campbell WG, Gordon SE, Carlson CJ, Pattison JS, Hamilton MT and Booth FW (2001). Differential global gene expression in red and white skeletal muscle. *Am J Physiol Cell Physiol* 280, C763–C768.

Cao Y, Yao Z, Sarkar D, Lawrence M, Sanchez GJ, Parker MH, et al. (2010). Genome-wide MyoD binding in skeletal muscle cells: a potential for broad cellular reprogramming. *Dev Cell* 18, 662–674.

Chin ER, Olson EN, Richardson JA, Yang Q, Humphries C, Shelton JM, et al. (1998). A calcineurin-dependent transcriptional pathway controls skeletal muscle fiber type. *Genes Dev* 12, 2499–2509.

Chomczynski P and Sacchi N (1987). Single-step method of RNA isolation by acid guanidinium thiocyanate-phenol-chloroform extraction. *Anal Biochem* 162, 156–159.

Cohen P (2002). The origins of protein phosphorylation. *Nat Cell Biol* 4, E127–E130.

Davidsen PK, Gallagher IJ, Hartman JW, Tarnopolsky MA, Dela F, Helge JW, et al. (2011). High responders to resistance exercise training demonstrate differential regulation of skeletal muscle microRNA expression. *J Appl Physiol* 110, 309–317.

Dunham I, Kundaje A, Aldred SF, Collins PJ, Davis CA, Doyle F, et al. (2012). An integrated encyclopedia of DNA elements in the human genome. *Nature* 489, 57–74.

Erce MA, Pang CN, Hart-Smith G and Wilkins MR (2012). The methylproteome and the intracellular methylation network. *Proteomics* 12, 564–586.

Fischer EH and Krebs EG (1955). Conversion of phosphorylase b to phosphorylase a in muscle extracts. *J Biol Chem* 216, 121–132.

Fuda NJ, Ardehali MB and Lis JT (2009). Defining mechanisms that regulate RNA polymerase II transcription in vivo. *Nature* 461, 186–192.

Glickman MH and Ciechanover A (2002). The ubiquitin-proteasome proteolytic pathway: destruction for the sake of construction. *Physiol Rev* 82, 373–428.

Gordon PM, Liu D, Sartor MA, IglayReger HB, Pistilli EE, Gutmann L, et al. (2012). Resistance exercise training influences skeletal muscle immune activation: a microarray analysis. *J Appl Physiol* 112, 443–453.

Hardie DG, Ross FA and Hawley SA (2012). AMPK: a nutrient and energy sensor that maintains energy homeostasis. *Nat Rev Mol Cell Biol* 13, 251–262.

He C, Sumpter R, Jr. and Levine B (2012). Exercise induces autophagy in peripheral tissues and in the brain. *Autophagy* 8, 1548–1551.

Henriksson J, Salmons S and Lowry OH (1989). Chronic stimulation of mammalian muscle: enzyme and metabolic changes in individual fibres. *Biomed Biochim Acta* 48, S445–S454.

Holloway KV, O'Gorman M, Woods P, Morton JP, Evans L, Cable NT, et al. (2009). Proteomic investigation of changes in human vastus lateralis muscle in response to interval-exercise training. *Proteomics* 9, 5155–5174.

Hornberger TA (2011). Mechanotransduction and the regulation of mTORC1 signaling in skeletal muscle. *Int J Biochem Cell Biol* 43, 1267–1276.

Ilsley JL, Sudol M and Winder SJ (2002). The WW domain: linking cell signalling to the membrane cytoskeleton. *Cell Signal* 14, 183–189.

Johnson ES (2004). Protein modification by SUMO. *Annu Rev Biochem* 73, 355–382.

Keller P, Vollaard NB, Gustafsson T, Gallagher IJ, Sundberg CJ, Rankinen T, et al. (2011). A transcriptional map of the impact of endurance exercise training on skeletal muscle phenotype. *J Appl Physiol* 110, 46–59.

Koutedakis Y, Metsios GS and Stavropoulos-Kalinoglou A (2006). Periodization of exercise training in sport. In *The physiology of training*, ed. MacLaren D, pp. 1–21. Elsevier, Edinburgh.

Laemmli UK (1970). Cleavage of structural proteins during the assembly of the head of bacteriophage T4. *Nature* 227, 680–685.

Li W, Bengtson MH, Ulbrich A, Matsuda A, Reddy VA, Orth A, et al. (2008). Genome-wide and functional annotation of human E3 ubiquitin ligases identifies MULAN, a mitochondrial E3 that regulates the organelle's dynamics and signaling. *PLoS ONE* 3, e1487.

Manning G, Whyte DB, Martinez R, Hunter T and Sudarsanam S (2002). The protein kinase complement of the human genome. *Science* 298, 1912–1934.

McCarthy JJ and Esser KA (2007). MicroRNA-1 and microRNA-133a expression are decreased during skeletal muscle hypertrophy. *J Appl Physiol* 102, 306–313.

McGee SL and Hargreaves M (2011). Histone modifications and exercise adaptations. *J Appl Physiol* 110, 258–263.

Mischerikow N and Heck AJ (2011). Targeted large-scale analysis of protein acetylation. *Proteomics* 11, 571–589.

Ohtsubo K and Marth JD (2006). Glycosylation in cellular mechanisms of health and disease. *Cell* 126, 855–867.

Park KD, Park J, Ko J, Kim BC, Kim HS, Ahn K, et al. (2012). Whole transcriptome analyses of six thoroughbred horses before and after exercise using RNA-Seq. *BMC Genomics* 13, 473.

Parker KC, Walsh RJ, Salajegheh M, Amato AA, Krastins B, Sarracino DA, et al. (2009). Characterization of human skeletal muscle biopsy samples using shotgun proteomics. *J Proteome Res* 8, 3265–3277.

Pearen MA, Myers SA, Raichur S, Ryall JG, Lynch GS and Muscat GE (2008). The orphan nuclear receptor, NOR-1, a target of beta-adrenergic signaling, regulates gene expression that controls oxidative metabolism in skeletal muscle. *Endocrinology* 149, 2853–2865.

Renart J, Reiser J and Stark GR (1979). Transfer of proteins from gels to diazobenzyloxymethyl-paper and detection with antisera: a method for studying antibody specificity and antigen structure. *Proc Natl Acad Sci U S A* 76, 3116–3120.

Resh MD (2006). Trafficking and signaling by fatty-acylated and prenylated proteins. *Nat Chem Biol* 2, 584–590.

Roff M, Thompson J, Rodriguez MS, Jacque JM, Baleux F, Arenzana-Seisdedos F, et al. (1996). Role of IkappaBalpha ubiquitination in signal-induced activation of NFkappaB in vivo. *J Biol Chem* 271, 7844–7850.

Safdar A, Abadi A, Akhtar M, Hettinga BP and Tarnopolsky MA (2009). miRNA in the regulation of skeletal muscle adaptation to acute endurance exercise in C57Bl/6J male mice. *PLoS ONE* 4, e5610.

Sanyal A, Lajoie BR, Jain G and Dekker J (2012). The long-range interaction landscape of gene promoters. *Nature* 489, 109–113.

Scarpulla RC (2002). Nuclear activators and coactivators in mammalian mitochondrial biogenesis. *Biochim Biophys Acta* 1576, 1–14.

Tintignac LA, Lagirand J, Batonnet S, Sirri V, Leibovitch MP and Leibovitch SA (2005). Degradation of MyoD mediated by the SCF (MAFbx) ubiquitin ligase. *J Biol Chem* 280, 2847–2856.

Toole BJ and Cohen PT (2007). The skeletal muscle-specific glycogen-targeted protein phosphatase 1 plays a major role in the regulation of glycogen metabolism by adrenaline in vivo. *Cell Signal* 19, 1044–1055.

Towbin H, Staehelin T and Gordon J (1979). Electrophoretic transfer of proteins from polyacrylamide gels to nitrocellulose sheets: procedure and some applications. *Proc Natl Acad Sci U S A* 76, 4350–4354.

Ulbricht A and Hohfeld J (2013). Tension-induced autophagy: May the chaperone be with you. *Autophagy* 9.

Valencia-Sanchez MA, Liu J, Hannon GJ and Parker R (2006). Control of translation and mRNA degradation by miRNAs and siRNAs. *Genes Dev* 20, 515–524.

Vaquerizas JM, Kummerfeld SK, Teichmann SA and Luscombe NM (2009). A census of human transcription factors: function, expression and evolution. *Nat Rev Genet* 10, 252–263.

Wolff J (1892). *Das Gesetz der Transformation der Knochen.* Springer, Berlin.

Yang S, Alnaqeeb M, Simpson H and Goldspink G (1996). Cloning and characterization of an IGF-1 isoform expressed in skeletal muscle subjected to stretch. *J Muscle Res Cell Motil* 17, 487–495.

4 Molecular adaptation to endurance exercise and skeletal muscle fibre plasticity

Keith Baar and Henning Wackerhage

LEARNING OBJECTIVES

At the end of the chapter you should be able to:

- Answer the question whether it is possible to prescribe endurance training for the general public and/or for athletes based on solid scientific evidence. Put another way: Will everyone respond the same to an endurance training programme?

- Explain the difference between physiological cardiac hypertrophy or the athlete's heart and pathological cardiac hypertrophy, also known as hypertrophic cardiomyopathy. Describe the signal transduction events that regulate both forms of hypertrophy.

- Describe the characteristics of slow type I, intermediate type IIa and fast type IIx and IIb muscle fibres. Answer the question how do fibres respond to endurance exercise, supra-physiological stimuli such as chronic electrical low-frequency stimulation and a lack of neronal input?

- Explain the regulation and function of calcineurin-NFAT signalling and the effect on the expression of 'fast' and 'slow' genes.

- Describe the genomic organization of myosin heavy chain genes. Explain how MyoMir-Sox6 and chromatin remodelling affects their expression.

- Explain how CaMK and AMPK via PGC-1α signalling and other events stimulate mitochondrial biogenesis in response to endurance exercise.

- Explain how CaMK and AMPK via PGC-1α, hypoxia and shear stress-NO signalling regulate the expression of angiogenic growth factors such as VEGF and metalloproteinases and how this promotes the extension of the capillary bed in skeletal muscle which is termed exercise-induced angiogenesis.

INTRODUCTION

Endurance exercise is not only a panacea to prevent and treat many lifestyle-related diseases but also the main tool to condition endurance athletes for events such as the Ironman Triathlon World Championships in Hawaii, the Tour de France, city marathons and winter climbs in Scotland. In this chapter, we will first explore whether common endurance training recommendations (for example how hard and how long one should train) can be backed up by sound scientific evidence. We show that there is little reliable, epidemiological evidence for many training recommendations and also highlight that the training response to a given training programme varies considerably among individuals. With this background, the rest of the chapter centres on the question 'What are the molecular mechanisms that mediate the adaptation of especially skeletal and cardiac muscle to endurance exercise?' First, we will discuss the mechanisms that are responsible for the development of an athlete's heart and compare the molecular events that drive the genesis of the athlete's heart with those associated with the disease state we know as hypertrophic cardiomyopathy. Second, we will describe type I, IIb, IIx and IIb muscle fibres. Even though this is a focus of much discussion in the training and performance world, we will show that the effect of endurance exercise on muscle fibre type percentages is small and usually only involves a reduction in type IIx fibres and increase in IIa fibres. We will also discuss how calcineurin-NFAT signalling differs between fast and slow muscle fibres and how this affects fibre type-specific gene expression. Third, we specifically look at the evolutionary conserved genomic organization of myosin heavy chain isoform genes in a fast and slow gene cluster. We review how MyoMir-Sox6 signalling contributes to the on/off signalling of these genes to ensure that most of the time only one myosin heavy chain isoform is expressed in a muscle fibre. As part of this discussion, we also show how the DNA of exercise-responsive genes is opened up or closed by epigenetic mechanisms such as methylation. Fourth, we discuss how increased energy turnover and other signals are sensed by CaMK, AMPK and SIRT1 and lead to the activation of these proteins. Active CaMK, AMPK and SIRT1 then activate the transcriptional co-factor PGC-1α via multiple mechanisms. PGC-1α regulates the expression of mitochondrial genes encoded in nuclear and of transcription factors that increase the expression of genes encoded in mitochondrial DNA (mtDNA). Finally we discuss how AMPK-PGC-1α, HIF-1 and nitric oxide (NO) regulate exercise-induced angiogenesis by affecting the expression of growth factors such as VEGF and of metalloproteinases that tunnel into the extracellular matrix so that endothelial cells can form capillaries in these tunnels.

ENDURANCE EXERCISE: CURRENT RECOMMENDATIONS, TRAINABILITY AND SCIENTIFIC EVIDENCE

Endurance performance depends on many limiting factors or, using the language of geneticists, quantitative traits (see Chapter 2). There are three major limiting factors (Bassett and Howley, 2000):

- **Maximal oxygen uptake** ($\dot{V}O_{2max}$): This is defined as the highest rate at which oxygen can be taken up and utilized by the body during severe exercise.
- **Percentage of $\dot{V}O_{2max}$ that can be sustained during endurance exercise** (%$\dot{V}O_{2max}$): This indicates the percentage of the $\dot{V}O_{2max}$ that can be sustained during, for example, a 10 km run, marathon run or Ironman triathlon. It is commonly measured as the %$\dot{V}O_{2max}$ at the lactate threshold.
- **Mechanical efficiency**: This is defined as the oxygen or energy cost (the two are linked: roughly 20 kJ of energy are used per 1 O_2 consumed) to sustain a power output or velocity.

These factors can be further broken down into a plethora of smaller limiting factors, which include cardiac output, stroke volume, fibre type proportions, connective tissue stiffness and mitochondrial density, whose adaptation to exercise will be discussed in the later part of this chapter. The endurance limiting factors depend both on genetic variation/talent (see Chapter 5) and on environmental factors, of which endurance training is the most important. Only those with a great deal of talent, or favourable genetic variations, that engage in the most effective training will become elite endurance athletes.

How should one train to increase endurance? As for strength exercise (see Chapter 6), there are many variables that determine an endurance training programme. These include **exercise intensity** (set as %$\dot{V}O_{2max}$, %HR_{max}, %$HR_{reserve}$, rate of perceived exertion or velocity/pace), **volume** (time, distance or step count per session, day or week), **type of session** (continuous exercise, interval training or high-intensity interval training (Gibala et al., 2012), Fartlek). Other factors are **training goals** (aerobic capacity, general fitness, risk factor reduction, health improvement), **progression/periodization** (how to change variables during a training programme; also tapering to prepare for competitions), **exercise mode** (running, cycling, swimming, etc.). Finally, the endurance training programme must be specific for the **individual** (elite athlete, disabled elite athlete, untrained individual, children, patient participating in an exercise therapy programme, etc.) as individuals differ in their trainability and may respond more to certain types of endurance training.

With so many variables it is impossible to recommend an 'ideal' endurance training programme for each individual based on sound scientific evidence. Also, as just mentioned, endurance exercise prescription is made difficult by the fact that the response to an identical endurance training programme or **endurance trainability** varies greatly in the human population. For example, some people do not improve their $\dot{V}O_{2max}$ with endurance training, whereas others may increase it by more than 1 L/min in response to an identical endurance training programme (Bouchard et al., 1999). Even more, a careful analysis of epidemiological data has demonstrated that a small minority ($\approx$10%) of individuals will experience worsened risk factors such as blood pressure, fasting insulin or blood lipids in response to an exercise programme (Bouchard et al., 2012). The worsening of one risk factor may not necessarily reduce health, as it may only be one risk factor that worsens while many others improve, but this still needs to be demonstrated. This variability can either be the result of the relative intensity of the training stimulus (McPhee et al., 2009) or a result of our genetic ability to respond to the exercise (Timmons et al., 2010; Bouchard et al., 2011). In the latest American College of Sports Medicine (ACSM) position stand on the 'quantity and quality of exercise for developing and maintaining cardiorespiratory, musculoskeletal, and neuromotor fitness in apparently healthy adults' (Garber et al., 2011) it is consequently recognized that 'there is considerable variability in individual responses to a standard dose of exercise'.

Given the variability of endurance trainability and the many variables, what recommendations can be given for prescribing endurance training programmes? In the 2011 ACSM position stand, the authors recommend the following for aerobic (endurance) exercise for apparently healthy adults (Garber et al., 2011):

- **Frequency**: ≥5 days per week of moderate exercise or ≥3 days per week of vigorous exercise, or a combination of moderate and vigorous exercise on ≥3–5 days per week is recommended.
- **Intensity**: Moderate and/or vigorous intensity is recommended for most adults.
- **Time**: 30–60 min per day (150 min per week) of purposeful moderate exercise or 20–60 min per day (75 min per week) of vigorous exercise, or a combination of moderate and vigorous exercise per day is recommended for most adults.
- **Type**: Regular, purposeful exercise that involves major muscle groups and is continuous and rhythmic in nature is recommended.

The authors suggest that the above recommendations are mostly supported by high-quality evidence. However, because of the variability of endurance trainability the outcomes of an endurance training programme should be verified periodically.

What about endurance training recommendations for endurance athletes? Specific endurance training recommendations are given in thousands of sports books, videos and Internet articles but almost all of this information is subjective. We know of no specific recommendations for endurance exercise for athletes that are based on a rich body of data coming from randomized controlled trials, which is seen as the highest level of scientific evidence (Garber et al., 2011) or from large-scale epidemiological studies. For example, there are no studies examining the relationship between training volume and marathon running time in relation to age, sex and training history, even though such data could quite easily obtained via questionnaire by targeting runners who complete city marathons. Thus we agree with Midgley and colleagues that 'there is insufficient direct scientific evidence to formulate [endurance] training recommendations based on the limited research' (Midgley et al., 2007). It is hoped that more reliable data will be generated in future, so that for example a 35-year-old female runner can determine the likelihood of running a marathon in 3 h 30 min if she runs a maximum volume of 40, 60 or 80 km per week during preparation. Until such data are available, athletes and coaches are best advised to copy endurance training programmes from successful endurance athletes. Also, while there is little reliable evidence coming from controlled trials or epidemiological studies, there is a strong literature describing the physiological and biochemical adaptations to endurance training in human and animals and the molecular mechanisms that cause such adaptations. Coaches and athletes can use such information to generate new ideas for endurance training programmes but should avoid over-interpreting such data.

In the remainder of this chapter we will discuss the molecular mechanisms that regulate adaptation to endurance exercise in the heart and skeletal muscles.

ATHLETE'S HEART VERSUS HYPERTROPHIC CARDIOMYOPATHY

The function of the cardiovascular system is to deliver O_2 and nutrients to muscles and other organs and to remove CO_2. The performance of the system depends on the **cardiac output**, which is the blood pumped by the heart per minute, and the O_2-carrying capacity of the blood, which is the haemoglobin mass partial pressure of oxygen in the air. Since **maximal cardiac output** (Q_{max}) correlates highly with **maximal oxygen uptake** ($\dot{V}O_{2max}$) (Ekblom, 1968) it is widely assumed that oxygen delivery to the working muscle is the primary determinant of the $\dot{V}O_{2max}$. Increases of the oxygen transport capacity further increase $\dot{V}O_{2max}$ (Bassett and Howley, 2000) but the adaptation of the oxygen transport capacity to endurance training at sea level is only small (Sawka et al., 2000). Therefore, the primary adaptation to endurance exercise training is an increase in the ability to supply oxygenated blood to the working muscle.

How do elite endurance athletes achieve a cardiac output that is in some cases twice as high as that measured in sedentary subjects? The maximal cardiac output (Q_{max}) is the maximal value for the product of **stroke volume** and **heart rate**. Table 4.1 summarizes some results for stroke volume, heart rate and the resultant cardiac output for control subjects before and after endurance training as well as for elite endurance athletes (Blomqvist and Saltin, 1983). The table shows that the high maximal cardiac output of >30 L/min of the elite endurance athlete can be attributed to a stroke volume of nearly 170 mL, which is more than 60 mL larger than that of untrained individuals. In contrast, the maximal heart rate differs little between untrained individuals and elite endurance athletes and thus the greater maximal cardiac output of elite endurance athletes is due to a greater stroke volume. The data also show that endurance training increases stroke volume.

Increased stroke volumes after endurance training and in elite endurance athletes are caused by exercise-induced physiological cardiac hypertrophy, also known as the **athlete's heart**. The athlete's heart has a greater left ventricle volume and a small increase in the thickness of the cardiac muscle (Scharhag et al., 2002). The thicker muscle allows the athlete's heart to pump the same percentage of blood out of the heart with each beat (ejection fraction) even with the greater stroke volume. However, large hearts are not always good news: one has to distinguish the beneficial athlete's heart (physiological, left ventricular hypertrophy) from **pathological cardiac hypertrophies** that are

Table 4.1 Examples of the maximal oxygen uptake, cardiac output, stroke volume and heart rate of normal subjects before and after endurance training and in typical elite endurance athletes

	NORMAL SUBJECTS		ELITE ENDURANCE ATHLETE
	BEFORE ENDURANCE TRAINING	AFTER ENDURANCE TRAINING	
Maximal oxygen uptake (L/min)	3.30	3.91	5.38
Maximal cardiac output (L/min)	20.0	22.8	30.4
Stroke volume (mL)	104	120	167
Maximal heart rate (per min)	192	190	182

From Blomqvist and Saltin (1983).

characterized by thick walls and an accumulation of non-contractile proteins such as collagen (Bernardo et al., 2010). The functional difference between the athlete's heart and a diseased heart is that in the diseased heart the muscle is not able to pump the blood returned to the heart, in other words the ejection fraction decreases. To summarize, the athlete's heart is a physiological adaptation to endurance exercise that increases the volume and thickness of the left ventricle without changing ejection fraction, resulting in increased stroke volume, cardiac output and maximal oxygen uptake. The athlete's heart is different from hypertrophic cardiomyopathy, a serious heart disease characterized by an increase in the volume and thickness of the left ventricle resulting in a decrease in ejection fraction.

For the molecular exercise physiologist there are two key questions related to the athlete's heart:

- What signal transduction events are responsible for the athlete's heart adaptation in response to endurance exercise?
- What are the differences in signal transduction between an athlete's heart and hypertrophic cardiomyopathy?

These important questions are difficult to answer for humans because it is impossible to obtain samples from living human hearts. Thus virtually all the research on cardiac hypertrophy has been performed using cultured cardiomyocytes, cultured hearts (so-called Langendorff-perfused hearts), wildtype and transgenic animals, and the hearts of human cadavers.

The arguably first landmark paper on the mechanisms that mediate cardiac hypertrophy was published by a team led by Eric Olson (Molkentin et al., 1998). Using a method that allows the identification of protein-binding partners they discovered that the transcription factor GATA-4 interacts with the Ca^{2+}-activated phosphatase calcineurin. This finding suggested that Ca^{2+}, the concentration of which increases in situations where hearts hypertrophy, might also activate a signal transduction pathway driving cardiac hypertrophy.

To test the hypothesis that calcineurin and its downstream transcription factor NFAT could promote cardiac hypertrophy, the scientists created two transgenic mouse strains where either calcineurin or NFAT were overexpressed in heart muscle cells. Both transgenic lines developed pathological cardiac hypertrophy and many mice suffered either serious cardiac problems and/or died prematurely. Figure 4.1 schematically shows the effect on calcineurin on the size of the heart and also that calcineurin inhibition with the drug cyclosporin A reverses this effect. These data demonstrate that increased calcineurin–NFAT signalling can induce pathological cardiac hypertrophy and that the drug cyclosporin A can prevent it.

In another experiment, Jeffrey Molkentin and his team identified that MEK1, a kinase in the mitogen-activated protein kinase (MAPK) signal transduction pathway, could promote physiological cardiac hypertrophy without signs of pathological decompensation or premature death (Bueno et al., 2000) The MEK1-overexpressing hearts had thicker walls and contracted more forcefully. However, in contrast to the endurance training-induced athlete's heart, left ventricle size was hardly changed. Thus MEK1 drives a concentric hypertrophy (Bueno et al., 2000) which is similar to the cardiac

Figure 4.1 Calcineurin and PI3K as regulators of pathological and physiological cardiac hypertrophy, respectively. (a) Calcineurin overexpression in transgenic mice causes pathological hypertrophic cardiomyopathy, which can be prevented by inhibiting calcineurin with cyclosporin A (CsA) (Molkentin et al., 1998). Mice in which calcineurin is overexpressed in the heart display hypertrophic cardiomyopathy and often die early. This suggests that calcineurin–NFAT signalling promotes pathological cardiac hypertrophy.(b) PI3K knockout prevents swimming-induced (swim) cardiac hypertrophy but not pathological cardiac hypertrophy induced by aortic banding (banding) (McMullen et al., 2003). This suggests that PI3K is required for physiological cardiac hypertrophy or 'athlete's heart'.

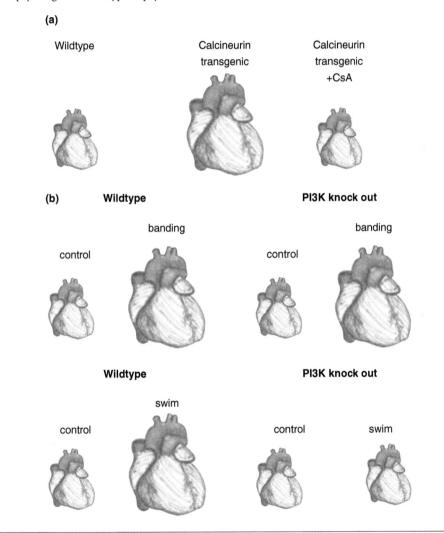

hypertrophy seen in strength athletes, where wall thickness but not left ventricular volume is increased. As a consequence, these hearts have only a small increase in stroke volume as a result of a higher ejection fraction, unlike endurance-trained hearts.

The Izumo laboratory studied the effect of the mTOR pathway regulators PI3K (Shioi et al., 2000) and PKB/Akt on the heart (Shioi et al., 2002). To investigate the function of PI3K, they knocked out PI3K in the heart (dnPI3K) and compared these mice

Figure 4.2 Signal transduction pathways that regulate cardiac hypertrophy. Schematic drawing of the signal transduction pathways that regulate physiological and pathological cardiac hypertrophy (Bernardo et al., 2010). (a) Endurance exercise induces cardiac hypertrophy via the increased PI3K and PKB/Akt and/or decreased C/EBPβ signalling which causes a physiological, left ventricular hypertrophy with maintained ejection fraction. As a consequence, stroke volume and thus maximal cardiac output and $\dot{V}O_{2max}$ increase and this is known as physiological cardiac hypertrophy or the athlete's heart. (b) Gene defects or diseases such as hypertension activate the calcineurin–NFAT pathway and other pathways. These pathways initially induce a compensatory hypertrophy that later decompensates. The decompensating heart shows signs of fibrosis, deranged gene expression, cell death and a decrease in ejection fraction.

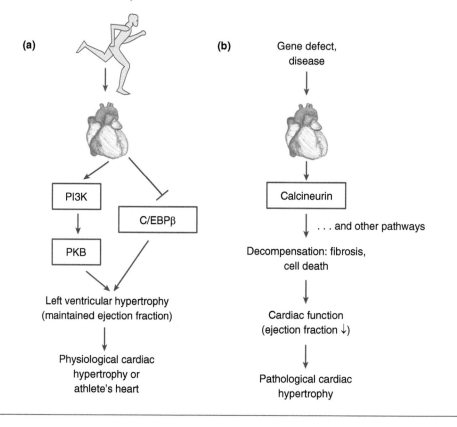

to wildtype mice (non-transgenic, NTG) (McMullen et al., 2003). The authors then used 'aortic banding' to induce pathological hypertrophic cardiomyopathy and swimming to induce physiological cardiac hypertrophy. They found that the PI3K knockout prevented physiological cardiac hypertrophy but not pathological cardiac hypertrophy (Figure 4.1). This suggests that PI3K is required for physiological but not pathological cardiac hypertrophy. Subsequently, it was shown that PKB/Akt, which lies downstream of PI3K, can similarly induce physiological cardiac hypertrophy (DeBosch et al., 2006). Taken together, these experiments suggest that PI3K and PKB/Akt, which increase protein synthesis via the mTOR pathway, can induce endurance exercise-induced cardiac hypertrophy (an athlete's heart) in mice.

In another paper, a team led by Bruce Spiegelman screened hearts for proteins with a regulatory function that change their expression in the heart in response to endurance exercise. They found that the transcription factor C/EBPβ was reduced in the hearts of mice that performed endurance exercise. They also observed that reduced C/EBPβ levels controlled the division of cardiomyocytes and increased their size (Bostrom et al., 2010). Taken together these data suggest that endurance exercise leads to a reduction of C/EBPβ, which contributes to the development of a physiological cardiac hypertrophy (Bostrom et al., 2010). It is unknown whether the C/EBPβ protein is linked to PI3K and PKB/Akt signalling in the heart or whether it regulates physiological cardiac hypertrophy or the athlete's heart independently.

To summarize, there are two major forms of cardiac hypertrophy. The first is endurance exercise-induced, physiological cardiac hypertrophy, which is an eccentric, left ventricular hypertrophy where ejection fraction is maintained. These hearts can generate a higher cardiac output during maximal exercise, a key adaptation to endurance training. The second form is pathological cardiac hypertrophy or hypertrophic cardiomyopathy where ejection fraction decreases as the disease progresses. Transgenic mouse studies suggest that physiological and cardiac hypertrophy are regulated by different signal transduction pathways: endurance exercise increases cardiac output, which is also known as volume overload and this drives cardiac hypertrophy via increased PI3K, PKB/Akt and decreased C/EBPβ signalling at least in mice. In contrast, pathological hypertrophy is mediated in part by the calcineurin–NFAT and/or other pathways that drive hypertrophy followed by decompensation. A simplified overview over the signalling that contributes to physiological and pathological cardiac hypertrophy is given in Figure 4.2.

REGULATION OF SKELETAL MUSCLE FIBRE PLASTICITY

A motor unit is defined as a motor neuron and all of the muscle fibres it innervates. There are different types of motor units, characterized by their α-motor neuron and muscle fibre type:

- **S (slow, fatigue resistant)** motor units comprise small α-motor neurons with a low excitation threshold and small axons that innervate slow type I fibres.
- **FR (fast, fatigue resistant)** motor units comprise α-motor neurons that innervate intermediate type IIa fibres. The α-motor neuron properties are likely in-between those of FF and S α-motor neurons.
- **FF (fast fatiguing)** motor units comprise large, hard-to-excite α-motor neurons with large, fast-conducting axons that innervate fast, but easily fatiguing, type IIb and/or IIx fibres.

The aim of this section in this chapter is to introduce the different types of muscle fibres and their response to increased or decreased contractile activity.

The muscle fibres within a muscle differ greatly in their contraction speed, force output and fatigability. This is a result of the proteins that regulate the Ca^{2+} concentration to control contraction and relaxation and myosin that determine contraction speed and force and proteins and enzymes that resynthesize ATP whenever it is broken down into ADP and P_i. Muscle fibres not only differ in the concentrations of these proteins but also express different isoforms of several of these proteins.

The key marker protein for a specific muscle fibre type are **myosin heavy chain** isoforms which are the engines of muscle fibres (Schiaffino, 2010). There are several reasons why myosin heavy chain isoforms are the proteins used to classify and name the different types of muscle fibres:

- Specific, high-quality antibodies against the different myosin heavy chain isoforms are available (Schiaffino, 2010). These antibodies allow a reliable determination of fibre type.
- Myosin heavy chain isoform determines the maximal shortening velocity and force per cross-sectional area of a muscle fibre (Barany, 1967; Harridge et al., 1996). Thus the expression of specific isoforms is linked to the function of muscle fibres.
- The concentration of myosin heavy chain isoforms is high within fibres and differs greatly between fibres. This makes their detection easier than the detection of proteins whose concentrations is low and/or differs only little between fibres.

Adult myosin heavy chains are divided into slow (type I) and fast (type II) isoforms. The type II isoforms are further subdivided into type IIa (intermediate), type IIx (fast) and type IIb isoforms. Note that the type IIb gene exists in the human genome but myosin heavy chain IIb protein is not expressed in the major human muscles (Smerdu et al., 1994). An overview of the functional characteristics of different muscle fibre types is given in Table 4.2.

Be aware that fibre type tables such as Table 4.2 are an oversimplification for several reasons. First, fibres have a limited ability change from one fibre type into another and

Table 4.2 Generalized fibre type characteristics of the major human and rodent fibre types

FIBRE TYPE (MAJOR MYOSIN HEAVY CHAIN EXPRESSED)	TYPE I	TYPE IIA	TYPE IIX/IIB*
Description	Slow, red, fatigue resistant	Intermediate, red, fatigue resistant	Fast, white, readily fatigued
Nuclei per mm of fibre (Tseng et al., 1994)	High	Medium?	Low
Maximal shortening speed when fully Ca^{2+} activated	Slow	Fast	Slightly faster still
Force build up and force relaxation before and after twitches and isometric contractions	Slow	Medium	Fast
ATPase activity under 'physiological' conditions, fully Ca^{2+} activated	Low	Medium	High
Glycolytic enzyme activity (for example phosphorylase, phosphofructokinase and lactate dehydrogenase)	Low–medium	Medium–very high	High
Myoglobin content	High	Medium–high	Low
Oxidative capacity, mitochondrial enzyme activity (NADH-TR, succinate dehydrogenase stains)	High	Highest	Low
Capillaries per fibre or mm²	High	Medium	Low

*Type IIb myosin heavy chain is only expressed in rodent skeletal muscle but not in most human muscles. Type IIx is also known as IId.
Modified after Jones et al. (2004) and Spurway and Wackerhage (2006).

during the transition phase fibres express more than one myosin heavy chain isoform. Such mixed fibres are termed **hybrid fibres**. Second, while there are large differences in the concentrations of myosin heavy chain isoform proteins, the concentrations of other proteins vary more and may overlap between fibres. Some researchers describe this as a **continuum** of fibre types to highlight that three or four fibre types are an oversimplification. Third, there are more myosin heavy chain genes and proteins than those shown in Table 4.2. In muscles that move the eye, for example, **extraocular** myosin heavy chain is expressed. **Embryonal** and **developmental** myosin heavy chains are expressed during embryonal development and in regenerating muscle fibres after injury but not in healthy, adult muscle.

Once researchers were able to determine the different muscle fibre types, they started to compare the fibre type composition between species, in the muscles of the human body and in the human population. This research revealed the following:

- **Inter-individual variation**: As a result of genetic variation and environmental factors, humans vary in their fibre type composition. For example, half of North American Caucasians have either less than 35% or more than 65% of type I fibres in their vastus lateralis muscle (Simoneau and Bouchard, 1995). Thus, the fibre type composition in any given skeletal muscle varies greatly in the human population.
- **Fibre type composition of athletes**: Sprint and endurance athletes commonly have a high percentage of fast type II or slow type I fibre percentages in a given muscle, respectively (Costill et al., 1976a, 1976b).
- **Intra-individual variation**: Researchers also found a large variation in the fibre type composition in human muscles. For example, the orbicularis oculi eye muscle has on average only 15% of slow type I fibres whereas the soleus has 89% type I fibres (Johnson et al., 1973). In general, postural muscles comprise more type I fibres, whereas those that contract only occasionally contain more type II fibres.
- **Inter-species variation**: Muscle fibre type proportions within a muscle differ within between species and even between different strains. A striking example is that type IIb muscle fibres can be found in many rodents, whereas humans have the type IIb gene in their genome but do not express myosin heavy chain IIb in their locomotory muscles (Smerdu et al., 1994). Another example is that the soleus of common mouse strains is 50% fast and 50% slow, whereas in a rat this muscle is 10% fast and 90% slow.

Exercise physiologists have also extensively studied whether endurance training can be used to change the fibre type composition of an individual. The first team to address this question comprehensively included the American biochemist Phil Gollnick and the Scandinavian Exercise Physiologist Bengt Saltin (Ingalls, 2004). They first compared the fibre type composition of trained and untrained men but then conducted an endurance training study. Six subjects exercised intensely for five months on a cycle ergometer for 1 h per day for four days per week (Gollnick et al., 1973). The subjects were biopsied before and after training. Histochemical assays were then used to determine the percentage of slow twitch (type I) and fast twitch (type II subgroups; the assays used at the time did not allow distinction between type II subgroups) fibres. They found that endurance training caused a non-significant increase of the percentage of slow twitch fibres from

32 to 36%. A limitation of the study was that only six subjects were investigated, but their data suggest that small, if any, changes from fast to slow occur with training. This finding has been confirmed a number of times, suggesting that training has a limited ability to shift myosin heavy chain from fast to slow in humans.

In model organisms, when extreme, supraphysiological stimuli are applied to skeletal muscle, greater changes in myosin heavy chain are possible. Over time, such stimuli can greatly change the fibre type proportions within a muscle. In general, profound increases in the contractile activity of muscle fibres result in a slower phenotype whereas any decrease in contractile activity will drive fibres towards a faster phenotype on the following scale: **I (slow)** ↔ **IIa (intermediate)** ↔ **IIx (fast)** ↔ (IIb; fast; not expressed in humans). The pathological, physiological and supraphysiological interventions that change the fibre composition of muscles are as follows:

- **Endurance exercise** (physiological) generally promotes a limited fibre type shift involving a decrease of type IIx and increase of type IIa fibres (Gollnick et al., 1973; Jansson and Kaijser, 1977). Over decades it seems that a significant type II-to-type I fibre type shift can occur especially in those that have a low percentage of type I fibres to start with (Trappe et al., 1995). However, it is difficult to disentangle the effects of long-term exercise from the effects of muscle ageing (Lexell et al., 1988).
- **Chronic electrical low frequency stimulation** (supra-physiological), a model for continuous low-intensity endurance exercise, can induce complete type II-to-type I fibre type transitions in rodent models (Pette and Vrbova, 1992).
- **Cross-reinnervation** (supra-physiological), i.e. connecting a fast muscle nerve to a slow muscle changes the muscle phenotype from fast to slow and vice versa (Buller et al., 1960) demonstrates the effects of different firing patterns of motor neurons on the innervated muscle fibres (Hennig and Lomo, 1985).
- **Denervation** (for example after spinal cord injury; pathological) stops contractile activity, induces a slow-to-fast phenotype shift with predominantly type IIx fibres years after injury (Biering-Sorensen et al., 2009).

To summarize, the fibre composition of the skeletal muscles within the human body and between humans varies greatly. Endurance training regimes over several months generally lower the proportion of type IIx fibres and increase the proportion of IIa fibres but will not convert a sprinter into a marathon runner. Although decades of endurance training may trigger type II-to-I conversions, it is impossible to distinguish this from ageing changes where muscle fibres are additionally lost. Finally, chronic low-frequency electrical stimulation induces a fast-to-slow fibre transformation, whereas a lack of innervation results in a slow-to-fast change.

FIBRE TYPE PLASTICITY AND THE CALCINEURIN–NFAT PATHWAY

The challenge for molecular exercise physiologists is to link the signals associated with increased or decreased muscle contraction to changes in skeletal muscle fibre phenotype. In this section we will discuss the calcineurin pathway as a major regulator of muscle fibre type.

Similar to cardiac hypertrophy the Ca^{2+}-activated phosphatase **calcineurin**, which dephosphorylates and activates the transcription factor NFAT, has been identified as a regulator of muscle fibre phenotype. In the cytosol of muscle fibres Ca^{2+} increases from a resting concentration of around 50 nM by $\approx$100 fold whenever an action potential triggers a release of Ca^{2+} from the sarcoplasmic reticulum (Berchtold et al., 2000). Since the calcium release, reuptake and binding protein concentrations and/or isoforms differ between slow type I and fast type II fibres (Berchtold et al., 2000) and since slow and fast motor neurons fire differently (Hennig and Lomo, 1985), both the frequency of Ca^{2+} waves and the shape of Ca^{2+} waves differ between slow type I and fast type II fibres (Carroll et al., 1997). Do the differences in Ca^{2+} concentrations between fibres affect the activity of the calcineurin pathway? A good readout of the activity of the calcineurin pathway is the localization of the calcineurin-regulated transcription factor NFAT, because when NFAT is dephosphorylated it moves to the nucleus. To investigate this, a team led by Stefano Schiaffino has overexpressed NFAT tagged with green fluorescent protein (GFP) in muscle fibres. The researchers observed that NFAT was largely in the nucleus in fibres taken from the slower soleus, whereas in the faster tibialis anterior NFAT was predominantly found in the cytoplasm (Tothova et al., 2006). Anaesthesia or inactivation of the muscle by denervation reduced NFAT in the nucleus, whereas increasing muscle activity through overload or chronic low-frequency (20 Hz) electrical stimulation increased NFAT in the nucleus (Tothova et al., 2006). Interestingly, short bouts of high-frequency (100 Hz) electrical stimulation did not increase nuclear localization. All this suggests that the Ca^{2+} patterns found in slow type I fibres and those that are induced by endurance exercise activate calcineurin. Calcineurin then dephosphorylates and drives NFAT into the nucleus in type I fibres and in response to low-frequency stimulation. In contrast, the lowering of Ca^{2+} or short bursts of Ca^{2+} do not activate calcineurin and as a consequence NFAT remains in the cytoplasm where it cannot regulate gene expression.

Thus, the activation of the calcineurin pathway is consistent with the hypothesis that calcineurin and NFAT activity contribute to the regulation of muscle fibre phenotype. But what is the evidence that calcineurin does actually regulate the phenotype of muscle fibres? In a landmark paper by a team led by Sanders Williams, the function of calcineurin in muscle was investigated (Chin et al., 1998). First, the authors found *in vitro* that overexpressed calcineurin induced the slow muscle fibre genes slow troponin and myoglobin, which was prevented by inhibiting calcineurin with cyclosporin A (Figure 4.3). They then used the calcineurin inhibitor cyclosporin A *in vivo* and found that this doubled the average percentage of fast type II fibres in the soleus muscle. This suggests that calcineurin reduces the number of type II fibres and increases the number of type I fibres when it is active. Later researchers over expressed or knocked out calcineurin in transgenic mice. This confirmed that high calcineurin activity increased the percentage of slow type I fibres *in vivo* (Naya et al., 2000;Parsons et al., 2003;Parsons et al., 2004). However, no intervention that targets the calcineurin-NFAT pathway so far converted all type II fibres into type I fibres. To summarize, the unique Ca^{2+} patterns in slow type I and fast type II muscle fibres differentially activate calcineurin in type I and type II fibres. Calcineurin then activates the transcription factor NFAT only in type I fibres and this drives the expression of 'slow' genes such as slow troponin and myoglobin. Also endurance exercise can activate this pathway and over time has a similar effect. The calcineurin-NFAT pathway is shown in Figure 4.4.

Figure 4.3 Calcineurin signalling in skeletal muscle. Schematic drawing depicting the localization of the transcription factor NFATc1 in slow and fast muscle fibres and its effect on innervation. (a) Mainly nuclear localization of green fluorescent protein-tagged NFATc1 in slow soleus and (b) mainly cytosolic localization of NFATc1 in the fast tibialis anterior muscle. (c) The effect of denervation (low Ca^{2+}), 100 Hz electrical stimulation (very high Ca^{2+}; intense contraction) and 10 Hz electrical stimulation (long-term elevated Ca^{2+}) in the soleus or tibialis anterior muscle. Redrawn after Tothova et al. (2006).

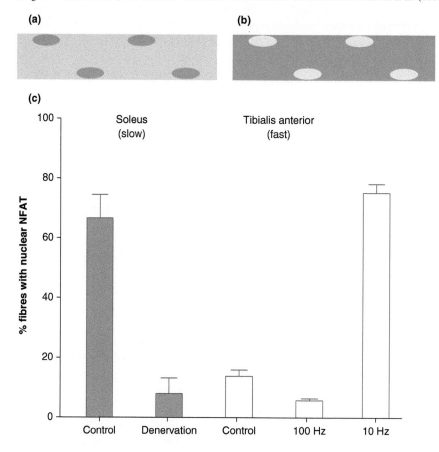

Is calcineurin the sole regulator of skeletal muscle fibre type identity? No. Many other signalling proteins, such as PGC-1α (Lin et al., 2002), PPARδ and ERRγ (Gan et al., 2011), members of the MAPK pathway (Murgia et al., 2000), Tead1 (Tsika et al., 2008) and the myogenic regulatory factor myogenin (Hughes et al., 1999) can also drive a slow muscle fibre phenotype. To summarize, the calcineurin–NFAT pathway is one potential regulator of fibre type but many other signalling proteins are also involved in the regulation of skeletal muscle fibre phenotype.

MYOSIN HEAVY CHAIN GENE CLUSTERS AND CHROMATIN REMODELLING

We will now focus on the developmental regulation of specific myosin heavy chain isoform genes. In mammalian genomes, the myosin heavy chain isoform genes are located

Figure 4.4 Calcineurin hypothesis. (a) Overexpression of activated calcineurin switches on the slow troponin I and myoglobin genes (MCK is a control) which can be prevented if the calcineurin inhibitor cyclosporin A (CsA) is added. (b) Calcineurin inhibition with CsA roughly doubles the percentage of fast type II fibres in the soleus muscle. Redrawn after Chin et al. (1998). (c) Schematic drawing depicting how Ca^{2+} via calmodulin activates calcineurin. Calcineurin then dephosphorylates the transcription factor NFAT, which enters the nucleus and, according to the hypothesis, increases the expression of slow muscle fibre genes.

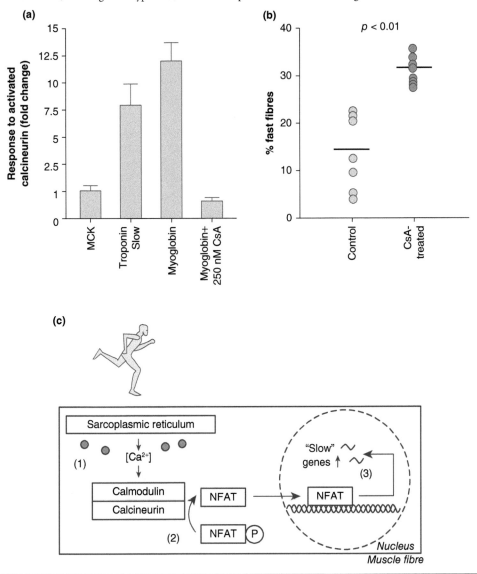

close together in either a slow/cardiac cluster in one genomic location or in a fast/developmental cluster in a second genomic location. The slow/cardiac cluster harbours myosin heavy chain types Iα (especially expressed in cardiac muscle) and Iβ and the fast/developmental cluster includes the type IIa/x/b subtypes as well as the perinatal (5′ of IIb, IIx/d and IIa) and embryonic (3′ of IIb, IIx/d and IIa; Table 4.3).

Table 4.3 Genomic locations (we first state the number of the chromosome followed by the location in megabases (Mb) on that chromosome) of skeletal and cardiac muscle human myosin heavy chain (MHC) genes and their orthologues in mouse

ISOFORM	MHC Iα (MYH6)	MHC Iβ (MYH7)	MHC IIa (MYH2)	MHC IIX/d (MYH1)	MHC IIb (MYH4)
Cluster	Slow/cardiac		Fast/developmental		
Human	14; 22.92 Mb	14; 22.95 Mb	17; 10.37 Mb	17; 10.34 Mb	17; 10.29 Mb
Mouse	14; 46.91 Mb	14; 46.91 Mb	11; 66.78 Mb	11; 66.83 Mb	11; 66.87 Mb

From Weiss et al. (1999).

It is unclear how aforementioned transcription factors such as NFAT, Tead1 and myogenin target myosin heavy chain clusters to regulate the expression of the genes in this cluster. However, in general these clusters are regulated by the packing of DNA. In recent years, it became clear that much of the DNA in the genome is tightly wrapped around histone proteins and needs to be opened up before transcription occurs. This process of DNA packaging and unpacking is termed chromatin remodelling or epigenetic regulation and it has been shown that such regulation is key for myogenesis and probably also plays an important role in the regulation of myosin heavy chain isoform expression and fibre type specification. An example of epigenetics and myogenesis can be seen if fibroblasts are treated with with 5-azacytidine, which opens up the DNA. After treating with 5-azacytidine, the fibroblasts can turn into myoblasts because the regions of the DNA containing muscle genes are now accessible for transcription (Lassar et al., 1986). This packaging and unravelling is a tightly regulated process controlled by DNA methylation and by histone acetylation and methylation on lysine residues, as described by the 'histone code' hypothesis (Jenuwein and Allis, 2001).

Over a decade ago, histone acetylases (HATs) and deacetylases (HDACs), which regulate the aforementioned acetylation of histones, were shown to be involved in the differentiaton of skeletal muscle (McKinsey et al., 2001). More recently authors have specifically analysed histone 3, lysine 27 and lysine 36 trimethylation (H3K27me3 and H3K36me3; 'K' stands for lysine) during the differentiation of C2C12 myoblasts into myotubes, which involves the upregulation of myosin heavy chain expression (Asp et al., 2011). The authors found that H3K27me3, which is associated with gene silencing, decreased, whereas H3K36me3, which is found on actively transcribed genes, generally increased during C2C12 differentiation. This finding is consistent with the hypothesis that such histone modifications contribute to the opening up of myosin heavy chain genes when myoblasts, which do not express myosin heavy chains, differentiate into myotubes that do express myosin heavy chains.

Another team investigated the general acetylation of histone 3 (H3ac) and the trimethylation of lysine 4 of histone 3 (H3K4me3), which are both associated with actively transcribed genes in myosin heavy chain genes in the rat plantaris (faster phenotype), soleus (slower phenotype), and hindlimb suspended soleus (slow-to-fast phenotype shift). The results (Figure 4.5) demonstrated that the activating H3ac and H3K4me3 of distinct myosin heavy chain isoforms matched the expression of these myosin heavy chain isoforms (Pandorf et al., 2009).

Figure 4.5 Histone regulation in skeletal muscle. (a) Schematic drawing showing how DNA is wrapped around histone complexes and the acetylation (ac) and methylation (me) of histone complexes. (b) Slow type I myosin heavy chain expression in the plantaris (Pla, fast phenotype), soleus (Sol, slow phenotype) and unloaded soleus in relation to (d) histone 3 lysine 4 trimethylation (H3K4me3) of histones in the area of the gene. (c) Fast type IIx myosin heavy chain expression and (e) H3K4me3 for the same conditions. These data suggest a relationship between histone 3 lysine 4 trimethylation and the expression of myosin heavy chain isoforms. Redrawn after Pandorf et al. (2009).

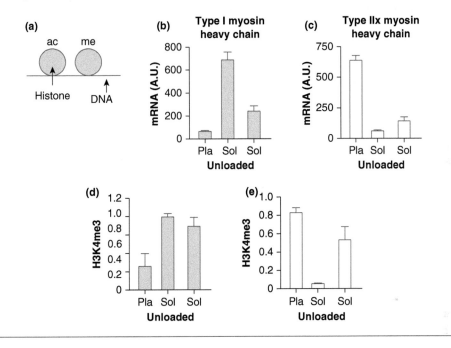

Together this suggests that chromatin remodelling opens up myosin heavy chain genes during differentiation, affects the expression of specific myosin heavy chain isoforms in type I and II fibre subtypes and changes in response to stimuli such as hindlimb unloading. Interestingly, the methylation of the promoters of key regulators of the adaptation to exercise, such as PGC-1α, is also affected by exercise (Barres et al., 2012), suggesting that such regulation is not restricted to myosin heavy chain genes. The challenge for the future is to link these chromatin remodelling events to the upstream signal transduction pathways and the exercise-related signals that control them.

MYOSIN HEAVY CHAIN ON/OFF REGULATION, MYOMIRS AND SOX6

One unique feature of myosin heavy chain isoform gene regulation is that their regulation is on/off, or binary, in contrast to many other genes where the regulation is graded. In other words, individual muscle fibres, apart from transitional hybrid fibres, express almost only one myosin heavy chain isoform, while the expression of all other myosin heavy chain isoforms is almost fully switched off. This is in stark contrast to, for example, glycolytic or oxidative enzymes which are always expressed but increase or decrease somewhat, for example, in response to endurance exercise. A team led by Eric Olson has identified one mechanism by which the binary regulation of myosin heavy chain isoforms is achieved. The research team discovered that the slow myosin heavy chain Iβ gene encodes two microRNAs (miRNA), miR-208b and miR-499 (termed **MyoMirs**), in its introns (van Rooij et al., 2009). These miRNAs bind to mRNAs and prevent their translation into protein. The researchers discovered that miR-208b and miR-499 were expressed in slow muscles whenever the myosin heavy chain Iβ gene is expressed. Moreover, they found that miR-208b and miR-499 blocked the translation of Sox6, a transcription factor that prevents the expression of slow genes and directly or indirectly upregulates fast genes (Figure 4.6, Quiat et al., 2011).

Figure 4.6 Regulation of muscle fibres by MyoMirs. MyoMirs (miR-208b and miR-499) are a key mechanism responsible for the binary (on/off) expression of myosin heavy chain (MHC) isoforms. (1) Developmental pathways via the calcineurin and other pathways upregulate the expression of slow type I myosin heavy chain (MHC; the genes names are *MYH7/7b*). (2) Two miRNAs, termed miR-208b and miR-499 or MyoMirs, are located in introns of the type I myosin heavy chain. Thus whenever the type I myosin heavy chain gene is expressed, these MyoMirs are also expressed. The MyoMirs then prevent the translation of the transcription factor Sox6 mRNA into protein. (3) Sox6 normally inhibits the expression of type I myosin heavy chain and increases the expression of fast myosin heavy chains. This MyoMir-dependent mechanism ensures that fast myosin heavy chains are not expressed in fibres where the type I myosin heavy chain is predominantly expressed (van Rooij et al., 2009; Quiat et al., 2011).

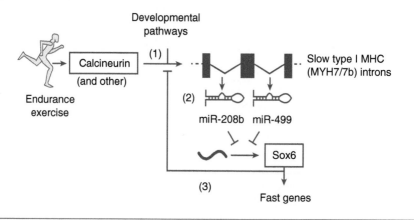

CAMK, AMPK, PGC-1α AND MITOCHONDRIAL BIOGENESIS

Even though the primary adaptation to endurance exercise is an increase in the delivery of oxygenated blood to the working muscle, skeletal muscles also increase their ability to extract oxygen and produce ATP. Here, two key exercise signals are the rise of the Ca^{2+} concentration, which signals increased contractile activity, and a rise of the AMP and ADP concentrations, which signals increased ATP usage, respectively. In this section we will discuss how these exercise signals trigger adaptations such as mitochondrial biogenesis.

As discussed above, Ca^{2+} is released every time a muscle contracts. While most of that calcium causes muscle contraction, some binds to a protein called calmodulin which has four Ca^{2+}-binding sites. In response to exercise, Ca^{2+}-calmodulin then activates the phosphatase calcineurin, which we have discussed above, and CaMK (calmodulin-activated protein kinase) isoforms, of which CaMKII is probably the most important in skel-etal muscle (Chin, 2005; Rose et al., 2006; Egan and Zierath, 2013). Early on it was shown that the overexpression of CaMK IV in skeletal muscle increased the expression of PGC-1α, a master regulator of mitochondrial biogenesis, and mitochondrial bio-genesis itself (Wu et al., 2002) demonstrating that CaMKs can induce mitochondrial biogenesis. Thus one mechanism by which exercise increases mitochondrial biogenesis is a Ca^{2+}-calmodulin–CaMK–PGC-1α signalling axis (see Figure 4.8).

A second mechanism is related to ATP turnover. During exercise, skeletal muscle converts the chemical energy in the ATP hydrolysis reaction (ATP→ADP + P_i) into mechanical energy for contraction and heat. The muscular ATP concentration is in the region of 3.5–8 mmol/kg in resting mammalian muscle and declines only little even during maximally fatiguing exercise (Figure 4.6). ATP is kept constant even though muscular ATP hydrolysis or usage can increase by more than 200-fold during maximal, short contractions when compared to a resting muscle (Meyer and Foley, 1996). This stability of the ATP concentration is achieved in the short term by ATP resynthesis via the very fast-acting Lohmann reaction (ADP + phosphocreatine→ATP + creatine), by the myokinase reaction (ADP + ADP→ATP + AMP), by glycolysis and by oxidative phosphorylation by mitochondria (Figure 4.7). Mitochondria are the main site for ATP resynthesis at rest and during endurance exercise. In this section we will review how an increased ATP turnover and contractile activity are sensed and how these signals stimu-late the production of additional mitochondria, termed 'mitochondrial biogenesis'.

How is an increased ATP turnover sensed? Whenever ATP is used, its millimolar concentration decreases only by a few micromoles, as the Lohmann reaction stabilizes the ATP concentration very quickly. The small decrease of ATP translates, however, to a large relative concentration of ADP and AMP, as these substances lie in the micromolar

Figure 4.7 Skeletal muscle metabolism. (a) Two ^{31}P-NMR spectra obtained from a resting (top) and fatigued, phosphocreatine (PCr)-depleted (bottom) human calf muscle (Wackerhage and Zange, unpublished). The peaks are proportional to the concentration of the metabolites within the muscle and show that PCr is depleted at the end of fatiguing exercise, whereas the concentration of ATP has changed little. (b) Electron microscopy image from a mouse tibialis anterior muscle showing sarcomeres and mitochondria. Scale bar 1 μm. From Greenhorn and Wackerhage (unpublished).

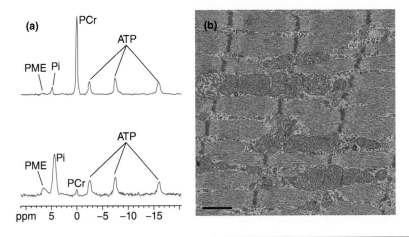

range. Thus a relative drop of ATP by less than 1% translates into a doubling or more of the ADP and AMP concentrations. The increases of the AMP and ADP concentrations are large but short during high-intensity exercise such as high-intensity interval training (Gibala et al., 2012) and moderate but long-lasting during low-intensity exercise. ADP, AMP and glycogen are all sensed and regulate the activity of AMP-activated protein kinase (AMPK). AMPK is one of the most important kinases in the human body, especially in relation to exercise, and thus an account by Grahame Hardie, who has led the discovery of AMPK is apposite (Box 4.1).

BOX 4.1 GRAHAME HARDIE: DISCOVERY OF AMP-ACTIVATED PROTEIN KINASE AND ITS ROLE DURING EXERCISE

At the beginning of 1977, at the age of only 26, I was lucky enough to get the opportunity to start my own research in Dundee. Philip Cohen had obtained a Fellowship that enabled him to focus full time on research for two years, and I was appointed as his temporary replacement on the teaching staff. Fortunately Philip never returned to his teaching post and I am still here! The head of department, Peter Garland, said that as the post was only temporary I would initially have to share space in Philip's lab, but since Philip was a great mentor this turned out to be a blessing in disguise. One evening Philip and I went to the bar below the Department to discuss what I might work on. I'd like to be able to say that I spent many hours carefully thinking about this, but in fact by the end of the second beer I'd come to a snap decision that profoundly affected my subsequent career!

Philip had been working on the regulation of glycogen metabolism, which in 1977 (strange as it may seem today) was the only process in which the role of protein kinases and protein phosphorylation was well understood. I decided to look at another pathway, fatty acid synthesis, and to follow up studies by Ki-Han Kim at Purdue University suggesting that the enzyme acetyl-CoA carboxylase (ACC) was inactivated by an unknown protein kinase. I had soon purified ACC and confirmed that it could be phosphorylated by various known kinases. However, none of these seemed to account for the large degree of ACC inactivation that occurred *in vivo*. Then a new PhD student, Dave Carling, partially purified a kinase from rat liver that seemed to fit the bill. Interestingly, it was activated both by AMP and by phosphorylation (we later showed that phosphorylation occurred at Thr172, now widely used as a biomarker for AMPK activation). In these respects it was similar to another kinase, first defined by David Gibson in Indianapolis, which inactivated HMG-CoA reductase, an enzyme involved in cholesterol synthesis. We soon obtained evidence that the same kinase phosphorylated and inactivated both targets, and I remember suggesting to Dave Carling one afternoon that we should change the name from ACC kinase to AMP-activated protein kinase. Another important step was that we identified the major site the kinase phosphorylated on ACC and made a peptide substrate based on it (the SAMS peptide), which made it much easier to assay the kinase activity.

Despite these advances, the real physiological role of AMPK remained unclear. One day in the mid 1990s I received a phone call from Will Winder who was at Brigham Young University, Utah. He said he would be passing by Dundee and asked if he could drop in. At the time I had no knowledge of the exercise physiology field and had no idea who he was, but I agreed to see him. He had found that ACC was inactivated during exercise in muscle (which, by lowering malonyl-CoA, would promote fat oxidation) and he wondered whether AMPK might be responsible for this. I had been thinking about the idea that AMPK might act as an energy sensor, and had already shown that mitochondrial poisons activated AMPK in liver cells. However, it was immediately obvious that exercise could be a more physiological situation where AMPK might be playing such a role. I therefore agreed to help Will with the AMPK assays, and within a few weeks he had shown that AMPK was activated in rat muscle during treadmill exercise. At that time, following the pioneering work of Georges van den Berghe, my lab had also started using the nucleoside AICAR to activate AMPK in intact cells. I suggested that Will tried this in perfused rat muscle, and he soon showed that AICAR not only stimulated fat oxidation as we had expected, but also glucose uptake. These experiments provided the first insights into the role of the AMPK system during exercise.

Grahame Hardie, University of Dundee, May 2013

AMPK is a protein kinase that forms a complex comprising three protein subunits:

- the α catalytic subunit (the actual kinase whose activity is regulated by Thr172 phosphorylation);
- the β subunit which holds the enzyme together and senses glycogen;
- the γ subunit which binds and therefore senses AMP/ADP/ATP.

AMPK is activated by ADP and AMP and inhibited by glycogen. The rise of ADP and AMP during intense exercise leads to an increase in the phosphorylation of AMPK at Thr172 due to a decrease in dephosphorylation and a constitutively active upstream kinase, LKB1 (Xiao et al., 2011; Carling et al., 2012). AMPK is also inhibited by glycogen, which is sensed by the β subunit (Towler and Hardie, 2007). Because of these events, the AMPK Thr172 phosphorylation and thus AMPK activity increases during endurance exercise and AMPK starts to regulate its downstream targets by phosphorylation (Figure 4.8). To summarize, AMPK is activated during endurance exercise via phosphorylation of Thr172 and this response is probably mainly dependent on the rise of AMP and ADP during exercise.

What happens when AMPK is active in muscle? AMPK is unusual as it regulates a whole host of adaptations in the short term (acutely) and long term (chronically). In a nutshell, active AMPK promotes ATP resynthesis in the short and long term and inhibits non-essential ATP consumption:

- **AMPK increases mitochondrial biogenesis** by especially increasing the expression of PGC-1α, but it may also increase its phosphorylation and reduce its acetylation (Canto et al., 2009; Philp et al., 2011). PGC-1α is, as we will show below, the master regulator of mitochondrial biogenesis.
- **AMPK increases carbohydrate catabolism**: AMPK promotes GLUT4-mediated glucose uptake into skeletal muscle and inhibits glycogen synthesis, which facilitates ATP resynthesis from carbohydrates (Hardie and Sakamoto, 2006).

Figure 4.8 Mitochondrial biogenesis. Schematic drawing depicting the regulation of mitochondrial biogenesis by endurance exercise. (1) Endurance exercise acutely increases Ca^{2+}, ADP, AMP and lowers glycogen in the contracting muscles. Especially elevated AMP and ADP trigger an increased phosphorylation of AMPK at Thr172 by its upstream kinase LKB1 (not shown) and the increased Ca^{2+} concentration via calmodulin causes CaMK autophosphorylation. Exercise-activated CaMK and AMPK then regulate transcription factors so that the PGC-1α gene expression increases, which is the key regulatory step of mitochondrial biogenesis. (2) The kinases AMPK and p38 and the acetylation modulators Gcn5 and Sirt1 then possibly phosphorylate PGC-1α and reduce the acetylation of PGC-1α, which increases its activity. (3) Thus endurance exercise results in more and more active PGC-1α. PGC-1α then binds to chromatin remodelling factors (CRF) which open up the chromatin for transcription. PGC-1α also binds to transcription factors such as NRF. (4) This increases the expression of mitochondrial proteins (P) and of mitochondrial transcription and replication factors (mtTF) encoded in nuclear DNA. (5) Once expressed, mtTF migrate to the mitochondrion, bind to mtDNA and drive the expression of proteins encoded in mtDNA (mtP) and the replication of mtDNA. Finally P and mtP form the mature complexes and mitochondrial fission occurs, which results in the production of new mitochondria.

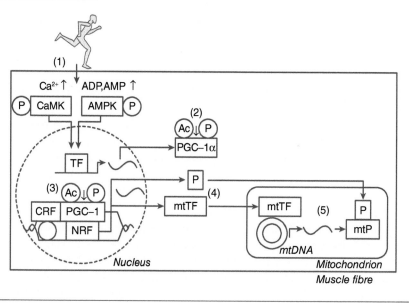

- **AMPK increases fat catabolism**: AMPK promotes fatty acid uptake into skeletal muscle and mitochondria, and as a consequence fatty acid oxidation increases (Hardie and Sakamoto, 2006).
- **AMPK promotes angiogenesis**: AMPK promotes the formation of additional capillaries, which is termed angiogenesis (Ouchi et al., 2005).

To summarize, rises in ADP and AMP are sensed by AMPK, which leads to the activation of AMPK via increased Thr172 phosphorylation. As an acute consequence, glycolysis becomes activated directly by ADP, carbohydrate uptake and catabolism via AMPK, fat catabolism via AMPK and oxidative phosphorylation, which is directly activated by ADP. In contrast, energy-consuming processes, such as protein synthesis, acutely decrease often due to the increased AMPK activity. AMPK also increases the activity and the transcription of PGC-1α, which in turn drives long-term changes in muscle, like mitochondrial biogenesis and angiogenesis.

Mitochondrial biogenesis in response to endurance exercise was the first positive effect of endurance exercise, demonstrated originally by Holloszy (1967). In this section we will review the mechanism by which active AMPK and PGC-1α promote mitochondrial biogenesis.

Active AMPK activates PGC-1α via several mechanisms:

- AMPK increases especially the expression of PGC-1α (Suwa et al., 2003).
- AMPK and p38 can phosphorylate PGC-1α (Knutti et al., 2001; Jager et al., 2007).
- AMPK reduces the acetylation of PGC-1α via the deacetylase SIRT1 in some situations (Canto et al., 2009). However, during endurance exercise the knockout of SIRT1 has no effect on PGC-1α deacetylation, which still occurs probably due to a reduced acetylation of PGC-1α by the acetyltransferase GCN5 (Philp et al., 2011).

We do not discuss the mechanisms in detail here as it is sufficient to know for the purpose of this chapter that endurance exercise activates AMPK in muscle (Baar et al., 2002; Pilegaard et al., 2003), which in turn activates and increases the expression of PGC-1α. Thus to summarize, endurance exercise activates AMPK-PGC-1α signalling. This also applies to a variant of endurance exercise, high-intensity interval training (HIT) (Gibala et al., 2012).

How does endurance exercise-activated PGC-1α function? PGC-1α is a transcriptional co-factor that does not bind DNA directly but co-activates other transcription factors which regulate mitochondrial biogenesis. Also, PGC-1α may contribute to regulate the expression of other genes that affect the muscle fibre phenotype (Lin et al., 2002). In the next sections we will first give an overview of mitochondria and mitochondrial biogenesis and then discuss in more detail how active CaMK/AMPK–PGC-1α signalling promotes mitochondrial biogenesis.

Mitochondria are unusual organelles since they have their own DNA, abbreviated as mtDNA, which encodes 13 mitochondrial protein-encoding genes, 22 transfer RNAs, and two ribosomal RNAs. The majority of the mitochondrial proteins, however, are encoded by the nuclear DNA. Table 4.4 shows that most electron transport chain complexes are built from proteins that are encoded in both mtDNA and nuclear DNA.

Two questions arise: First, why do mitochondria have their own DNA and second, how does AMPK–PGC-1α signalling switch on the expression of the proteins encoded in both the nuclear and mtDNA, as these genes all need to be expressed in order to build new mitochondria?

Table 4.4 Numbers of subunits of electron transfer chain complexes encoded in either mtDNA or nuclear DNA

ELECTRON TRANSFER CHAIN COMPLEX	MTDNA	NUCLEAR DNA
I NADH dehydrogenase complex	7	>25
II Succinate dehydrogenase	–	4
III Cytochrome bc1 complex	1	10
IV Cytochrome c oxidase	3	10
(V) F_0F_1 ATP synthase	2	11

From Poyton and McEwen (1996).

Mitochondria have probably evolved from bacteria that infected nucleus-containing cells millions of years ago. According to the endosymbiotic hypothesis, these bacteria and their ability to use oxygen to create energy provided their host cell with an evolutionary advantage; combining for the first time glycolytic and oxidative metabolism. Evolution over millions of years improved upon this system and led to the development of regulatory systems of which CaMK/AMPK–PGC-1α signalling is the most important. So how does endurance exercise via CaMK/AMPK–PGC-1α signalling lead to the formation of new mitochondria? It involves the following steps (Figure 4.8):

1   CaMK and AMPK can increase the expression of PGC-1α through the regulation of transcription factor binding to the PGC-1α promoter.

2   The increase in PGC-1α mRNA results in more protein and this protein becomes deacetylated and phosphorylated resulting in more PGC-1α in the nucleus where it (a) interacts with proteins that open up the chromatin for transcription and (b) co-activates transcription factors that increase the expression oxidative phospho-rylation and fat oxidation genes encoded in **nuclear DNA**.

3   Activated PGC-1α also increases the expression of the TFAM and TFBM tran-scription factors that are encoded in nuclear DNA. TFAM and TFBM proteins then migrate to mitochondria where they activate **mtDNA gene expression** and **mtDNA replication**. The latter ensures that daughter mitochondria have their own mtDNA (Scarpulla, 2010).

4   Finally nuclear proteins are imported by mitochondria, protein complexes are assembled followed by **mitochondrial fission**.

To summarize, endurance exercise results in a rise of the concentrations of ADP, AMP and Ca^{2+}, which results in an activation of AMPK and CaMK. Active AMPK and CaMK increase the expression of PGC-1α by directly regulating transcription factor activity. PGC-1α is also deacetylated and possibly more phosphorylated which increases its activity. PGC-1α binds to a variety of transcription factors and increases their activ-ity, resulting in increased expression of mitochondrial genes encoded in the nucleus but also of transcription factors that migrate to the mitochondrion where they increase the expression of genes encoded by the mtDNA and promote the replication of mtDNA. With all the building blocks now present, the functional mitochondrial mass increases.

EXERCISE-INDUCED ANGIOGENESIS

Endurance training not only increases the capacity for cardiac output or maximal blood flow due to triggering the athlete's heart but also increases the capillary density in the trained muscles, termed **angiogenesis**. Capillaries can be demonstrated with electron microscopy as shown in Figure 4.9.

Classical exercise physiologists have demonstrated that slow type I fibres have a greater capillary density than faster type II fibres, that endurance athletes have a greater capillary-to-muscle fibre ratio in their muscles than untrained subjects (Brodal et al., 1977) and that endurance training increases the capillary density in skeletal muscle (Andersen and Henriksson, 1977; Ingjer, 1979).

Figure 4.9 Electron microscopy image of a capillary within a mouse soleus muscle. An endothelial cell and its nucleus is visible. The capillary is surrounded by sarcomeres and mitochondria. Inside the capillary an erythrocyte, or red blood cell, can be seen. Scale bar = 0.5 μm. From Wettstein and Greenhorn (unpublished).

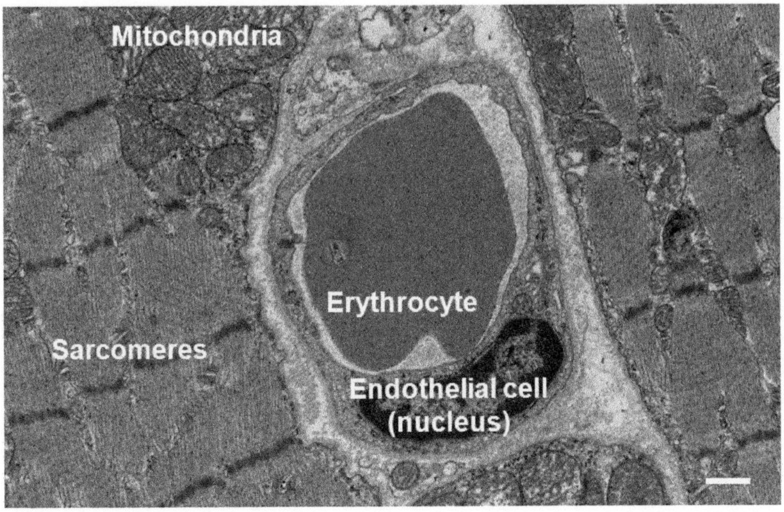

In the previous section we described how endurance exercise activates CaMK, AMPK and PGC-1α. PGC-1α then binds and co-activates several transcription factors including ERRα. Together, PGC-1α and ERRα upregulate the expression of VEGF and other angiogenic growth factors (Arany et al., 2008). A second pathway that induces exercise-induced angiogenesis senses the oxygen tension, or PO_2. It has been estimated that the PO_2 decreases from ≈30 mmHg at rest to ≈3–4 mmHg during moderate to heavy exercise (Wagner, 2011). Given that oxygen is the key molecule for aerobic metabolism, it is no surprise that this drop in PO_2 is sensed by cells and that adaptations such as angiogenesis are triggered. Oxygen is sensed by the van Hippel–Lindau (VHL) protein, which in turn regulates the breakdown of the hypoxia-sensing transcription factor HIF-1. Whenever the PO_2 is high, VHL is more bound to HIF-1. VHL binding to HIF-1 targets HIF-1 for breakdown by the proteasome, and as a consequence the concentration of HIF-1 is low in normoxic or resting muscle. During exercise or at high altitude the PO_2 drops and this causes VHL to detach from HIF-1. This reduces the degradation of HIF-1 and as a consequence the concentration of HIF-1 increases (Ameln et al., 2005). Similar to PGC-1α–ERRα complexes, HIF-1 also increases the expression of VEGF by binding to a DNA-binding site which is termed hypoxia response element (HRE) (Xie et al., 2004; Wagner, 2011). In order to test the importance of HIF-1, researchers developed a transgenic mouse where HIF-1 was knocked down. Surprisingly, the capillary density was not significantly decreased. However, VEGF mRNA increased in response to exercise only by 13%, whereas it increased by 73% in wildtype animals. This suggests that HIF-1 does not determine the capillary density under resting conditions but that HIF-1 contributes to the angiogenesis signalling response in response to exercise (Mason et al., 2004).

Shear stress and wall tension within capillaries are other signals that may contribute to the regulation angiogenesis. Blood flow has been estimated to increase ≈100-fold from rest to maximum in exercising knee-extensor muscles (Saltin et al., 1998). Muscle capillaries that receive this flow will experience greater shear stress across the wall of the blood vessel. The increased wall stress leads the lining cells, termed endothelial cells, to produce nitric oxide (NO) which also stimulates angiogenesis. Evidence for this hypothesis stems from experiments where NO synthase was inhibited with N(G)-nitro-1-arginine (1-NNA) in rats. 1-NNA treatment abolished the increase in capillary-to-fibre ratio that is induced by electrical stimulation (Hudlicka et al., 2000). These data suggest that shear stress-dependent release of NO may also contribute to exercise-induced angiogenesis.

To summarize, Ca^{2+}, ADP, AMP, PO_2 and shear stress activate CaMK/AMPK–PGC-1α, HIF-1 and NO signalling which leads to the increased expression of VEGF and other growth factors that stimulate angiogenesis. VEGF appears to be the master regulator of angiogenesis and as such regulates the proliferation, migration, elongation, network formation, branching and leakiness of endothelial cells. Both VEGF and VEGF receptor mice die during development with defects of angiogenesis (Yancopoulos et al., 2000; Wagner, 2011). Additionally, the concentration of the angiopoietins Ang1 and its inhibitor Ang2 regulate the leakiness of capillaries (Yancopoulos et al., 2000). Furthermore, proteins that tunnel through the extracellular matrix, metalloproteinases, are expressed in response to VEGF so that endothelial cells then can form a tunnel which will mature into a capillary. Inhibiting these proteases with the inhibitor GM6001 prevents capillary growth (Haas et al., 2000), demonstrating the importance of metalloproteinases for angiogenesis. Together these angiogenesis-regulating growth factors and metalloproteinases change their expression during exercise and this leads to an expansion of the capillary network in skeletal muscle, mainly by sprouting (Figure 4.10).

SUMMARY

The major limiting factors or quantitative traits for endurance performance are the maximal oxygen uptake ($\dot{V}O_{2max}$), the percentage of $\dot{V}O_{2max}$ that can be sustained during endurance exercise (%$\dot{V}O_{2max}$) and mechanical efficiency. Prescribing an endurance training programme requires setting variables such as exercise intensity and volume. There is good scientific evidence that endurance training is effective for the general population, but limited evidence to give specific endurance training recommendations. Also, individuals will respond differently to an endurance training programme and although the vast majority of individuals will improve endurance-related variables such as $\dot{V}O_{2max}$ and health, a small minority may not respond and may even worsen some risk factors.

A key adaptation to endurance exercise is an increased maximal cardiac output. The increase in cardiac output is dependent on an increased stroke volume which depends on the development of a physiological cardiac hypertrophy or an athlete's heart. Physiological hypertrophy relies at least partially on increased PI3K, PKB/Akt and decreased C/EBPβ signalling. In contrast, increased calcineurin signalling stimulates pathological cardiac hypertrophy.

Figure 4.10 Exercise-induced angiogenesis. Schematic drawing depicting the events that increase angiogenesis in response to endurance exercise and training. (1) Endurance exercise activates CaMK/AMPK and PGC-1α, which in turn binds and co-activates ERRα. ERRα promotes the expression of VEGF. (2) From rest to moderate to heavy exercise the PO_2 drops from ≈30 mmHg to ≈3–4 mmHg. This leads to less degradation of HIF-1 in a process that involves VHL. HIF-1 also increases the expression of VEGF. (3) Shear stress leads to NO production by endothelial cells. All the aforementioned events regulate the expression of angiogenic factors, of which VEGF appears to be the most important one. Additionally the expression of metalloproteinases (MMP) also increases and the MMPs cut a path through the extracellular matrix that surrounds all muscle cells allowing endothelial cells to grow into these tunnels to form capillaries.

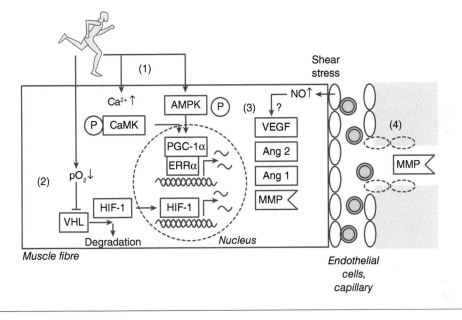

Muscle fibres can be sub-divided into slow type I fibres, intermediate type IIa fibres, fast type IIx fibres and even faster type IIb fibres. The name derives from the myosin heavy chain that is predominantly expressed in a muscle fibre. For example, type IIx fibres mainly express type IIx myosin heavy chain. Importantly, while humans have a myosin heavy chain IIb gene in their genomes, the myosin heavy chain IIb protein is not expressed in major muscles. Calcineurin–NFAT signalling is active in type I fibres and increases in response to prolonged electrical stimulation in model organisms. When active, calcineurin–NFAT signalling promotes the expression of 'slow fibre' genes, but there are exceptions. In response to endurance training programmes over several months, type IIx fibres decrease and IIa fibres increase. More pronounced fibre type conversions may occur in response to decades of endurance training but the evidence is limited. Myosin heavy chain genes are organized in a slow/cardiac cluster and a fast/developmental cluster and this is evolutionary conserved. Expression of myosin heavy chain I also increases the expression of MyoMirs which, via the transcription factor Sox6, prevent the expression of faster myosin heavy chains. This can in part explain the on/off regulation of myosin heavy chains. Also the chromatin, or the wrapping of DNA around histones, is opened up in myosin heavy chain isoforms that are actively transcribed in a muscle fibre. This is also known as epigenetic regulation.

Endurance exercise increases the number of mitochondria, which is termed exercise-induced mitochondrial biogenesis. Here, the major players are CaMK and AMPK. CaMK is activated by calcium release during exercise and likely explains the effect of long-slow endurance training. AMPK senses the concentrations of AMP and ADP, which increase during acute high intensity endurance exercise. AMPK also senses a drop of glycogen. During exercise, AMPK is activated by phosphorylation of Thr172 as a result of decreased phosphatase activity and a constitutively active upstream kinase. CaMK and AMPK then increase the expression of the transcriptional co-factor PGC-1α. PGC-1α can further be modulated by deacetylation and increased phosphorylation, which increases its greater activity. PGC-1α can then co-activate transcription factors to increase the expression of mitochondrial genes encoded by the nuclear DNA and of transcription and replication factors that promote the expression of genes encoded in, and the replication of, mitochondrial DNA (mtDNA). CaMK/AMPK–PGC-1α, hypoxia–HIF-1 and shear stress–NO regulate the expression of angiogenic growth factors of which VEGF is the major factor. Metalloproteinases expression also increases and active metalloproteinases cut a tunnel through the extracellular matrix to promote the formation of new capillaries. This process is termed exercise-induced angiogenesis.

REVIEW QUESTIONS

1   What strategy do you recommend for the prescription of an endurance training programme which takes the variation in trainability into account?

2   Give an example of and explain a signal transduction pathway that regulates either physiological (athlete's heart) or pathological cardiac hypertrophy (hypertrophic cardiomyopathy).

3   Compare and contrast type I, IIa, IIx and IIb muscle fibres. What is special about IIb fibres? What is the effect of endurance exercise on fibre type percentages?

4   Explain mechanisms that may contribute to the on/off regulation of myosin heavy chain isoforms in skeletal muscle.

5   How do the rises of the Ca^{2+}, AMP and ADP concentrations during endurance exercise stimulate mitochondrial biogenesis?

6   Explain how Ca^{2+}, an increased energy turnover, changed oxygen levels and increased blood flow may stimulate capillary growth in response to endurance exercise.

FURTHER READING

Egan B and Zierath JR (2013). Exercise metabolism and the molecular regulation of skeletal muscle adaptation. *Cell Metab* 17, 162–184.

Hardie DG (2011). Energy sensing by the AMP-activated protein kinase and its effects on muscle metabolism. *Proc Nutr Soc* 70, 92–99.

Schiaffino S (2010). Fibre types in skeletal muscle: a personal account. *Acta Physiol (Oxford)* 199, 451–463.

REFERENCES

Ameln H, Gustafsson T, Sundberg CJ, Okamoto K, Jansson E, Poellinger L, et al. (2005). Physiological activation of hypoxia inducible factor-1 in human skeletal muscle. *FASEB J* 19, 1009–1011.

Andersen P and Henriksson J (1977). Capillary supply of the quadriceps femoris muscle of man: adaptive response to exercise. *J Physiol* 270, 677–690.

Arany Z, Foo SY, Ma Y, Ruas JL, Bommi-Reddy A, Girnun G, et al. (2008). HIF-independent regulation of VEGF and angiogenesis by the transcriptional coactivator PGC-1alpha. *Nature* 451, 1008–1012.

Asp P, Blum R, Vethantham V, Parisi F, Micsinai M, Cheng J, et al. (2011). Genome-wide remodeling of the epigenetic landscape during myogenic differentiation. *Proc Natl Acad Sci U S A* 108, E149–E158.

Baar K, Wende AR, Jones TE, Marison M, Nolte LA, Chen M, et al. (2002). Adaptations of skeletal muscle to exercise: rapid increase in the transcriptional coactivator PGC-1. *FASEB J* 16, 1879–1886.

Barany M (1967). ATPase activity of myosin correlated with speed of muscle shortening. *J Gen Physiol* 50, Suppl 218.

Barres R, Yan J, Egan B, Treebak JT, Rasmussen M, Fritz T, et al. (2012). Acute exercise remodels promoter methylation in human skeletal muscle. *Cell Metab* 15, 405–411.

Bassett DR, Jr. and Howley ET (2000). Limiting factors for maximum oxygen uptake and determinants of endurance performance. *Med Sci Sports Exerc* 32, 70–84.

Berchtold MW, Brinkmeier H and Muntener M (2000). Calcium ion in skeletal muscle: its crucial role for muscle function, plasticity, and disease. *Physiol Rev* 80, 1215–1265.

Bernardo BC, Weeks KL, Pretorius L and McMullen JR (2010). Molecular distinction between physiological and pathological cardiac hypertrophy: experimental findings and therapeutic strategies. *Pharmacol Ther* 128, 191–227.

Biering-Sorensen B, Kristensen IB, Kjaer M and Biering-Sorensen F (2009). Muscle after spinal cord injury. *Muscle Nerve* 40, 499–519.

Blomqvist CG and Saltin B (1983). Cardiovascular adaptations to physical training. *Annu Rev Physiol* 45, 169–189.

Bostrom P, Mann N, Wu J, Quintero PA, Plovie ER, Panakova D, et al. (2010). C/EBPbeta controls exercise-induced cardiac growth and protects against pathological cardiac remodeling. *Cell* 143, 1072–1083.

Bouchard C, An P, Rice T, Skinner JS, Wilmore JH, Gagnon J, et al. (1999). Familial aggregation of $VO_{(2max)}$ response to exercise training: results from the HERITAGE Family Study. *J Appl Physiol* 87, 1003–1008.

Bouchard C, Sarzynski MA, Rice TK, Kraus WE, Church TS, Sung YJ, et al. (2011). Genomic predictors of the maximal O uptake response to standardized exercise training programs. *J Appl Physiol* 110, 1160–1170.

Bouchard C, Blair SN, Church TS, Earnest CP, Hagberg JM, Hakkinen K, et al. (2012). Adverse metabolic response to regular exercise: is it a rare or common occurrence? *PLoS ONE* 7, e37887.

Brodal P, Ingjer F and Hermansen L (1977). Capillary supply of skeletal muscle fibers in untrained and endurance-trained men. *Am J Physiol* 232, H705–H712.

Bueno OF, De Windt LJ, Tymitz KM, Witt SA, Kimball TR, Klevitsky R, et al. (2000). The MEK1-ERK1/2 signaling pathway promotes compensated cardiac hypertrophy in transgenic mice. *EMBO J* 19, 6341–6350.

Buller AJ, Eccles JC and Eccles RM (1960). Interactions between motoneurones and muscles in respect of the characteristic speeds of their responses. *J Physiol* 150, 417–439.

Canto C, Gerhart-Hines Z, Feige JN, Lagouge M, Noriega L, Milne JC, et al. (2009). AMPK regulates energy expenditure by modulating NAD+ metabolism and SIRT1 activity. *Nature* 458, 1056–1060.

Carling D, Thornton C, Woods A and Sanders MJ (2012). AMP-activated protein kinase: new regulation, new roles? *Biochem J* 445, 11–27.

Carroll SL, Klein MG and Schneider MF (1997). Decay of calcium transients after electrical stimulation in rat fast- and slow-twitch skeletal muscle fibres. *J Physiol* 501 (Pt 3), 573–588.

Chin ER (2005). Role of Ca^{2+}/calmodulin-dependent kinases in skeletal muscle plasticity. *J Appl Physiol* 99, 414–423.

Chin ER, Olson EN, Richardson JA, Yang Q, Humphries C, Shelton JM, et al. (1998). A calcineurin-dependent transcriptional pathway controls skeletal muscle fiber type. *Genes Dev* 12, 2499–2509.

Costill DL, Daniels J, Evans W, Fink W, Krahenbuhl G and Saltin B (1976a). Skeletal muscle enzymes and fiber composition in male and female track athletes. *J Appl Physiol* 40, 149–154.

Costill DL, Fink WJ and Pollock ML (1976b). Muscle fiber composition and enzyme activities of elite distance runners. *Med Sci Sports* 8, 96–100.

DeBosch B, Treskov I, Lupu TS, Weinheimer C, Kovacs A, Courtois M, et al. (2006). Akt1 is required for physiological cardiac growth. *Circulation* 113, 2097–2104.

Egan B and Zierath JR (2013). Exercise metabolism and the molecular regulation of skeletal muscle adaptation. *Cell Metab* 17, 162–184.

Ekblom B (1968). Effect of physical training on oxygen transport system in man. *Acta Physiol Scand Suppl* 328, 1–45.

Gan Z, Burkart-Hartman EM, Han DH, Finck B, Leone TC, Smith EY, et al. (2011). The nuclear receptor PPARbeta/delta programs muscle glucose metabolism in cooperation with AMPK and MEF2. *Genes Dev* 25, 2619–2630.

Garber CE, Blissmer B, Deschenes MR, Franklin BA, Lamonte MJ, Lee IM, et al. (2011). American College of Sports Medicine position stand. Quantity and quality of exercise for developing and maintaining cardiorespiratory, musculoskeletal, and neuromotor fitness in apparently healthy adults: guidance for prescribing exercise. *Med Sci Sports Exerc* 43, 1334–1359.

Gibala MJ, Little JP, MacDonald MJ and Hawley JA (2012). Physiological adaptations to low-volume, high-intensity interval training in health and disease. *J Physiol* 590, 1077–1084.

Gollnick PD, Armstrong RB, Saltin B, Saubert CW, Sembrowich WL and Shepherd RE (1973). Effect of training on enzyme activity and fiber composition of human skeletal muscle. *J Appl Physiol* 34, 107–111.

Haas TL, Milkiewicz M, Davis SJ, Zhou AL, Egginton S, Brown MD, et al. (2000). Matrix metalloproteinase activity is required for activity-induced angiogenesis in rat skeletal muscle. *Am J Physiol Heart Circ Physiol* 279, H1540–H1547.

Hardie DG and Sakamoto K (2006). AMPK: a key sensor of fuel and energy status in skeletal muscle. *Physiology (Bethesda)* 21, 48–60.

Harridge SD, Bottinelli R, Canepari M, Pellegrino MA, Reggiani C, Esbjornsson M, et al. (1996). Whole-muscle and single-fibre contractile properties and myosin heavy chain isoforms in humans. *Pflugers Arch* 432, 913–920.

Hennig R and Lomo T (1985). Firing patterns of motor units in normal rats. *Nature* 314, 164–166.

Holloszy JO (1967). Biochemical adaptations in muscle. Effects of exercise on mitochondrial oxygen uptake and respiratory enzyme activity in skeletal muscle. *J Biol Chem* 242, 2278–2282.

Hudlicka O, Brown MD and Silgram H (2000). Inhibition of capillary growth in chronically stimulated rat muscles by N(G)-nitro-1-arginine, nitric oxide synthase inhibitor. *Microvasc Res* 59, 45–51.

Hughes SM, Chi MM, Lowry OH and Gundersen K (1999). Myogenin induces a shift of enzyme activity from glycolytic to oxidative metabolism in muscles of transgenic mice. *J Cell Biol* 145, 633–642.

Ingalls CP (2004). Nature vs. nurture: can exercise really alter fiber type composition in human skeletal muscle? *J Appl Physiol* 97, 1591–1592.

Ingjer F (1979). Effects of endurance training on muscle fibre ATP-ase activity, capillary supply and mitochondrial content in man. *J Physiol* 294, 419–432.

Jager S, Handschin C, St-Pierre J and Spiegelman BM (2007). AMP-activated protein kinase (AMPK) action in skeletal muscle via direct phosphorylation of PGC-1alpha. *Proc Natl Acad Sci U S A* 104, 12017–12022.

Jansson E and Kaijser L (1977). Muscle adaptation to extreme endurance training in man. *Acta Physiol Scand* 100, 315–324.

Jenuwein T and Allis CD (2001). Translating the histone code. *Science* 293, 1074–1080.

Johnson MA, Polgar J, Weightman D and Appleton D (1973). Data on the distribution of fibre types in thirty-six human muscles. An autopsy study. *J Neurol Sci* 18, 111–129.

Jones D, Round J and de Haan A (2004). *Skeletal muscle. From molecules to movement.* Churchill Livingstone, Edinburgh.

Knutti D, Kressler D and Kralli A (2001). Regulation of the transcriptional coactivator PGC-1 via MAPK-sensitive interaction with a repressor. *Proc Natl Acad Sci U S A* 98, 9713–9718.

Lassar AB, Paterson BM and Weintraub H (1986). Transfection of a DNA locus that mediates the conversion of 10T1/2 fibroblasts to myoblasts. *Cell* 47, 649–656.

Lexell J, Taylor CC and Sjostrom M (1988). What is the cause of the ageing atrophy? Total number, size and proportion of different fiber types studied in whole vastus lateralis muscle from 15- to 83-year-old men. *J Neurol Sci* 84, 275–294.

Lin J, Wu H, Tarr PT, Zhang CY, Wu Z, Boss O, et al. (2002). Transcriptional co-activator PGC-1 alpha drives the formation of slow-twitch muscle fibres. *Nature* 418, 797–801.

Mason SD, Howlett RA, Kim MJ, Olfert IM, Hogan MC, McNulty W, et al. (2004). Loss of skeletal muscle HIF-1alpha results in altered exercise endurance. *PLoS Biol* 2, e288.

McKinsey TA, Zhang CL and Olson EN (2001). Control of muscle development by dueling HATs and HDACs. *Curr Opin Genet Dev* 11, 497–504.

McMullen JR, Shioi T, Zhang L, Tarnavski O, Sherwood MC, Kang PM, et al. (2003). Phospho-inositide 3-kinase(p110alpha) plays a critical role for the induction of physiological, but not patho-logical, cardiac hypertrophy. *Proc Natl Acad Sci U S A* 100, 12355–12360.

McPhee JS, Williams AG, Stewart C, Baar K, Schindler JP, Aldred S, et al. (2009). The training stimu-lus experienced by the leg muscles during cycling in humans. *Exp Physiol* 94, 684–694.

Meyer RA and Foley JM (1996). Cellular processes integrating the metabolic response to exercise. In *Handbook of Physiology. Section 12. Exercise: Regulation and Integration of multiple Systems,* eds. Rowell LB and Shepherd JT, pp. 841–869. Oxford University Press, Oxford.

Midgley AW, McNaughton LR and Jones AM (2007). Training to enhance the physiological determi-nants of long-distance running performance: can valid recommendations be given to runners and coaches based on current scientific knowledge? *Sports Med* 37, 857–880.

Molkentin JD, Lu JR, Antos CL, Markham B, Richardson J, Robbins J, et al. (1998). A calcineurin-dependent transcriptional pathway for cardiac hypertrophy. *Cell* 93, 215–228.

Murgia M, Serrano AL, Calabria E, Pallafacchina G, Lomo T and Schiaffino S (2000). Ras is involved in nerve-activity-dependent regulation of muscle genes. *Nat Cell Biol* 2, 142–147.

Naya FJ, Mercer B, Shelton J, Richardson JA, Williams RS and Olson EN (2000). Stimulation of slow skeletal muscle fiber gene expression by calcineurin in vivo. *J Biol Chem* 275, 4545–4548.

Ouchi N, Shibata R and Walsh K (2005). AMP-activated protein kinase signaling stimulates VEGF expression and angiogenesis in skeletal muscle. *Circ Res* 96, 838–846.

Pandorf CE, Haddad F, Wright C, Bodell PW and Baldwin KM (2009). Differential epigenetic modi-fications of histones at the myosin heavy chain genes in fast and slow skeletal muscle fibers and in response to muscle unloading. *Am J Physiol Cell Physiol* 297, C6–16.

Parsons SA, Wilkins BJ, Bueno OF and Molkentin JD (2003). Altered skeletal muscle phenotypes in calcineurin Aalpha and Abeta gene-targeted mice. *Mol Cell Biol* 23, 4331–4343.

Parsons SA, Millay DP, Wilkins BJ, Bueno OF, Tsika GL, Neilson JR, et al. (2004). Genetic loss of calcineurin blocks mechanical overload-induced skeletal muscle fiber-type switching but not hypertrophy. *J Biol Chem* 279, 26192–26220.

Pette D and Vrbova G (1992). Adaptation of mammalian skeletal muscle fibers to chronic electrical stimulation. *Rev Physiol Biochem Pharmacol* 120, 115–202.

Philp A, Chen A, Lan D, Meyer GA, Murphy AN, Knapp AE, et al. (2011). Sirtuin 1 (SIRT1) dea-cetylase activity is not required for mitochondrial biogenesis or peroxisome proliferator-activated receptor-gamma coactivator-1alpha (PGC-1alpha) deacetylation following endurance exercise. *J Biol Chem* 286, 30561–30570.

Pilegaard H, Saltin B and Neufer PD (2003). Exercise induces transient transcriptional activation of the PGC-1alpha gene in human skeletal muscle. *J Physiol* 546, 851–858.

Poyton RO and McEwen JE (1996). Crosstalk between nuclear and mitochondrial genomes. *Annu Rev Biochem* 65, 563–607.

Quiat D, Voelker KA, Pei J, Grishin NV, Grange RW, Bassel-Duby R, et al. (2011). Concerted regulation of myofiber-specific gene expression and muscle performance by the transcriptional repressor Sox6. *Proc Natl Acad Sci U S A* 108, 10196–10201.

Rose AJ, Kiens B and Richter EA (2006). Ca^{2+}-calmodulin-dependent protein kinase expression and signalling in skeletal muscle during exercise. *J Physiol* 574, 889–903.

Saltin B, Radegran G, Koskolou MD and Roach RC (1998). Skeletal muscle blood flow in humans and its regulation during exercise. *Acta Physiol Scand* 162, 421–436.

Sawka MN, Convertino VA, Eichner ER, Schnieder SM and Young AJ (2000). Blood volume: importance and adaptations to exercise training, environmental stresses, and trauma/sickness. *Med Sci Sports Exerc* 32, 332–348.

Scarpulla RC (2010). Metabolic control of mitochondrial biogenesis through the PGC-1 family regulatory network. *Biochim Biophys Acta* 1813, 1269–1278.

Scharhag J, Schneider G, Urhausen A, Rochette V, Kramann B and Kindermann W (2002). Athlete's heart: right and left ventricular mass and function in male endurance athletes and untrained individuals determined by magnetic resonance imaging. *J Am Coll Cardiol* 40, 1856–1863.

Schiaffino S (2010). Fibre types in skeletal muscle: a personal account. *Acta Physiol (Oxford)* 199, 451–463.

Shioi T, Kang PM, Douglas PS, Hampe J, Yballe CM, Lawitts J, et al. (2000). The conserved phosphoinositide 3-kinase pathway determines heart size in mice. *EMBO J* 19, 2537–2548.

Shioi T, McMullen JR, Kang PM, Douglas PS, Obata T, Franke TF, et al. (2002). Akt/protein kinase B promotes organ growth in transgenic mice. *Mol Cell Biol* 22, 2799–2809.

Simoneau JA and Bouchard C (1995). Genetic determinism of fiber type proportion in human skeletal muscle. *FASEB J* 9, 1091–1095.

Smerdu V, Karsch-Mizrachi I, Campione M, Leinwand L and Schiaffino S (1994). Type IIx myosin heavy chain transcripts are expressed in type IIb fibers of human skeletal muscle. *Am J Physiol* 267, C1723–C1728.

Spurway NC and Wackerhage H (2006). *Genetics and molecular biology of muscle adaptation*. Elsevier/Churchill Livingstone, Edinburgh.

Suwa M, Nakano H and Kumagai S (2003). Effects of chronic AICAR treatment on fiber composition, enzyme activity, UCP3, and PGC-1 in rat muscles. *J Appl Physiol* 95, 960–968.

Timmons JA, Knudsen S, Rankinen T, Koch LG, Sarzynski M, Jensen T, et al. (2010). Using molecular classification to predict gains in maximal aerobic capacity following endurance exercise training in humans. *J Appl Physiol* 108, 1487–1496.

Tothova J, Blaauw B, Pallafacchina G, Rudolf R, Argentini C, Reggiani C, et al. (2006). NFATc1 nucleocytoplasmic shuttling is controlled by nerve activity in skeletal muscle. *J Cell Sci* 119, 1604–1611.

Towler MC and Hardie DG (2007). AMP-activated protein kinase in metabolic control and insulin signaling. *Circ Res* 100, 328–341.

Trappe SW, Costill DL, Fink WJ and Pearson DR (1995). Skeletal muscle characteristics among distance runners: a 20-yr follow-up study. *J Appl Physiol* 78, 823–829.

Tseng BS, Kasper CE and Edgerton VR (1994). Cytoplasm-to-myonucleus ratios and succinate dehydrogenase activities in adult rat slow and fast muscle fibers. *Cell Tissue Res* 275, 39–49.

Tsika RW, Schramm C, Simmer G, Fitzsimons DP, Moss RL and Ji J (2008). Overexpression of TEAD-1 in transgenic mouse striated muscles produces a slower skeletal muscle contractile phenotype. *J Biol Chem* 283, 36154–36167.

van Rooij E., Quiat D, Johnson BA, Sutherland LB, Qi X, Richardson JA, et al. (2009). A family of microRNAs encoded by myosin genes governs myosin expression and muscle performance. *Dev Cell* 17, 662–673.

Wagner PD (2011). The critical role of VEGF in skeletal muscle angiogenesis and blood flow. *Biochem Soc Trans* 39, 1556–1559.

Weiss A, McDonough D, Wertman B, Acakpo-Satchivi L, Montgomery K, Kucherlapati R, et al. (1999). Organization of human and mouse skeletal myosin heavy chain gene clusters is highly conserved. *Proc Natl Acad Sci U S A* 96, 2958–2963.

Wu H, Kanatous SB, Thurmond FA, Gallardo T, Isotani E, Bassel-Duby R, et al. (2002). Regulation of mitochondrial biogenesis in skeletal muscle by CaMK. *Science* 296, 349–352.

Xiao B, Sanders MJ, Underwood E, Heath R, Mayer FV, Carmena D, et al. (2011). Structure of mammalian AMPK and its regulation by ADP. *Nature* 472(7342), 230–233.

Xie K, Wei D, Shi Q and Huang S (2004). Constitutive and inducible expression and regulation of vascular endothelial growth factor. *Cytokine Growth Factor Rev* 15, 297–324.

Yancopoulos GD, Davis S, Gale NW, Rudge JS, Wiegand SJ and Holash J (2000). Vascular-specific growth factors and blood vessel formation. *Nature* 407, 242–248.

5 Genetics and endurance sports

Stephen M Roth and Henning Wackerhage

INTRODUCTION

Before reading this chapter you should have read Chapter 2, which introduces sport and exercise genetics. Chapter 4 is also useful as it covers the adaptations to endurance training.

Research questions related to endurance exercise were a major research focus of many exercise physiologists from the time when exercise physiology emerged as a sub-discipline of physiology. Maximal oxygen uptake ($\dot{V}O_{2max}$), for example, had already been measured in the 1920s by Archibald Vivian Hill (Hill and Lupton, 1923) and others. Classical endurance research is focused on the function of the cardiovascular and muscular organ systems during exercise and on the effect of environmental factors such as training and nutrition. However, while $\dot{V}O_{2max}$ and $\dot{V}O_{2max}$ trainability are both $\approx 50\%$ inherited (Bouchard et al., 1999), only a small proportion of $\dot{V}O_{2max}$ and endurance training research has been directed at identifying the genetic variations that are responsible for the large variation in endurance capacity and trainability in the human population. However, the tide is changing and the number of exercise physiologists that engage in genetic research on endurance is increasing.

We start the present chapter by reviewing factors such as maximal cardiac output, blood oxygen transport capacity, skeletal muscle fibre type variability and efficiency that limit performance in endurance events. Endurance capacity overall and probably most limiting factors are significantly inherited and polygenic. The variation of endurance capacity seen in the human population depends partially on common DNA sequence variations or polymorphisms. In this context we discuss the *ACE I/D* genotype which was the first polymorphism linked to endurance-related traits. After that we will discuss recent genome-wide association studies (GWAS) which have identified common DNA sequence variations that explain a large proportion of the variation of endurance trainability (Timmons et al., 2010; Bouchard et al., 2011). The results of these studies offer the real prospect of genetic tests for personalized endurance exercise prescription (Roth, 2008). After that we will discuss a rare DNA sequence variation of the erythropoietin (EPO) receptor found in the family of the cross-country skier and triple Olympic gold medal winner Eero Antero Mäntyranta (de la Chapelle et al., 1993). This suggests that some rare DNA sequence variations may have a large effect size on endurance-related traits.

After that we will review transgenic mouse models that have an increased endurance capacity, perhaps most strikingly in the case of the PEPCK mouse (Hakimi et al., 2007). Such transgenic mice highlight candidate genes where common or rare DNA sequence mutations may have an effect on endurance-related traits. At the end we will discuss selective breeding for endurance performance in rodents and race horses and discuss the variation of endurance-related traits in inbred mouse strains.

ENDURANCE: A POLYGENIC TRAIT

Endurance performances such as a marathon run depend on many limiting factors or quantitative traits, such as $\dot{V}O_{2max}$, the percentage of $\dot{V}O_{2max}$ at the lactate threshold (% $\dot{V}O_{2max}$ at LT) and movement efficiency (summarized in Figure 5.1) (Bassett and Howley, 2000). The variations of these factors are explained by both environmental factors, such as endurance training and nutrition and by DNA sequence variation. Because endurance is dependent on many limiting factors that are affected by genetic variation, it is clear that endurance itself is, as we will show throughout this chapter, a polygenic trait.

HERITABILITY OF $\dot{V}O_{2MAX}$, $\dot{V}O_{2MAX}$ TRAINABILITY AND FIBRE TYPE PERCENTAGES

The earliest studies on the genetics of human endurance performance were twin and family studies. These studies estimated the heritability especially of $\dot{V}O_{2max}$ and fibre type

Figure 5.1 Endurance performance depends on many limiting factors, which in turn are influenced by common and rare DNA sequence variations or quantitative trait loci and environmental factors such as training and nutrition. QTL, quantitative trait locus; HR_{max}, maximal heart rate; SV_{max}, maximal stroke volume; Q_{max}, maximal cardiac output; $C_a-C_vO_{2max}$, maximal difference in the oxygen content of arterial and venous blood; LT, lactate threshold; ACE, angiotensin-converting enzyme; EpoR, erythropoietin receptor.

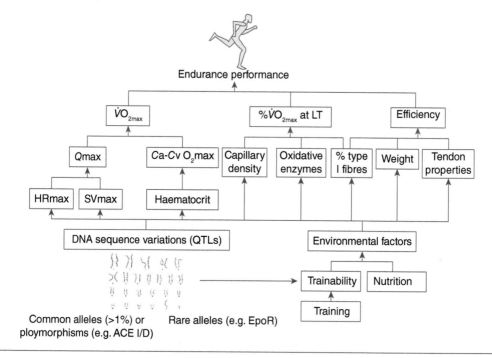

proportions. The heritability of $\dot{V}O_{2max}$ was estimated to be between 40% (Bouchard et al., 1986a) and 93% (Klissouras, 1971). More recently, the heritability of $\dot{V}O_{2max}$ has been estimated to be 50% (Bouchard et al., 1998), while the heritability of $\dot{V}O_{2max}$ trainability has been estimated to be 47% (Bouchard et al., 1999). Thus keeping the limitations of the methods of measuring heritability in mind, it is close enough to assume that $\dot{V}O_{2max}$ and $\dot{V}O_{2max}$ trainability are probably both ≈50% inherited.

After $\dot{V}O_{2max}$, the fibre type proportion is the second major factor to affect endurance performance. In humans, adult skeletal muscle is composed of slow type I, intermediate type IIa and fast type IIx fibres, as discussed in Chapter 4. Humans carry the gene for myosin heavy chain IIb but it is not normally expressed in human skeletal muscle (Smerdu et al., 1994). Within individuals, the proportions of fibre types differ from muscle to muscle. They range from 15% of slow type I fibres in eye muscles to 89% in the soleus (Johnson et al., 1973). From a genetics perspective the most important finding is that the fibre proportions in a given muscle differs greatly in the population. For example, a quarter of North American Caucasians have either less than 35% or more than 65% type I fibres (Simoneau and Bouchard, 1995). Extremes of fibre type proportions for locomotor muscles are found in speed/power and endurance athletes, with speed/power athletes having a high percentages of fast type II and endurance athletes having a high percentage of slow type I fibres (Costill et al., 1976a, 1976b). In the

1970s, Costill and Saltin concluded that 'these measurements confirm earlier reports which suggest that the athlete's preference for strength, speed, and/or endurance events is in part a matter of genetic endowment' (Costill et al., 1976a). The heritability estimates for fibre types range from no significant genetic effect (Bouchard et al., 1986b) to 92.8–99.5% for females and males, respectively (Komi et al., 1977), which highlights the limitations of such studies. Today the consensus estimate is a heritability of ≈45% (Simoneau and Bouchard, 1995).

Taken together, the key endurance limiting factors of $\dot{V}O_{2max}$, $\dot{V}O_{2max}$ trainability and fibre type proportions are ≈50% inherited. In the following text we will now discuss common and rare DNA sequence variations that contribute to the variation of endurance traits seen in the human population.

ASSOCIATION STUDIES AND THE *ACE I/D* POLYMORPHISM

Until recently, updates on genes associated with sport and exercise-related traits were published in journal articles in *Medicine and Science in Sports and Exercise* (Bray et al., 2009). These articles give a good overview of the results from association and other genetic studies in relation to endurance performance and other sport and exercise-related traits. In this section we will first focus in detail on the angiotensin-converting enzyme (ACE) insertion/deletion (I/D) polymorphism as a putative endurance-related polymorphism and then list other polymorphisms that have been linked to endurance related traits. However, sport and exercise genetics is changing quickly and so single-gene polymorphism studies will increasingly become a study type of the past.

The *ACE* gene arose as a possible candidate for endurance performance given its central role in the renin–angiotensin system and the regulation of blood pressure. The *ACE* gene contains a 287-base pair insertion/deletion (*I/D*) in intron 16. Both alleles are common in many populations, which is why this DNA sequence variation is considered to be a polymorphism (Rigat et al., 1990). A 1997 report showed that the *ACE I/D* genotype was associated with growth of the left ventricle in response to physical training in military recruits, with *D/D* genotype carriers having greater cardiac growth than *I/I* carriers (Montgomery et al., 1997). This led to a flood of studies aimed at determining the importance of this DNA sequence variation on endurance-related traits. Cross-sectional studies of athletes followed, several of which showed a greater proportion of high-level endurance athletes carrying the *I/I* genotype than would be expected in the general population. However, the results were not all consistent, as several studies found no or only little association with endurance-related traits, especially $\dot{V}O_{2max}$ (Rankinen et al., 2000a) (Figure 5.2).

There are two main explanations for these inconsistent results. First, the effect size of the *ACE I/D* polymorphism is relatively small and for this reason hundreds or even thousands of subjects are required for sufficient statistical power to detect true associations (Altmuller et al., 2001; Chanock et al., 2007). As in other fields, this has not been achieved in most studies on the *ACE I/D* polymorphism. Second, the *ACE I/D* polymorphism may not be associated with $\dot{V}O_{2max}$ but instead with skeletal muscle endurance. Skeletal muscle has a local, tissue-specific renin–angiotensin system whose function might be altered by the *ACE I/D* polymorphism. Studies emerged examining metabolic efficiency and skeletal

Figure 5.2 *ACE I/D* polymorphism and exercise. (a) Frequency of the *ACE I/D* polymorphism in 25 British mountaineers/climbers (black bars) and 1906 healthy British men (white bars), who served as controls. (b) Improvement of duration of repetitive elbow flexion after 10 weeks of physical training among British army recruits in relation to the *ACE I/D* polymorphism. (a, b) Redrawn after Montgomery et al. (1998). (c) No association between the *ACE I/D* genotype and $\dot{V}O_{2max}$ in a study of elite endurance athletes. Redrawn after Rankinen et al. (2000a).

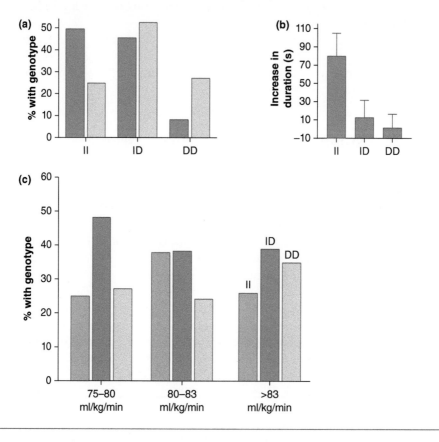

muscle fibre type and here researchers found more success. The *ACE I/I* genotype has been associated with both enhanced metabolic efficiency and a higher proportion of type I muscle fibres in comparison to the *D/D* genotype, both of which could contribute to improved endurance performance. Interestingly, the *D/D* genotype has been associated with strength and power-related traits but we will not discuss this further in this chapter.

Overall, the *ACE I/D* polymorphism appears to have a small influence on endurance performance such that *I/I* individuals have a slightly greater advantage. But this advantage must be put in context: there are elite endurance athletes without the *I/I* genotype and many endurance athletes with the *I/I* genotype never reach elite status as the effect size of the *ACE I/D* polymorphism is small. In other words, the *ACE I/D* polymorphism is just one of many contributors to endurance performance and should be considered within this context.

Over the years, many other polymorphisms have been linked to endurance-related traits. We give a brief overview in Table 5.1 of some key genes studied in relation to endurance performance traits.

Table 5.1 DNA sequence variations affecting endurance-related traits in humans (a partial listing of key studies)			
GENE	TRAIT AFFECTED AND STUDY DESIGN	NUMBER OF SUBJECTS	REFERENCE
PPARGC1A	Predicted $\dot{V}O_{2max}$	599	Franks et al., 2003
PPARGC1A	Case control of endurance athletes versus controls	204	Lucia et al., 2005
PPARGC1A	Case control of endurance athletes versus controls	395	Eynon et al., 2010
ADRB2	$\dot{V}O_{2max}$	63	Moore et al., 2001
ADRB2	Case control of endurance athletes versus controls	600	Wolfarth et al., 2007
ADRB2	$\dot{V}O_{2max}$	62	McCole et al., 2004
VEGFA	$\dot{V}O_{2max}$	146	Prior et al., 2006
NOS3	Triathlon competition performance	443	Saunders et al., 2006
ADRB1	$\dot{V}O_{2peak}$, exercise time	263	Wagoner et al., 2002
ADRB1	$\dot{V}O_{2peak}$	892	Defoor et al., 2006
HIF1A	$\dot{V}O_{2max}$ and training response of $\dot{V}O_{2max}$	125	Prior et al., 2003
BDKRB2	Muscle efficiency	115	Williams et al., 2004
BDKRB2	Case–control of endurance athletes versus controls	346	Saunders et al., 2006
CKM	$\dot{V}O_{2max}$ and training response of $\dot{V}O_{2max}$	240	Rivera et al., 1997
AMPD1	Training response of $\dot{V}O_{2max}$	400	Rico-Sanz et al., 2003
ATP1A2	Training response of $\dot{V}O_{2max}$	472	Rankinen et al., 2000b
PPARD	Training response of $\dot{V}O_{2max}$	264	Hautala et al., 2007

Overall, these studies demonstrate the presence of common DNA sequence variations or polymorphisms that explain generally a small proportion of the variability of endurance-related traits such as $\dot{V}O_{2max}$ or endurance trainability. They do not, however, explain all the heritability of such traits and it is currently unclear whether the remaining heritability is related to currently unknown polymorphisms or whether rare DNA sequence variations have a major effect. The whole-genome sequencing of large cohorts that are reliably pheno-typed for endurance-related traits will eventually yield an answer to this question.

MITOCHONDRIAL HAPLOGROUPS – A MARKER FOR MATERNAL INHERITANCE AND ENDURANCE-RELATED TRAITS

In this section we will discuss **mitochondrial haplogroups**, which are common mitochondrial DNA sequence variations that are passed on only by mothers. The analysis of such haplogroups allows researchers to trace maternal heritage and to test whether this is linked to endurance-related traits.

Mitochondria have their own, 16.6-kb-long DNA which is abbreviated mtDNA. Human mtDNA was sequenced in the mid-1980s as the first larger scale human DNA sequencing project (Anderson et al., 1981). MtDNA is unique for several reasons: First it is located, unlike all the other DNA, in mitochondria, and second, an offspring's mitochondria and their mtDNA are direct progeny of the mother's mitochondria in the ovum. Thus, if there is a DNA sequence variation in the mother's mtDNA then all the children will be carriers. Also, mtDNA is haploid as it is only inherited from the mother and it does not recombine. Because of that, new mutations in maternal mtDNA will stay in the mtDNA and will be passed on for many generations. Researchers can thus determine the maternal inheritance based on specific mitochondrial mutations, which makes it a powerful tool for studying human migration and ancestry.

So-called mitochondrial haplogroups are polymorphisms on a haploid DNA molecule which can be measured by polymerase chain reaction (PCR)-based assays, often in combination with mtDNA sequencing. The first haplogroups were termed A–D and all subsequent haplogroups were named using combinations of letters and numbers (van Oven and Kayser, 2009). The labelling of haplogroups and their ancestral geographical locations are shown in Figure 5.3.

Figure 5.3 Labelling of mtDNA haplogroups, their migration and the ancestral location of the dominant haplogroups. Redrawn after Shriver and Kittles (2004).

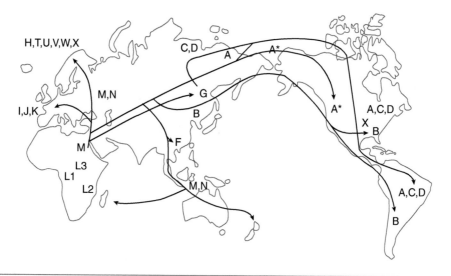

For molecular exercise physiologists, the question in relation to mtDNA haplogroups is: Are mtDNA haplogroups associated with sport and exercise-related traits such as endurance? Or, to give a more specific example, could a specific haplogroup be defining the African Rift Valley, where so many elite endurance athletes arise? Several groups have sought answers to these intriguing questions, but disappointingly there are no clear findings so far (Eynon et al., 2011). For example, Scott and colleagues examined Kenyan endurance athletes and controls and found an excess of L0 and lower L3* haplogroups in the elite athletes compared to a control population (Scott et al., 2009). This could indicate that the L0 haplogroup provides an advantage (and L3* a disadvantage) for endurance performance. An alternative interpretation is that the geographic ancestries of the athletes could be slightly different so that these haplogroup frequency differences are simply a byproduct of population stratification rather than a true genetic association (Scott et al., 2009). Other studies have reported higher $\dot{V}O_{2max}$ values in specific haplogroups. A Spanish research group has reported the highest $\dot{V}O_{2max}$ values associated with haplogroup H and the lowest with haplogroup J (Martinez-Redondo et al., 2010). Another report found fewer T haplogroups than expected in a group of elite Spanish endurance athletes compared to controls (Castro et al., 2007). However, none of these findings have been replicated or conclusively confirmed. A recent review article gives an overview of the state of research in this area (Eynon et al., 2011).

MULTIPLE POLYMORPHISMS: GENE SCORE AND GENOME-WIDE ASSOCIATION STUDIES

Association studies for single-gene polymorphisms with small effect size are not very informative, and testing athletes for one polymorphism generally yields little information about an individual's genetic potential for endurance sport. For these reasons, researchers have started to analyse several polymorphisms together to calculate 'genetic scores' in order to better predict the genetic endurance potential of an individual (Williams and Folland, 2008; Ahmetov et al., 2009). More recently, GWAS have been performed in order to search in a non-biased way for multiple polymorphisms that are associated with sport and exercise-related traits (Bouchard et al., 2011).

In one 'genetic score' study 1423 Russian athletes and 1132 controls were genotyped for multiple endurance-related polymorphisms (Ahmetov et al., 2009). The authors analysed their data for multigene effects and found that individuals who carry ≥ 9 endurance alleles are significantly more likely to be endurance athletes, suggesting that endurance-associated polymorphisms contribute to talent for endurance sports. However, the practical usefulness of this genetic score test is still limited because 38% of the controls also had ≥ 9 endurance alleles, suggesting that the test does not allow researchers to reliably identify endurance athletes based on the presence of several alleles associated with endurance (Figure 5.4).

The next step up from multigene or 'genetic score' studies is to search the whole genome for common DNA sequence variations or polymorphisms that are associated with endurance-related traits. Here, we discuss two such studies. In the first, Timmons et al. (2010) first took muscle biopsies from individuals in whom the response of the $\dot{V}O_{2max}$ to a standard endurance training programme had been measured; this is known as the $\dot{V}O_{2max}$ trainability. The researchers then used a complicated strategy in order to find single-nucleotide polymorphisms (SNPs) that predicted $\dot{V}O_{2max}$ trainability and a so-called expression quantitative trait locus (QTL) strategy. In a nutshell, their strategy was as follows (Timmons et al., 2010):

1 Use microarrays to measure the expression of mRNAs in resting muscle.
2 Use statistics to determine those mRNAs whose expression correlates best with $\dot{V}O_{2max}$ trainability. These mRNAs were termed **quantitative molecular classifiers**.
3 In another experiment measure SNPs within the gene sequence of the quantitative molecular classifiers and analyse the SNP data for association with $\dot{V}O_{2max}$ trainability.

The important findinag of this experimental strategy was that the researchers were able to identify 11 SNPs that predict 23% of the total variance of $\dot{V}O_{2max}$ trainability (Figure 5.5).

Figure 5.4 Endurance alleles and sporting performance. Percentage of subjects with a high number of 'endurance alleles'. The study suggests that more endurance athletes carry a high number of endurance alleles (i.e. ≥9 endurance alleles). Redrawn after Ahmetov et al. (2009).

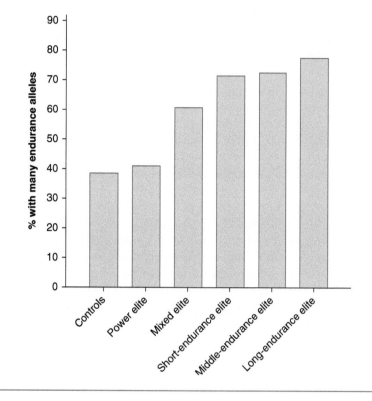

Given that $\dot{V}O_{2max}$ trainability is ≈50% inherited, (Bouchard et al., 1999), the described method predicts roughly half of the inherited variation of $\dot{V}O_{2max}$ trainability based on common DNA sequence variations (Timmons et al., 2010). Explaining such a large proportion of $\dot{V}O_{2max}$ trainability would be useful for identifying likely low and high responders prior to an endurance training programme.

In another $\dot{V}O_{2max}$ trainability study, the researchers performed a GWAS (Bouchard et al., 2011). In this study gene chips were used to measure 324 611 SNPs spread over all chromosomes in 473 individuals from 99 families. The association between all SNPs and $\dot{V}O_{2max}$ trainability was then calculated using a sophisticated statistical analysis. At the end of the procedure the research team identified 21 SNPs that predicted 49% of the variation of $\dot{V}O_{2max}$ trainability (Bouchard et al., 2011), which explains all of the genetic variation! Also, given that these gene chips measure common SNPs it implies that nearly all of the genetic variation of $\dot{V}O_{2max}$ trainability is due to common as opposed to rare alleles. Clearly these results will need to be replicated, but both studies show how newer approaches can improve our ability to find DNA sequence variations that explain much of the variability of endurance-related traits.

Sport and exercise genetics has come a long way and the recent GWAS studies by Timmons et al. (2010) and especially Bouchard et al. (2011) provide the basis for

Figure 5.5 Genome-wide studies and endurance trainability. (a) The authors used mRNA expression analysis to identify mRNAs of so-called classifier genes, whose expression predicts $\dot{V}O_{2max}$ trainability. The DNA of the classifier genes was then checked for SNPs that also predict $\dot{V}O_{2max}$ trainability. Eleven of these SNPs were used to calculate a gene predictor score. (b) Relationship between the gene predictor score and $\dot{V}O_{2max}$ trainability. According to this study, the gene predictor score roughly explains 23% of $\dot{V}O_{2max}$ trainability. Redrawn after Timmons et al. (2010). (c) GWAS study of SNPs (each dot in this schematic drawing represents one SNP; in reality many more SNPs were measured) that are associated with $\dot{V}O_{2max}$ trainability. The authors found that variation in 39 SNPs accounted for 49% of the variance of $\dot{V}O_{2max}$ trainability. (d) A predictor SNP score calculated from 21 SNPs predicts $\dot{V}O_{2max}$ trainability well. Redrawn after Bouchard et al. (2011).

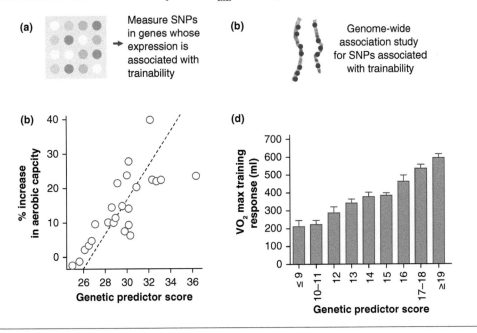

genetic tests that can potentially predict $\dot{V}O_{2max}$ trainability to a large extent. However, as with all studies the results must be confirmed in other populations with similar and different ethnic backgrounds before the full value of these results is known. But for the first time sport and exercise geneticists seem to be close to developing tests which will be useful for personalized exercise prescription by distinguishing those who will benefit little from an endurance training programme from those who are likely to benefit a lot from such an intervention (Roth, 2008).

EERO ANTERO MÄNTYRANTA AND THE EPO RECEPTOR MUTATION AS AN EXAMPLE OF A RARE DNA SEQUENCE VARIATION WITH A LARGE EFFECT SIZE

In the previous section we discussed common endurance alleles or polymorphisms. However, exceptional endurance performance may also partially depend on the presence of one or more rare DNA sequence variations with a large effect on endurance performance. The best example of such a rare DNA sequence variation is a variation in the erythropoietin receptor (*EPOR*) gene which was found in a Finnish family. In this section we review this study in some detail.

Many studies have demonstrated that removal or re-infusion of red blood cells (erythrocytes) immediately reduces or increases both haematocrit and $\dot{V}O_{2max}$, respectively (Cooper, 2008). Thus blood doping or the administration of EPO, which stimulates erythrocyte production by the bone marrow and thus increases haematocrit, can have a large effect on $\dot{V}O_{2max}$ and endurance performance (Cooper, 2008). To control blood doping and EPO, the cycling world governing body (Union Cycliste Internationale (UCI)) has stated that a haematocrit >50% and >170 g/L haemoglobin are defined as abnormal levels and athletes who exceed these thresholds will be declared temporarily unfit for at least 15 days (UCI regulation 13.2.012). This is a good point at which to start the story of the Finnish cross-country skier and triple Olympic gold medal winner Eero Antero Mäntyranta and his family. Eero was a member of a family in whom a rare erythrocyte-related phenotype was segregating. In this family the affected males had haemoglobin concentrations of between 183 and 231 g/L (de la Chapelle et al., 1993), which are abnormal levels according to the UCI. The fact that this trait was segregating in a family suggested that it was inherited. Furthermore, the researchers found normal bone marrow, low or normal EPO values and the plasma of the individuals did not stimulate erythroid progenitor cells. Thus the problem was unlikely to be a problem of the bone marrow or of EPO or another serum hormone that had an effect on the production of red blood cells.

What could be the cause of the high haematocrit? The researchers found that bone marrow and blood cells from affected family members formed more erythrocyte colonies at low EPO concentrations and even in the absence of EPO, where controls do not form colonies. This suggested that something within the cells was responsible for the increased erythrocyte production. The researchers reasoned that the EPO receptor (*EPOR*) might be a candidate gene. They designed primers to cover the *EPOR* gene and used the genomic DNA of affected family members and controls to amplify the *EPOR* gene. The amplified *EPOR* gene was then sequenced and a G-to-A mutation was found in position 6002, which was abbreviated as *EPOR* 6002 G→A. The mutation results in a premature stop codon in the DNA which gives rise to a shortened *EPOR* mRNA and protein. Normally, incompletely transcribed proteins are degraded or function less, but in this case it seemed as if the truncated version of the EPOR protein was more active. The most likely explanation is that the last bit of the EPOR protein must somehow inhibit EPOR activity.

Because DNA sequencing is expensive, the authors used the restriction enzymes *Nco*I and *Sty*I to digest the PCR product of the *EPOR* gene. In this assay the PCR product of control subjects is cut by *Nco*I and *Sty*I into two bands (Figure 5.6) but not in the affected family members (one band; heterozygous individuals display three bands) (de la Chapelle et al., 1993). The figure shows that there are no individuals with just one band, suggesting that there are no homozygous carriers. As there was no individual homozygous for the *EPOR* mutation in the whole study it seems likely that the mutation is lethal if homozygous.

The researchers then also checked DNA from 50 random Finnish blood donors to ensure that this mutation was not present in the normal population, which was confirmed by this experiment.

Figure 5.6 EPO receptor rare mutation. Schematic drawing showing a pedigree for the Finnish family affected by the rare *EPOR* DNA sequence variation and the result of a genetic test on agarose gel. Black symbols and '+' represent affected family members, white symbols and '–' unaffected members. Unaffected family members, who do not carry the mutation, have a 233 bp and a 100 bp fragment when their *EPOR* PCR product is cut into two pieces by *Nco*I or *Sty*I restriction enzymes. Heterozygous affected family members have three bands: 233 bp and 100 bp for the normal *EPOR* allele and 333 bp for the mutated allele. Family members who are homozygous for the mutated form would have just one 333 bp band but none were present, suggesting that being homozygous for this genotype is lethal. Redrawn after de la Chapelle et al. (1993).

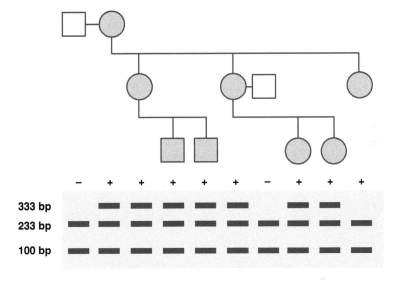

Now back to Eero Antero Mäntyranta. As a heterozygous carrier of the mutation his haematocrit and haemoglobin were so high that he would probably be banned from competing in cycling today. Essentially he had the advantage of what is equivalent to EPO or blood doping, which would have increased his $\dot{V}O_{2max}$ and endurance performance. Given that he was competing at a time when EPO doping was unknown and blood doping was at least unlikely, he had a genetic advantage over his competitors which partially explains why he was able to win one gold medal in 1960 in Squaw Valley and two gold medals in 1964 in Innsbruck.

However, this poses an important question: If he was a cyclist today, should he be banned from competing as his haematocrit is too high? Of course not, because talent for sport is all about being a carrier of advantageous common and rare DNA sequence variations (Williams and Folland, 2008). Thus individuals who carry a DNA sequence variation that results in a conflict with anti-doping rules should be allowed to use genetic testing in order to prove that such a mutation is natural and be allowed to compete if it can be shown that an abnormal result is due the mutation. To summarize, the *EPOR* 6002 G→A DNA sequence variation demonstrates as a proof-of-principle the presence of rare DNA sequence variations with a large effect on endurance performance. Such rare DNA sequence variations are currently difficult to identify but as we enter the age of whole-genome sequencing (Wheeler et al., 2008), it is likely that more rare DNA sequence variations will be identified and that the relative contributions of common and rare DNA sequence variations to endurance-related traits can be quantified.

TRANSGENIC MICE WITH AN ENDURANCE PHENOTYPE

The identification of polymorphisms or rare mutations depends on previous knowledge of genes that affect endurance-related traits. Genes where this has been demonstrated are termed **candidate genes**. Such candidate genes are commonly identified via gain or loss-of-function experiments in cultured muscle cells *in vitro* or in transgenic mice or other species *in vivo*.

Much information on endurance-related candidate genes was obtained from transgenic mouse studies. In these studies genes were either knocked out or knocked in and we have described several of these mouse models in Chapter 3, as frequently a gain or loss-of-function of a signalling molecule changes endurance-related traits. In Table 5.2 we have also stated whether the knock out or in is global or whether a tissue-specific promoter such as muscle creatine kinase (MCK) or human α-actin has been used to knock in or knock out a gene.

Possibly the most striking experiment from an exercise physiology viewpoint is the study in which the α-actin promoter was used to drive the overexpression of the gluconeogenic enzyme PEPCK-C in skeletal muscle. The authors found that the transgenic

Table 5.2 Examples of genetically altered mice with an endurance-related phenotype

PROMOTER AND TRANSGENE	PHENOTYPE	REFERENCE
Cardiac specific, α-MHC promoter-driven overexpression of constitutively active MEK1	Cardiac hypertrophy with improved function (more strength athlete's heart)	Bueno et al., 2000
Plasmids to overexpress constitutively active Ras in regenerating skeletal muscle	Fast-to-slow muscle phenotype shift in regenerating skeletal muscle	Murgia et al., 2000
Knockout of calcineurin Aα and Aβ	Slow-to-fast phenotype shift	Parsons et al., 2003
Muscle promoter-driven knockout of calcineurin isoforms	Impaired fast-to-slow fibre type switching after overload	Parsons et al., 2004
MCK promoter-driven PGC-1α overexpression in skeletal muscle	Mitochondrial biogenesis and increased percentage of fibres with a 'slower' phenotype	Lin et al., 2002
Human α-skeletal actin promoter driven PPARδ overexpression in skeletal muscle	Increased % of type I fibres and improved endurance running performance	Wang et al., 2004
Human α-skeletal actin promoter driven PEPCK-C overexpression in skeletal muscle	Improved endurance running performance and metabolic changes	Hakimi et al., 2007
miR-208B knockout, miR-409 knockout	Slow-to-fast fibre switch via Sox6	van Rooij E. et al., 2009
MCK promoter-driven muscle Sox6 knockout in skeletal muscle	Fast-to-slow skeletal muscle fibre phenotype changes	Quiat et al., 2011
Global ACTN3 knockout	Reduced force generation and fast fibre diameter in skeletal muscle and a slower phenotype	MacArthur et al., 2008
MCK promoter-driven glycogen synthase overexpression in skeletal muscle	Higher muscle glycogen content in skeletal muscle	Manchester et al., 1996

MEK1, mitogen-activated protein kinase kinase 1; Ras, rat sarcoma; PGC-1, peroxisome proliferator activator protein co-activator-1; PPARδ, peroxisome proliferator-activated receptor δ; IGF-1, insulin-like growth factor 1; PKB/Akt, protein kinase B; Pax7, paired box gene 7; ACTN3, alpha actinin 3; MCK, muscle creatine kinase; Sox6, sex determining region Y gene.

Figure 5.7 *PEPCK* skeletal muscle overexpression and performance. Effect of skeletal muscle-specific expression of the *PEPCK* gene in skeletal muscle on running distance at 20 m/min. *PEPCK* muscle mice can run more than 10 times the distance at 20 m/min when compared to wildtype controls. Redrawn after Hakimi et al. (2007).

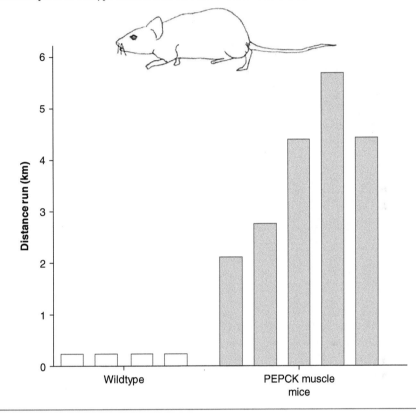

mice ran up to 6 km at 20 m/min, while the wildtype controls ran only ≈200 m at that speed to fatigue (Figure 5.7).

The *PEPCK* mouse demonstrates the large effect which some transgenes can have on endurance-related traits. It seems likely that some of the genes listed in Table 5.2 are also affected by naturally occurring common or rare DNA sequence variations in humans.

SELECTIVE BREEDING FOR ENDURANCE AND ENDURANCE-RELATED TRAITS OF INBRED MOUSE STRAINS

We will now discuss the results of selective breeding and inbred mouse strain studies as a non-biased strategy to identify genes that affect endurance-related traits. Selective breeding for endurance-related traits has been used for a long time to breed race dogs and horses. The aim of selective breeding for endurance is to accumulate DNA sequence variations found in the founder population that increase endurance capacity. In a scientific experiment, researchers have selectively bred rats for low and high running ability

Figure 5.8 Selective breeding for high (high-capacity runners, HCR) and low (low-capacity runners, LCR) exercise capacity in rats over 11 generations. The study shows that such breeding selects, among others, for high concentrations of known regulators of mitochondrial biogenesis and fibre type (i.e. PGC-1α and PPARγ). Redrawn after Wisloff et al. (2005).

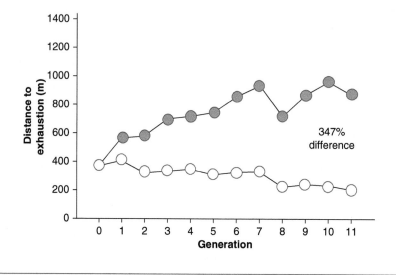

(Wisloff et al., 2005). After 11 generations the running distance to exhaustion differed by 347% between the high- and low-capacity runners (Figure 5.8).

Genetically, this has led to the enrichment of gene variants that are favourable for running capacity. The researchers additionally measured the expression of known regulatory factors of endurance and found that, for example, PGC-1α and PPARγ are expressed at significantly higher levels in the high-capacity than in the low-capacity runners. This could be due to DNA sequence variations in the promoter or enhancer regions of these genes or due to variations that affect the expression or activity of factors which increase the expression of PGC-1α and PPARγ.

A different model is inbred mice. In this case the genetic variations are fixed in a species by interbreeding family members for at least 20 consecutive generations. Via this procedure, the animals become genetically more or less identical and have similar phenotypes and thus exercise-related traits, such as muscularity or fibre type percentages. Such inbred mice strains are a powerful tool for molecular sport and exercise geneticists because the genomes of many of these inbred mouse strains have now been DNA sequenced. Researchers can thus measure the phenotype or quantitative traits (QTs), such as voluntary running, muscle size, number of muscle fibres for a given muscle or fibre type proportions, and then use linkage analysis in order to identify the regions that harbour variations in the DNA sequence or QTLs that are responsible for the variation of QTs between inbred strains.

Inbred mouse strains have been used, for example, to search for QTLs that affect voluntary wheel running, which also depends on psychological factors. Figure 5.9b shows the large difference in average wheel running between the 41 inbred mouse strains studied and the small variation within each strain. This directly suggests two things: First, the large variation between strains suggests that there are DNA sequence variations in the heterogeneous founder population that affect voluntary wheel running. Second, the small variation within a group suggests that the effect of these DNA sequence variations is large. Figure 5.9a then shows QTLs in the mouse genome that are associated with difference in voluntary wheel running distance. The highest QTL is located on chromosome 12.

Figure 5.9 Genome-wide search for loci that affect running. (a) The *x*-axis shows the location on the chromosomes of mice and the *y*-axis shows a measure of the likelihood that variation in that genomic location has an influence on average running distance. (b) Average running wheel distance (mean ± SD) per day for different inbred mouse strains. Redrawn after Lightfoot et al. (2010).

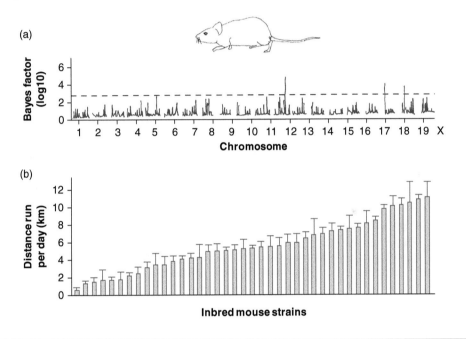

Researchers can now search more selectively for the exact DNA sequence variations which cause the differences in voluntary wheel running that are seen in the inbred mouse strains studied.

SUMMARY

Endurance performance as a complex physiological endpoint is a polygenic trait. The $\dot{V}O_{2max}$ is a key variable and both $\dot{V}O_{2max}$ and $\dot{V}O_{2max}$ trainability are ≈50% inherited (Bouchard et al., 1999) while fibre type proportions are probably ≈45% inherited (Simoneau and Bouchard, 1995). Earlier association studies have suggested that common DNA sequence variations or polymorphisms such as the *ACE I/D* polymorphism are associated with endurance-related traits (Montgomery et al., 1998).

However, many of these polymorphisms have a small effect size on endurance and thus large sample numbers are needed in order to conclusively demonstrate the effect of such polymorphisms (Altmuller et al., 2001; Chanock et al., 2007). In many of the published studies the sample sizes are too small, which explains the divergent published results. The trend is towards GWAS for larger cohorts and other sophisticated study designs show that such studies can be used to predict a practically useful amount of variation of $\dot{V}O_{2max}$ trainability based on polymorphisms (Timmons et al., 2010; Bouchard et al., 2011). Thus it seems likely that genetic tests will be developed on the basis of such studies that will be used to designed personalized exercise programmes (Roth, 2008). However, rare DNA sequence variations also contribute to the variation of endurance-related traits. A proof-of-principle example is the *EPOR* 6002 G→A allele which results in a premature stop codon and a hyperactive EPOR and increased haematocrit. Only heterozygous individuals are born, suggesting that homozygous carriers are not viable. The importance of rare DNA sequence variations for endurance-related traits is currently unknown and only whole-genome sequencing (Wheeler et al., 2008) of individuals phenotyped for endurance will eventually yield an answer. Transgenic mice display changes in cardiac performance, muscle fibre type percentages, mitochondrial content and actual running performance. The most striking example is a mouse where the PEPCK enzyme has been overexpressed in skeletal muscles. These mice run >10 times longer to exhaustion than wildtype controls. Such transgenic mouse studies highlight candidate genes where DNA sequence variations may affect endurance-related traits. Selective breeding has been used to breed race dogs and horses but also in scientific studies for high and low running capacity (Wisloff et al., 2005). Inbred mouse strains differ among each other in endurance capacity but within an inbred mouse strain the difference in capacity is small, suggesting that underlying genetic variants explain these differences (Lightfoot et al., 2010). As the genomes of many inbred mouse strains have been sequenced, genetic variations that are responsible for high- or low-endurance capacity can be identified (Lightfoot et al., 2010).

REVIEW QUESTIONS

1   Draw a diagram to illustrate factors that limit endurance performance. Ensure common and rare DNA sequence variations are included as causative factors.

2   What do we know about the heritability of key traits underlying endurance performance?

3   Explain one polymorphism that has been linked to endurance performance. What are the limitations of polymorphism studies?

4   What genetic advantage contributed to the success of Eero Mäntyranta and what specific trait was impacted and how?

5   How can transgenic mice and inbred mouse strains be used to improve our understanding of the genetic underpinnings of endurance performance and endurance-related traits?

FURTHER READING

Bouchard C and Hoffman EP (2011). Genetic and molecular aspects of sports performance, Vol. 18 of Encyclopedia of sports medicine. John Wiley & Sons, Chichester.

Bray MS, Hagberg JM, Perusse L, Rankinen T, Roth SM, Wolfarth B, et al. (2009). The human gene map for performance and health-related fitness phenotypes: the 2006–2007 update. Med Sci Sports Exerc 41, 35–73.

de la Chapelle A, Traskelin AL and Juvonen E (1993). Truncated erythropoietin receptor causes dominantly inherited benign human erythrocytosis. Proc Natl Acad Sci U S A 90, 4495–4499.

Hagberg JM, Rankinen T, Loos RJ, Perusse L, Roth SM, Wolfarth B, et al. (2011). Advances in exercise, fitness, and performance genomics in 2010. Med Sci Sports Exerc 43, 743–752.

Peeters MW, Thomis MA, Beunen GP and Malina RM (2009). Genetics and sports: an overview of the pre-molecular biology era. Med Sport Sci 54, 28–42.

Pescatello LS and Roth SM (2011). Exercise genomics. Springer, New York.

REFERENCES

Ahmetov II, Williams AG, Popov DV, Lyubaeva EV, Hakimullina AM, Fedotovskaya ON, et al. (2009). The combined impact of metabolic gene polymorphisms on elite endurance athlete status and related phenotypes. *Hum Genet* 126, 751–761.

Altmuller J, Palmer LJ, Fischer G, Scherb H and Wjst M (2001). Genomewide scans of complex human diseases: true linkage is hard to find. *Am J Hum Genet* 69, 936–950.

Anderson S, Bankier AT, Barrell BG, de Bruijn MH, Coulson AR, Drouin J, et al. (1981). Sequence and organization of the human mitochondrial genome. *Nature* 290, 457–465.

Bassett DR, Jr. and Howley ET (2000). Limiting factors for maximum oxygen uptake and determinants of endurance performance. *Med Sci Sports Exerc* 32, 70–84.

Bouchard C, Lesage R, Lortie G, Simoneau JA, Hamel P, Boulay MR, et al. (1986a). Aerobic performance in brothers, dizygotic and monozygotic twins. *Med Sci Sports Exerc* 18, 639–646.

Bouchard C, Simoneau JA, Lortie G, Boulay MR, Marcotte M and Thibault MC (1986b). Genetic effects in human skeletal muscle fiber type distribution and enzyme activities. *Can J Physiol Pharmacol* 64, 1245–1251.

Bouchard C, Daw EW, Rice T, Perusse L, Gagnon J, Province MA, et al. (1998). Familial resemblance for $\dot{V}O_{2max}$ in the sedentary state: the HERITAGE family study. *Med Sci Sports Exerc* 30, 252–258.

Bouchard C, An P, Rice T, Skinner JS, Wilmore JH, Gagnon J, et al. (1999). Familial aggregation of $\dot{V}O_{(2max)}$ response to exercise training: results from the HERITAGE Family Study. *J Appl Physiol* 87, 1003–1008.

Bouchard C, Sarzynski MA, Rice TK, Kraus WE, Church TS, Sung YJ, et al. (2011). Genomic predictors of the maximal O uptake response to standardized exercise training programs. *J Appl Physiol* 110, 1160–1170.

Bray MS, Hagberg JM, Perusse L, Rankinen T, Roth SM, Wolfarth B, et al. (2009). The human gene map for performance and health-related fitness phenotypes: the 2006–2007 update. *Med Sci Sports Exerc* 41, 35–73.

Bueno OF, De Windt LJ, Tymitz KM, Witt SA, Kimball TR, Klevitsky R, et al. (2000). The MEK1-ERK1/2 signaling pathway promotes compensated cardiac hypertrophy in transgenic mice. *EMBO J* 19, 6341–6350.

Castro MG, Terrados N, Reguero JR, Alvarez V and Coto E (2007). Mitochondrial haplogroup T is negatively associated with the status of elite endurance athlete. *Mitochondrion* 7, 354–357.

Chanock SJ, Manolio T, Boehnke M, Boerwinkle E, Hunter DJ, Thomas G, et al. (2007). Replicating genotype-phenotype associations. *Nature* 447, 655–660.

Cooper CE (2008). The biochemistry of drugs and doping methods used to enhance aerobic sport performance. *Essays Biochem* 44, 63–83.

Costill DL, Daniels J, Evans W, Fink W, Krahenbuhl G and Saltin B (1976a). Skeletal muscle enzymes and fiber composition in male and female track athletes. *J Appl Physiol* 40, 149–154.

Costill DL, Fink WJ and Pollock ML (1976b). Muscle fiber composition and enzyme activities of elite distance runners. *Med Sci Sports* 8, 96–100.

de la Chapelle A, Traskelin AL and Juvonen E (1993). Truncated erythropoietin receptor causes dominantly inherited benign human erythrocytosis. *Proc Natl Acad Sci U S A* 90, 4495–4499.

Defoor J, Martens K, Zielinska D, Matthijs G, Van NH, Schepers D, et al. (2006). The CAREGENE study: polymorphisms of the beta1-adrenoceptor gene and aerobic power in coronary artery disease. *Eur Heart J* 27, 808–816.

Eynon N, Meckel Y, Sagiv M, Yamin C, Amir R, Sagiv M, et al. (2010). Do PPARGC1A and PPARalpha polymorphisms influence sprint or endurance phenotypes? *Scand J Med Sci Sports* 20, e145–e150.

Eynon N, Moran M, Birk R and Lucia A (2011). The champions' mitochondria: is it genetically determined? A review on mitochondrial DNA and elite athletic performance. *Physiol Genomics* 43, 789–798.

Franks PW, Barroso I, Luan J, Ekelund U, Crowley VE, Brage S, et al. (2003). PGC-1alpha genotype modifies the association of volitional energy expenditure with [OV0312]O2max. *Med Sci Sports Exerc* 35, 1998–2004.

Hakimi P, Yang J, Casadesus G, Massillon D, Tolentino-Silva F, Nye CK, et al. (2007). Overexpression of the cytosolic form of phosphoenolpyruvate carboxykinase (GTP) in skeletal muscle repatterns energy metabolism in the mouse. *J Biol Chem* 282, 32844–32855.

Hautala AJ, Leon AS, Skinner JS, Rao DC, Bouchard C and Rankinen T (2007). Peroxisome proliferator-activated receptor-delta polymorphisms are associated with physical performance and plasma lipids: the HERITAGE Family Study. *Am J Physiol Heart Circ Physiol* 292, H2498–H2505.

Hill AV and Lupton H (1923). Muscular exercise, lactic acid and the supply and utilization of oxygen. *Q J Med* 16, 135–171.

Johnson MA, Polgar J, Weightman D and Appleton D (1973). Data on the distribution of fibre types in thirty-six human muscles. An autopsy study. *J Neurol Sci* 18, 111–129.

Klissouras V (1971). Heritability of adaptive variation. *J Appl Physiol* 31, 338–344.

Komi PV, Viitasalo JH, Havu M, Thorstensson A, Sjodin B and Karlsson J (1977). Skeletal muscle fibres and muscle enzyme activities in monozygous and dizygous twins of both sexes. *Acta Physiol Scand* 100, 385–392.

Lightfoot JT, Leamy L, Pomp D, Turner MJ, Fodor AA, Knab A, et al. (2010). Strain screen and haplotype association mapping of wheel running in inbred mouse strains. *J Appl Physiol* 109, 623–634.

Lin J, Wu H, Tarr PT, Zhang CY, Wu Z, Boss O, et al. (2002). Transcriptional co-activator PGC-1 alpha drives the formation of slow-twitch muscle fibres. *Nature* 418, 797–801.

Lucia A, Gomez-Gallego F, Barroso I, Rabadan M, Bandres F, San Juan AF, et al. (2005). PPARGC1A genotype (Gly482Ser) predicts exceptional endurance capacity in European men. *J Appl Physiol* 99, 344–348.

MacArthur DG, Seto JT, Chan S, Quinlan KG, Raftery JM, Turner N, et al. (2008). An Actn3 knockout mouse provides mechanistic insights into the association between alpha-actinin-3 deficiency and human athletic performance. *Hum Mol Genet* 17, 1076–1086.

Manchester J, Skurat AV, Roach P, Hauschka SD and Lawrence JC, Jr. (1996). Increased glycogen accumulation in transgenic mice overexpressing glycogen synthase in skeletal muscle. *Proc Natl Acad Sci U S A* 93, 10707–10711.

Martinez-Redondo D, Marcuello A, Casajus JA, Ara I, Dahmani Y, Montoya J, et al. (2010). Human mitochondrial haplogroup H: the highest VO_{2max} consumer – is it a paradox? *Mitochondrion* 10, 102–107.

McCole SD, Shuldiner AR, Brown MD, Moore GE, Ferrell RE, Wilund KR, et al. (2004). Beta2- and beta3-adrenergic receptor polymorphisms and exercise hemodynamics in postmenopausal women. *J Appl Physiol* 96, 526–530.

Montgomery HE, Clarkson P, Dollery CM, Prasad K, Losi MA, Hemingway H, et al. (1997). Association of angiotensin-converting enzyme gene I/D polymorphism with change in left ventricular mass in response to physical training. *Circulation* 96, 741–747.

Montgomery HE, Marshall R, Hemingway H, Myerson S, Clarkson P, Dollery C, et al. (1998). Human gene for physical performance. *Nature* 393, 221–222.

Moore GE, Shuldiner AR, Zmuda JM, Ferrell RE, McCole SD and Hagberg JM (2001). Obesity gene variant and elite endurance performance. *Metabolism* 50, 1391–1392.

Murgia M, Serrano AL, Calabria E, Pallafacchina G, Lomo T and Schiaffino S (2000). Ras is involved in nerve-activity-dependent regulation of muscle genes. *Nat Cell Biol* 2, 142–147.

Parsons SA, Wilkins BJ, Bueno OF and Molkentin JD (2003). Altered skeletal muscle phenotypes in calcineurin Aalpha and Abeta gene-targeted mice. *Mol Cell Biol* 23, 4331–4343.

Parsons SA, Millay DP, Wilkins BJ, Bueno OF, Tsika GL, Neilson JR, et al. (2004). Genetic loss of calcineurin blocks mechanical overload-induced skeletal muscle fiber-type switching but not hypertrophy. *J Biol Chem* 279, 26192–26200.

Prior SJ, Hagberg JM, Phares DA, Brown MD, Fairfull L, Ferrell RE, et al. (2003). Sequence variation in hypoxia-inducible factor 1alpha (HIF1A): association with maximal oxygen consumption. *Physiol Genomics* 15, 20–26.

Prior SJ, Hagberg JM, Paton CM, Douglass LW, Brown MD, McLenithan JC, et al. (2006). DNA sequence variation in the promoter region of the VEGF gene impacts VEGF gene expression and maximal oxygen consumption. *Am J Physiol Heart Circ Physiol* 290, H1848–H1855.

Quiat D, Voelker KA, Pei J, Grishin NV, Grange RW, Bassel-Duby R and Olson EN (2011). Concerted regulation of myofiber-specific gene expression and muscle performance by the transcriptional repressor Sox6. *Proc Natl Acad Sci U S A* 108, 10196–10201.

Rankinen T, Perusse L, Gagnon J, Chagnon YC, Leon AS, Skinner JS, et al. (2000a). Angiotensin-converting enzyme ID polymorphism and fitness phenotype in the HERITAGE Family Study. *J Appl Physiol* 88, 1029–1035.

Rankinen T, Wolfarth B, Simoneau JA, Maier-Lenz D, Rauramaa R, Rivera MA, et al. (2000b). No association between the angiotensin-converting enzyme ID polymorphism and elite endurance athlete status. *J Appl Physiol* 88, 1571–1575.

Rico-Sanz J, Rankinen T, Joanisse DR, Leon AS, Skinner JS, Wilmore JH, et al. (2003). Associations between cardiorespiratory responses to exercise and the C34T AMPD1 gene polymorphism in the HERITAGE Family Study. *Physiol Genomics* 14, 161–166.

Rigat B, Hubert C, Ahenc-Gelas F, Cambien F, Corvol P and Soubrier F (1990). An insertion/deletion polymorphism in the angiotensin I-converting enzyme gene accounting for half the variance of serum enzyme levels. *J Clin Invest* 86, 1343–1346.

Rivera MA, Dionne FT, Simoneau JA, Perusse L, Chagnon M, Chagnon Y, et al. (1997). Muscle-specific creatine kinase gene polymorphism and $\dot{V}O_{2max}$ in the HERITAGE Family Study. *Med Sci Sports Exerc* 29, 1311–1317.

Roth SM (2008). Perspective on the future use of genomics in exercise prescription. *J Appl Physiol* 104, 1243–1245.

Saunders CJ, Xenophontos SL, Cariolou MA, Anastassiades LC, Noakes TD and Collins M (2006). The bradykinin beta 2 receptor (BDKRB2) and endothelial nitric oxide synthase 3 (NOS3) genes and endurance performance during Ironman Triathlons. *Hum Mol Genet* 15, 979–987.

Scott RA, Fuku N, Onywera VO, Boit M, Wilson RH, Tanaka M, et al. (2009). Mitochondrial haplogroups associated with elite Kenyan athlete status. *Med Sci Sports Exerc* 41, 123–128.

Shriver MD and Kittles RA (2004). Genetic ancestry and the search for personalized genetic histories. *Nat Rev Genet* 5, 611–618.

Simoneau JA and Bouchard C (1995). Genetic determinism of fiber type proportion in human skeletal muscle. *FASEB J* 9, 1091–1095.

Smerdu V, Karsch-Mizrachi I, Campione M, Leinwand L and Schiaffino S (1994). Type IIx myosin heavy chain transcripts are expressed in type IIb fibers of human skeletal muscle. *Am J Physiol* 267, C1723–C1728.

Timmons JA, Knudsen S, Rankinen T, Koch LG, Sarzynski M, Jensen T, et al. (2010). Using molecular classification to predict gains in maximal aerobic capacity following endurance exercise training in humans. *J Appl Physiol* 108, 1487–1496.

van Oven M and Kayser M (2009). Updated comprehensive phylogenetic tree of global human mitochondrial DNA variation. *Hum Mutat* 30, E386–E394.

van Rooij E., Quiat D, Johnson BA, Sutherland LB, Qi X, Richardson JA, et al. (2009). A family of microRNAs encoded by myosin genes governs myosin expression and muscle performance. *Dev Cell* 17, 662–673.

Wagoner LE, Craft LL, Zengel P, McGuire N, Rathz DA, Dorn GW and Liggett SB (2002). Polymorphisms of the beta1-adrenergic receptor predict exercise capacity in heart failure. *Am Heart J* 144, 840–846.

Wang YX, Zhang CL, Yu RT, Cho HK, Nelson MC, Bayuga-Ocampo CR, et al. (2004). Regulation of muscle fiber type and running endurance by PPARdelta. *PLoS Biol* 2, e294.

Wheeler DA, Srinivasan M, Egholm M, Shen Y, Chen L, McGuire A, et al. (2008). The complete genome of an individual by massively parallel DNA sequencing. *Nature* 452, 872–876.

Williams AG and Folland JP (2008). Similarity of polygenic profiles limits the potential for elite human physical performance. *J Physiol* 586, 113–121.

Williams AG, Dhamrait SS, Wootton PT, Day SH, Hawe E, Payne JR, et al. (2004). Bradykinin receptor gene variant and human physical performance. *J Appl Physiol* 96, 938–942.

Wisloff U, Najjar SM, Ellingsen O, Haram PM, Swoap S, Al Share Q, et al. (2005). Cardiovascular risk factors emerge after artificial selection for low aerobic capacity. *Science* 307, 418–420.

Wolfarth B, Rankinen T, Muhlbauer S, Scherr J, Boulay MR, Perusse L, et al. (2007). Association between a beta2-adrenergic receptor polymorphism and elite endurance performance. *Metabolism* 56, 1649–1651.

6 Molecular adaptation to resistance exercise

Keith Baar and Henning Wackerhage

LEARNING OBJECTIVES

At the end of the chapter you should be able to:

- Discuss training strategies to increase muscle mass and strength and the evidence that support these.

- Explain how the mTOR pathway increases protein synthesis and causes hypertrophy in response to resistance exercise.

- Explain the effect of the myostatin–Smad pathway on muscle mass.

- Define satellite cells and explain their function during muscle adaptation to overload and regeneration after injury.

- Explain the molecular mechanisms that regulate the identity of satellite cells and their differentiation into muscle fibres.

INTRODUCTION

Resistance or **strength exercise** is defined as skeletal muscles working against high loads over a short time. When resistance exercise is repeated at a sufficient frequency, intensity and duration this type of training improves neural activation and increases muscle size, strength and/or power. We start this chapter with practical resistance training principles and review the hyperplasia and hypertrophy adaptations to resistance exercise. After discussing the classical exercise physiology we will then review the molecular mechanisms that cause adaptation to resistance exercise.

Resistance exercise activates signal transduction pathways that regulate muscle protein synthesis, protein degradation, transcription and satellite cell behaviour. We first review research aimed at elucidating how our muscles turn the mechanical signal of resistance exercise into a chemical signal that triggers the growth response, mainly in the form of increased protein synthesis. This discussion will focus on the mTOR pathway, the central node within our muscles that integrates multiple inputs and determines the degree to which protein synthesis and thus muscle hypertrophy occur. An opponent of mTOR and muscle growth is the myostatin–Smad signalling pathway. Genetic defects that decrease functional myostatin or Smad signalling have a large effect on muscle size but the exact role of this pathway in load-induced muscle growth is at this time poorly understood. Next, we will shift the focus to muscle stem cells and show how the key muscle stem cell, the satellite cell, can both self-renew and differentiate to assist in muscle repair. We review the function of satellite cells in response to resistance training and injury. We also use the satellite cells to introduce some basic cell physiology concepts such as the immortal strand. To prepare for this chapter you should first read Chapter 3 as it covers the basics of signal transduction and adaptation. In addition, we cover the genetics of the muscle growth-regulating mTOR and myostatin–Smad pathways in Chapter 7.

RESISTANCE EXERCISE: CURRENT RECOMMENDATIONS, TRAINABILITY AND SCIENTIFIC EVIDENCE

At the core of a resistance or strength training programme is the overload principle which states that strength gains occur as a result of systematic and progressive exercise of sufficient frequency, intensity and duration to cause adaptation. Therefore, planning a resistance training programme requires decisions about: **loading** (% of 1 repetition maximum, abbreviated as RM), **volume** (sessions per week, sets, repetitions), **rest** (time between sets) and **progression** (how to change variables during a training programme). This planning should also consider **training goals** (strength, power, hypertrophy), the **type of exercise/equipment** available (machines, free weights, own body weight), and the **muscle action** (eccentric, concentric, isometric, fast/slow) to be used. Moreover, the resistance training programme needs to be specific for the **individual** as different programmes will be needed for patients rehabilitating following trauma, untrained individuals, athletes, children, etc.

Even if an optimized programme is designed and executed, the magnitude of the adaptation in muscle mass and strength will vary dramatically between individuals. This is best shown by a study by Hubal et al. (2005), who studied the response of individuals to a given strength training programme. Hubal et al. found that some individuals did not respond with any gain in muscle size (cross-sectional area, CSA) or strength (1 RM), whereas others increased size and strength by >40% and >100%, respectively (Figure 6.1). You will also note that the figure showing the change in strength looks very different from that showing the change in muscle size, suggesting that even though the two parameters are associated, the link between size and strength is not absolute. Even though these values may have some variance associated with the precision of measurement and subject motivation, the data clearly show that there is a large variation in resistance exercise trainability in the human population and thus even a very good training programme will not work well for all subjects.

Keeping this high variation of trainability in mind, what training variables can be recommended for effective resistance training? In 2002, the American College of Sports Medicine (ACSM) (Kraemer et al., 2002) produced a position stand essentially for how to set the frequency, intensity and duration during a basic progressive resistance training programme. However, whether the recommendations in the ACSM position stand were

Figure 6.1 Resistance training trainability. The increase in cross-sectional area (CSA) in response to a resistance training programme differs dramatically in males and females. Redrawn after Hubal et al. (2005).

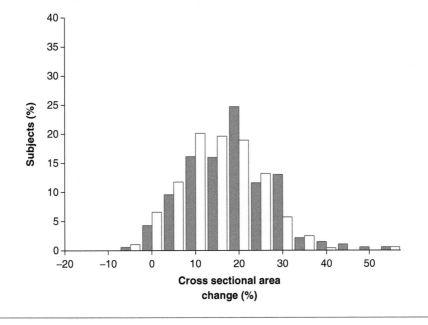

supported by solid scientific evidence or largely relying on expert opinion was questioned. A critical review of the ACSM position stand came to the conclusion that many recommendations within the position stand are not supported by scientific evidence (Carpinelli et al., 2004). Faced with such criticism, the ACSM panel went on to issue a revised position stand in 2009 where the evidence for recommendations was rated on an A–D scale. On this scale 'A' was the rating for sound evidence (randomized control trials, rich body of data) and 'D' the rating for poor scientific evidence (panel consensus judgement) (ACSM, 2009). A critical review of the new 2009 position stand states that 'the new ACSM position stand [. . .] is very similar to the 2002 position stand' and criticizes that 'the authors cited references that failed to support their opinions and recommendations' (Carpinelli, 2009).

So with all this confusion, how should one train for strength or power? First, an 'optimal' resistance training programme, fully based on reliable scientific data, does not exist. Our practical recommendation is that a programme using anywhere from 1 to 5 sets of 6–12 repetitions, if the individual works to failure, significantly increases size and strength in most people (ACSM, 2009). However, be aware that differences in motivation, genetics, nutrition and goals of the individual will affect the response to training and accept that the effects of a resistance training programme will therefore vary greatly between individuals (Hubal et al., 2005).

The initial increase of strength in response to resistance training is due to an increase in neuromuscular activation, and increases in muscle size only contribute to strength gains at a later stage (Sale, 1988). Specifically, it has been shown that resistance training increases the maximal firing frequency of motor units and action potential doublets can be detected post training (Van et al., 1998). However, even though neural adaptations predominate over the initial period of adaptation, protein synthesis also increases immediately after a bout of resistance exercise in both untrained and trained subjects (Kim et al., 2005). This apparent discrepancy that hypertrophy occurs in untrained individuals even though it contributes little to strength is due to the fact that the neural adaptations allow us to recruit more motor units to perform a given movement, making us appear stronger even though our individual muscle fibres have yet to grow very much. At this early stage, hypertrophy cannot be detected even with the most sensitive techniques such as MRI.

HYPERPLASIA

Human muscles contain thousands to hundreds of thousands of muscle fibres. For example, the vastus lateralis of young accident victims with a mean age of 19 ± 3 years contained between 393 000 and 903 000 muscle fibres (Lexell et al., 1988). This shows not only the high number but also the large variability of fibre numbers within a given muscle. This variation in muscle fibre number at least partially explains why some untrained individuals have large muscles whereas others have small muscles. Human muscle fibres can be up to 20 cm long in long muscles such as the sartorius or gracilis muscle (Heron and Richmond, 1993) and are thus among the largest cells of the human body.

Does resistance exercise increase the number of muscle fibres (termed **hyperplasia**)? Generally, it is assumed that hyperplasia contributes little to the increase in muscle mass after resistance training. However, it is clear that extreme growth stimuli can increase muscle fibre number. Furthermore, a meta-analysis showed that muscle fibre mass can increase by over 50% in animal models in response to synergist ablation, with an average of approximately 7% of that increase in mammals, due to an increase in fibre number. In several of the cited studies fibre numbers increased by more than 30%, which demonstrates that hyperplasia can be a significant factor for muscle hypertrophy in response to extreme growth stimulation (Kelley, 1996). Thus hyperplasia can occur, especially in animal models after synergist ablation. It is unclear, however, whether long-term muscle growth stimulation in humans following resistance training and/or steroid use results in significant hyperplasia.

HYPERTROPHY

Before discussing muscle hypertrophy, which depends on adding protein to muscle, we will first review several facts related to skeletal muscle and the proteins therein. In men ≈38% and in women ≈31% of the body mass is skeletal muscle (Janssen et al., 2000). However, this depends on the individual as lean subjects have a higher percentage whereas obese subjects have a lower percentage. Depending on hydration status, muscle comprises ≈70% water (700 mL/kg muscle) and ≈30% solids (300 g per kg muscle). Of the solids ≈70% (215 g/kg muscle) is protein (Forsberg et al., 1991). In muscle, thousands of different proteins are expressed but the most abundant proteins are the **myofibrillar proteins** of the sarcomere. It has been estimated that ≈20–40% of overall protein (in absolute terms 45–85 g per each kg of muscle) is myosin heavy chain and that ≈15% (30 g/kg muscle) is actin (Carroll et al., 2004). We hope that this gives an idea of the composition of skeletal muscle and of the amount of protein therein. We will now discuss the effect of resistance exercise on muscle protein.

Muscle growth occurs mainly by muscle fibre **hypertrophy**, which refers to an increase of the volume or mass of muscle fibres, or more generally cells, without an increase in cell number. Muscle hypertrophy occurs if protein is added to a muscle fibre or, using more scientific terms, if the protein balance is positive. The protein balance is defined as the difference between **protein synthesis** and **protein breakdown**. Skeletal muscle protein synthesis, and even more so protein breakdown, is challenging to measure in humans but a few laboratories worldwide have mastered this technique.

In simple terms, human muscle protein synthesis is measured by infusing **stable isotope**-labelled amino acids into the blood of an individual. For example to ≈99% of carbon naturally occurs as ^{12}C, whereas the other ≈1% is the heavier ^{13}C (i.e. non-radioactive, stable) isotope. For stable isotope protein synthesis measurements, these low-percentage isotopes (^{13}C or ^{2}H) are added to amino acids in greater quantities to generate a **tracer** for the experiment. The amino acid tracer is then infused into the body, where it is incorporated into protein as a result of protein synthesis. To measure muscle protein synthesis, muscle samples are taken and the enrichment of the amino acid tracer into muscle protein is measured by separating the ^{13}C from the naturally occurring ^{12}C by mass spectroscopy. The greater the increase in the stable isotope within muscle proteins over time, the greater the rate of muscle protein synthesis.

What is clear from these studies is that a single bout of resistance exercise can increase protein synthesis for up to 72 h (Figure 6.2a (Miller et al., 2005)). This phase of elevated protein synthesis is much longer than the one that occurs after a meal (Bohe et al., 2001). However, if resistance exercise is performed in the fasted state there will be an even greater increase in protein degradation which results in a negative protein balance (Phillips et al., 1997). The protein balance only becomes positive, and thus hypertrophy only occurs in the presence of amino acids, which increase when proteins are digested after a meal (Tipton et al., 1999) (Figure 6.2b). The fact that consuming essential amino acids decreases protein breakdown following resistance exercise suggests that fasting protein breakdown increases following resistance exercise in order to supply essential amino acids for protein synthesis. Since essential amino acids only come from the diet or stored protein, they must be supplied by breaking down existing protein if no protein was ingested.

To conclude, resistance exercise against a heavy load to failure in the presence of nutrients, specifically proteins, leads to an increase in the protein balance largely due to an increase in synthesis. Over time, this increase in protein synthesis causes muscle fibre and whole-muscle hypertrophy. In small mammals, pronounced muscle growth is accompanied by an increase in the number of muscle fibres termed hyperplasia. During short-term resistance exercise in humans, there is no or at best only a negligible increase in the number of muscle fibres and thus the resultant whole-muscle hypertrophy is almost solely due to muscle fibre hypertrophy. It is unclear, however, whether muscle fibre numbers can increase in response to many years of resistance training in humans.

Figure 6.2 Resistance training and muscle protein synthesis. (a) Resistance exercise increases myofibrillar protein synthesis for up to 72 h. Redrawn after Miller et al. (2005). (b) Only resistance exercise and amino acids (or protein, as amino acids are digested protein) results in a positive net protein balance. Redrawn after Tipton et al. (1999). AA amino acids.

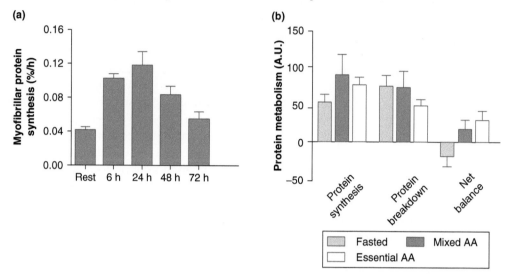

MOLECULAR MECHANISMS OF RESISTANCE TRAINING ADAPTATION

Molecular exercise physiologists have made several advances in our understanding of how a mechanical signal such as the load across a muscle is converted to chemical signals that trigger muscle growth. From this work, there appear to be three major players:

- The anabolic **mTOR** signal transduction pathway increases protein synthesis resulting in hypertrophy.
- The catabolic **myostatin–Smad** signal transduction pathway inhibits muscle growth by poorly understood mechanisms.
- **Satellite cell pathways** aid in the repair of damaged fibres but this is not essential for short-term, modest, hypertrophy.

In the text below we will introduce these three signal transduction pathways or networks and then briefly mention other mechanisms that can also regulate muscle size and strength.

THE mTOR PATHWAY REGULATES PROTEIN SYNTHESIS

The mammalian/mechanistic target of rapamycin (mTOR) is a serine/threonine kinase that increases **protein synthesis** (i.e. the translation of mRNA into protein by the ribosome) and **ribosome biogenesis** (i.e. the capacity of a cell for protein synthesis) and in some cells inhibits a form of protein breakdown termed **autophagy**. In addition, mTOR promotes cell division and the transcription of certain genes. mTOR got its name from the fact that it is selectively inhibited by the macrolide antibiotic rapamycin (Davies et al., 2000). As a result, rapamycin has become a powerful research tool in the identification of the function of mTOR. As mTOR is the master growth regulator of all cells, a personal account of the discovery of mTOR by Michael N Hall is included here (Box 6.1).

BOX 6.1 MICHAEL N HALL: DISCOVERY OF TOR

It was the late 1980s and I was a fresh assistant professor at the Biozentrum of the University of Basel, Switzerland. We were studying nuclear protein import as an extension of my postdoctoral work at the University of California, San Francisco, during which I described the nuclear localization signal. Our work on nuclear import was not going very well and we were getting desperate, so desperate that (as later described by the journalist Karen Hopkin) we turned to drugs – in this case the immunosuppressive drugs cyclosporin A (CsA) and FK506. Joe Heitman had just joined the lab as a postdoc after finishing the PhD part of his MD–PhD studies in NYC and, given his medical background, was interested in how drugs worked. Another very fortunate circumstance was an ongoing collaboration with Rao Movva who was a group leader at the Basel pharmaceutical company Sandoz (now Novartis). CsA was a blockbuster drug for Sandoz and Rao was interested in determining its mode of action. Little was known about how CsA and FK506 worked other than they blocked nuclear import of a signal, possibly a protein, downstream of the T cell receptor. Thus, we thought we could use the drugs to probe novel signaling pathways or the nuclear import process. Something interesting had to come out, particularly since at the time no complete signaling pathway was known. What was unusual about our approach was that we used yeast genetics to study the drugs that were, of course, developed for use on humans.

In our first experiments, CsA and FK506 had little to no effect on yeast cells. Rao then told us about rapamycin, a brand new FK506-like compound that was not yet approved for use in the clinic and was not even available commercially. At the time, Sandoz was one of the few places in the world where one could obtain rapamycin. Unlike CsA and FK506, rapamycin blocked growth of yeast cells, and Joe quickly selected rapamycin-resistant mutants. Most of the yeast mutants were defective in the *FPR1* gene, which Joe had already characterized in the context of his FK506 work (*FPR1* stands for FK506-binding proline rotamase, also known as FKBP). The few remaining mutants were altered in one of two new genes which we named *TOR1* and *TOR2* (target of rapamycin). We published Joe's findings in 1991, shortly after he returned to NYC to finish the MD part of his studies.

The rapamycin study was taken over by Jeannette Kunz, a Swiss PhD student, who was later joined on the project by Stephen Helliwell, a British PhD student. Jeannette and Stephen cloned and characterized the two *TOR* genes to discover that they encode two highly homologous proteins that resemble lipid kinases. However, the TORs turned out to be the founding members of a novel class of atypical protein kinases known as PIKKs (PI kinase-related protein kinases). Jeannette knocked out the two *TOR* genes to reproduce the effect of rapamycin treatment, confirming that the TOR proteins were indeed the target of rapamycin *in vivo*. Mutations in

(Continued)

BOX 6.1 CONTINUED

the *FPR1* and *TOR* genes conferred resistance because rapamycin acts by binding FKBP and the FKBP–rapamycin complex then binds and inhibits *TOR*. Jeannette published her work in 1993. Shortly thereafter, the Schreiber, Snyder, Berlin and Abraham groups independently described the mammalian TOR orthologue, giving it the names FRAP, RAFT, RAPT and mTOR, respectively. The field eventually chose the name mTOR based on the yeast precedent.

The race to discover TOR was extremely competitive. We were fortunate to win this race, in large part because of the talented postdocs and students working on the project, but also because we worked with genetically tractable yeast. We made the assumption that the target of rapamycin was conserved, an assumption based on the fact that rapamycin is a natural product secreted by a soil bacterium to inhibit other microbes such as yeast. Ironically, in the late 1980s, this was not widely appreciated and many looked at us askance for using yeast to study a compound better known as a drug to treat allograft rejection in humans.

By the mid-1990s, it was clear that TOR was a highly conserved kinase and the *in vivo* target of rapamycin. The next step was to understand its physiological role in the cell. Work from our and others' labs showed that TOR activates several anabolic processes and inhibits catabolic processes to control cell growth (increase in cell size or mass) in response to nutrients. In mammalian cells, mTOR is controlled also by growth factors such as insulin (via PI3K) and cellular energy (via AMPK). The fact that the so-called metabolic tissues, including muscle, are the most responsive to the three inputs (nutrients, insulin, energy) that control mTOR led us to focus on mTOR in these tissues. The work on mTOR in muscle was performed by our collaborator Markus Rüegg, David Glass and many others. These studies established that mTOR controls muscle mass, metabolism, and performance.

Michael N Hall, University of Basel, July 2013

The mTOR pathway was first identified as being activated by resistance exercise in rats (Baar and Esser, 1999). In this experiment, the researchers showed that the amount of muscle hypertrophy following six weeks of training was directly related to the degree of phosphorylation of p70 S6K1 (also known as S6K1), a known target of mTOR. Even though this study suggested a relationship between mTOR and hypertrophy, it took a number of other experiments to show that mTOR was required for load-induced muscle growth. In the first, Bodine et al. (2001) showed in mice that rapamycin could prevent load-induced skeletal muscle hypertrophy. In the second, Drummond et al. (2009) showed that volunteers receiving 12 mg of rapamycin prior to performing 11 sets of 10 repetitions of leg extension at 70% of their 1 RM (Drummond et al., 2009) did not show the classic increase in muscle protein synthesis after exercise, whereas those who got a saline injection did (Figure 6.3). In the third and most conclusive experiment, Goodman et al (2011) used mice in which they had mutated mTOR so that it was no longer sensitive to rapamycin to show that mTOR was required for muscle growth. In this experiment, wildtype and rapamycin-resistant mice had their gastrocnemius and soleus muscles removed in a procedure called a

Figure 6.3 mTOR is required for the increase of human muscle protein synthesis after resistance exercise. The mTOR inhibitor rapamycin blocks resistance exercise-induced increase of protein synthesis in human muscle. Redrawn after Drummond et al. (2009).

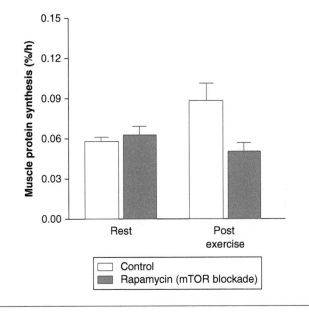

synergist ablation. Following removal of these two muscles, the third muscle in the area, named the plantaris, must take on the extra load and as a result hypertrophies. In the wildtype mice, ablation resulted in a 42% larger muscle in 14 days and this was completely prevented by daily injections of rapamycin. In the rapamycin-resistant mTOR mice, ablation led to a similar increase in muscle mass that rapamycin was unable to prevent, showing that mTOR is required for load-induced skeletal muscle hypertrophy.

Even though mTOR is required for muscle hypertrophy, this does not mean that it is sufficient on its own to cause it. The demonstration that activating mTOR on its own was enough to cause muscle hypertrophy was provided in 2010 by Goodman and his colleagues (2010). In this study, the scientists injected DNA into muscle to genetically activate mTOR. Seven days later, they removed the muscles and found that the muscle fibres where they had activated mTOR had grown ≈40%, whereas a control injection had no effect on muscle fibre size. Taken together, these data show that resistance exercise activates mTOR and that the activation of mTOR is sufficient and required for load-induced skeletal muscle hypertrophy.

Below we will now better describe mTOR, introduce the two protein complexes that contain mTOR and review the activation of mTOR. After that we will provide a brief overview of the continuing search for the mechanoreceptor that senses resistance exercise and activates the mTOR pathway.

HOW DOES THE INFORMATION FLOW FROM RESISTANCE EXERCISE TO PROTEIN SYNTHESIS VIA THE MTOR PATHWAY?

Before discussing the elements of the mTOR pathway, it is important to understand that mTOR exists in two protein complexes, termed mTOR complex 1 and mTOR complex 2 (**mTORC1, mTORC2**). These two protein complexes differ in the proteins that associate with mTOR, their location within the cell and the proteins that each complex phosphorylates:

- **mTORC1** includes mTOR, raptor, mLST8/GβL, PRAS40 and deptor.
- **mTORC2** includes mTOR, rictor, mLST8/GβL, protor, deptor and mSIN1.

From the perspective of resistance exercise, mTORC1 is the centre of the mTOR pathway (Figure 6.4). If instead, we were interested in insulin signalling, mTORC2 would be far more important since it regulates insulin-stimulated glucose uptake. There are a number of independent ways that mTORC1 can be activated, but all of these pathways end with mTOR being activated by a small G-protein named Rheb (Ras homologue enriched in brain). It is thought that Rheb activates mTOR by recruiting phospholipase D (PLD) and increasing the amount of phosphatidic acid bound to the so-called FRB domain of the mTOR. Even though the last step in the activation of mTOR is the same, there are several different ways that this activation can occur:

- **Growth factors such as IGF-1 and insulin** activate mTOR via a cascade involving the IGF-1 or insulin receptor, insulin receptor substrates (IRS), PI3K, PDK1/mTORC2, PKB/Akt, PRAS40, TSC2 and Rheb. The hypertrophic effect of this pathway has been demonstrated by overexpressing IGF-1 (Coleman et al., 1995) or constitutively active PKB/Akt in skeletal muscle (Bodine et al., 2001). Both interventions result in skeletal muscle hypertrophy (Pallafacchina et al., 2002; Lai et al., 2004).
- **Amino acids** via the Rag small G-proteins and the ragulator bring mTOR to Rheb and thereby activate mTOR (Kim and Guan, 2009).
- **Resistance exercise** via an as yet unidentified mechanoreceptor inhibits TSC2, which results in the activation of Rheb and mTOR.
- **G-protein coupled receptors (GPCR)** via Gα.

The activation of mTOR is further complicated by the fact that **endurance exercise**, possibly through high AMP or NAD^+ and low glycogen, can potentially inhibit mTOR activation in part through AMPK, TSC2 and Rheb (Inoki et al., 2003). This mechanism ensures that muscle is not built when energy is needed elsewhere.

The combined, aforementioned signalling then either activates or inhibits mTORC1 activity, respectively. We do not further discuss all the signalling interactions within the pathway. Most are protein phosphorylation events, but phospholipids and G-proteins also play a role. An overview is given in Figure 6.4.

The activity of mTORC1 depends on the combined activity of the aforementioned four inputs. Activated mTORC1 then regulates several cellular functions. After resistance exercise, the most important of these is the increase of protein synthesis, or more technically **translation initiation** and **elongation**. Proteins are synthesized by

Figure 6.4 Regulation of muscle protein synthesis by the mTOR pathway. Current model of the regulation of mTORC1 by (1) resistance exercise, (2) amino acids, (3) growth hormone via IGF-1, muscle secreted IGF-1 and insulin and (4) endurance exercise via AMPK. (5) Active mTORC1 then phosphorylates 4E-BP1 and p70 S6k, and 4E-BP1 detaches from eIF4E which increases protein synthesis (translation). In reality many more proteins are involved and mTORC1 regulates functions other than protein synthesis.

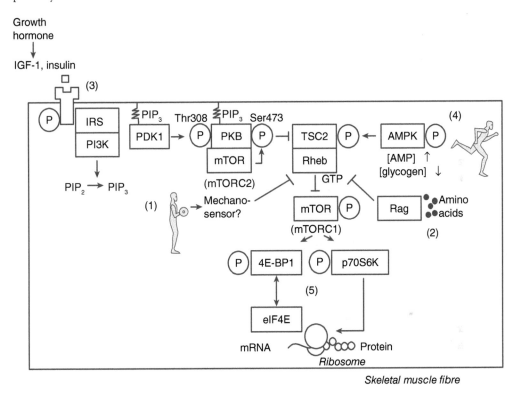

the ribosome, an organelle made from ribosomal RNAs and 79 proteins. In 2009, Venkatraman Ramakrishnan, Thomas A Steitz and Ada E Yonath won the Nobel Prize in Chemistry for their 'Studies of the structure and function of the ribosome', which demonstrated what a ribosome looks like and how it works. mTORC1 signalling activates translation via the phosphorylation of the regulatory proteins 4E-BP1 p70 S6K1. Phosphorylation of 4E-BP1 detaches it from eIF4E (Holz et al., 2005) and this allows the ribosome to bind to the mRNA to initiate translation. How p70 S6K regulates initiation is less clear but no less important. mTORC1-dependent signalling also regulates the movement of the ribosome along the mRNA, which is termed translation elongation. The result is the synthesis of a peptide chain on the basis of the mRNA blueprint. To summarize, mTORC1 is at the centre of the mTOR pathway. It is activated by resistance exercise, amino acids and hormones but can be inhibited by acute endurance exercise. Active mTORC1 promotes translation initiation and elongation, resulting in an increase in protein synthesis. When protein synthesis is chronically higher than protein breakdown as it is after resistance exercise in fed individuals, then muscle fibres hypertrophy.

THE HUNT FOR THE ELUSIVE MECHANOSENSOR

Initially many researchers assumed that resistance exercise and other forms of muscle overload activate the mTOR pathway by changing the expression of the growth factor IGF-1. However, hypertrophy can occur in response to overload in transgenic mice where the IGF-1 receptor was knocked out (Spangenburg et al., 2008) or where the signalling between IGF-1 and PKB/Akt was otherwise inhibited. Moreover, the time course of IGF-1 expression and mTOR activity does not match. Other results against a requirement of IGF-1 followed, combined with a point–counterpoint debate on whether IGF-1 is required for overload-induced mTOR activation and hypertrophy in the *Journal of Applied Physiology*. What is clear from this work is that tension directly stimulates mTOR-dependent protein synthesis via a growth factor-independent mechanism (Philp et al., 2011). The alternative candidate mechanisms include stretch-activated calcium channels, amino acids and PLD (phospholipase D), and the phosphorylation and movement of TSC2 (Hornberger review 2011). However, the protein or molecule that senses mechanical stress is still unknown. This is still an area of intense research and arguably the most important unanswered question in this field.

MYOSTATIN AND ADAPTATION TO RESISTANCE EXERCISE

The mTOR and myostatin–Smad pathways are the accelerator and brake or the Yin and Yang of muscle growth, respectively. Thus in contrast to mTOR, activating the myostatin–Smad signalling pathway inhibits skeletal muscle growth. This is most obvious in animal models where myostatin activity has been removed. Mice, dogs or cows where myostatin has been knocked out or mutated have muscles that are twice the size of animals with normal myostatin (McPherron et al., 1997). Combining myostatin knockout with overexpression of FLRG, which is an inhibitor of myostatin and other proteins, even quadruples muscle mass (Lee, 2007). Inhibiting the pathway further down, by blocking Smad transcriptional activity with the transcriptional repressor *ski* also results in greater than 3-fold increase in muscle mass (Sutrave et al., 1990). Therefore, the myostatin–Smad pathway clearly regulates skeletal muscle size, but does the increase in muscle mass lead to an increase in muscle strength? The answer to this question is: It depends on how and where it is measured. In the myostatin knockout mouse, there is an increase in the absolute amount force the muscle produces, but the specific force, which is the force in relation to muscle size, is lower (Mendias et al., 2006). The reports on the boy who is a myostatin knockout and his family (Schuelke et al., 2004) and the heterozygous myostatin race dog (Mosher et al., 2007) suggest that the myostatin loss-induced increase in muscle mass can in some cases not only increase muscle mass but also strength and speed. Therefore, it appears that blocking myostatin increases the absolute strength of muscle, but for some reason the muscle is, at least in some cases, unable to produce the same relative force as normal muscle. Put another way, myostatin mutants are stronger, but in some cases not as strong as they should be considering their greater size.

It is clear that inhibiting the myostatin–Smad pathway increases muscle size, but is the myostatin–Smad pathway also responsive to resistance exercise? Also, what are the mechanisms that link less myostatin to more muscle? Figure 6.5 gives an introduction into the signalling of the myostatin pathway (Lee, 2004, 2010).

Figure 6.5 Myostatin-Smad pathway. Schematic drawing depicting the role of the myostatin–Smad pathway in skeletal muscle size inhibition. (1) Glucocorticoids and immobilization commonly increase myostatin mRNA, while resistance exercises in some, but not all, studies decrease myostatin. (2) Transcribed and translated myostatin is then cleaved by proteases, forms a dimer and is secreted. (3) Extracellular myostatin can alternatively form inactive complexes with inhibitor proteins such as follistatin, FLRG and GASP-1 (Lee, 2010). Only when active does myostatin bind to the TGFβ type II receptor, which phosphorylates and binds to the type I receptor which then phosphorylates Smad2/3 at its C-terminal end. Phosphorylated Smad2/3 together with Smad4 enters the nucleus, where they regulate genes that control muscle mass.

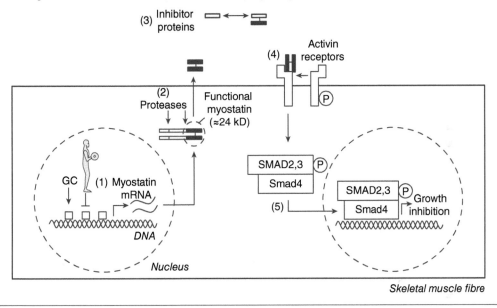

Even though it is clear that genetically inhibiting myostatin results in huge muscle growth, how resistance exercise affects this pathway is more controversial. Louis et al. (2007) measured myostatin mRNA using reverse transcriptase polymerase chain reaction (RT-PCR) from 0 to 24 h after resistance and endurance exercise. They observed a significant decrease in myostatin mRNA after resistance exercise, peaking 8 h after exercise. However, they also observed a decrease in myostatin mRNA after endurance exercise, which does not increase muscle mass. Interestingly, Kim et al. also saw a decrease in myostatin mRNA following resistance exercise, but this effect was not related to the change in muscle mass that occurred with training (Kim et al., 2007). Therefore, there is no clear relationship between myostatin mRNA after resistance exercise and changes in muscle mass.

However, when you consider the pathway shown in Figure 6.6 it is perhaps not surprising that myostatin mRNA does not always relate to muscle growth. After myostatin mRNA is made in a muscle, it has to be translated, cleaved and shipped out of the muscle where it can be bound up by inhibitors. When myostatin protein does bind to its receptor it has to activate the Smad protein and can still be inhibited by inhibitors of Smads such as ski or sno. Therefore, all that we can say at the moment is that we do not know whether/how myostatin signalling is affected by resistance exercise in a way that suggests that myostatin contributes to mediating the growth response to resistance exercise.

Even though it is unclear how or whether myostatin is affected by resistance exercise, the genetic models show that it can control muscle size, so how does it work? If myostatin is knocked out or mutated then muscle cells lack myostatin from the moment they are created during development. If this is the case, muscle cells divide more during development and we end up with hyperplasia as well as hypertrophy. If myostatin is reduced in the adult, using myostatin antibodies or follistatin, then fibre hypertrophy is the main effect (Whittemore et al., 2003; Welle et al., 2007). In both cases, skeletal muscle hypertrophy occurs because protein synthesis is greater in muscles where myostatin is either knocked out or reduced than in the wildtype (Welle et al., 2006, 2011). It is currently unclear how this effect on protein synthesis is achieved. Myostatin is thought to act primarily through Smad transcription factors. Interestingly, a secondary effect of Smad3 appears to be the regulation of PKB/Akt (Winbanks et al., 2012). When activated by myostatin, Smad3 can no longer bind to PKB/Akt and activate mTOR. The result is a decrease in basal mTOR activity and muscle size. Whether the Smads also regulate genes that affect protein synthesis or whether myostatin can modulate ribosome activity in some other way is currently unknown. Thus despite well over 10 years of myostatin research, the mechanisms by which the myostatin–Smad system increases muscle mass up to 4-fold are currently incompletely understood.

SATELLITE CELLS, REPAIR AFTER MUSCLE INJURY AND ADAPTATION TO RESISTANCE EXERCISE

Muscle fibres within the human sartorius and gracilis muscles can be up to 20 cm long (Heron and Richmond, 1993). They are thus the second largest cells in the human body after α-motor neurons. Each muscle fibre is a cell that contains many nuclei termed myonuclei. In rats, there are between 44 and 116 myonuclei per mm of fibre, with more nuclei being found in the slow type I fibres than in the type II fibres (Tseng et al., 1994). Applying these nuclei-per-mm of fibre values to human muscle fibres suggests that the largest human muscle fibres could contain between 9700 and 23 000 nuclei. This would make muscle fibres by far the most nucleated cells of the human body. The unique feature of myonuclei is that they cannot divide. The scientific term for their inability to divide is to say that they are **post-mitotic**. Thus if additional myonuclei are required during post-natal growth to regenerate damaged or dead muscle fibres after injury or to avoid the dilution of nuclei during muscle fibre hypertrophy after resistance exercise, where do the additional myonuclei come from?

In skeletal muscle the production of new myonuclei is outsourced to so-called **satellite cells**. Satellite cells are the resident stem cells of skeletal muscle. In human muscle it has been estimated that between 1.4 and 7.3% of all nuclei are satellite cells, with higher frequencies in muscles or around muscle fibres with a slower phenotype (Kadi et al., 2005). Satellite cells were discovered by Alexandra Mauro and Bernhard Katz who used electron microscopes to study frog muscle fibres (Katz, 1961; Mauro, 1961). In his paper, Mauro described satellite cells as a subset of cells 'wedged' between the plasma membrane, also known as plasmalemma, and basement membrane or basal lamina. Because of their position he named these cells 'satellite cells' (Mauro, 1961). Mauro noted that satellite cells were almost all nucleus with very little cytoplasm, and he hypothesized that their function might be to repair skeletal muscle. This is clear in the electron microscope image in Figure 6.6.

Figure 6.6 A satellite cell of a mouse tibialis anterior muscle near the myotendinous junction. Above the satellite cells are the sarcomeres and mitochondria of a muscle fibre. Note the membrane surrounding the satellite cell. Scale bar 2 μm.

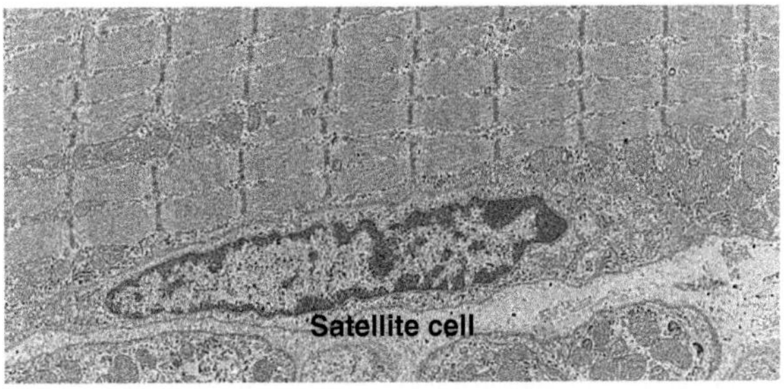

What is the evidence that satellite cells regenerate damaged muscle? The ability of adult muscle to repair itself after injury was well studied already in the nineteenth century (Scharner and Zammit, 2011). The regenerative capacity of skeletal muscle was most impressively demonstrated by Studitsky, who minced whole muscles from small animals such as chick, pigeon, rat and mouse using scissors and reinserted the homogenized muscle (Studitsky, 1964). Surprisingly, the homogenates regenerated at least partially to form innervated muscles. These experiments highlighted that muscle fibres can regenerate extremely well and satellite cells were the obvious candidates for the cells that permit this renewal.

However, in 1998 a paper by an Italian team threw a spanner into the works. The Italian team demonstrated conclusively that bone marrow-derived cells could migrate to sites of muscle injury and regenerate the damaged fibres (Ferrari et al., 1998). So were bone marrow-derived cells the true myogenic (i.e. muscle making) stem cells that regenerated skeletal muscle? Were satellite cells just minor players?

These questions were answered by several experiments that showed:

- the regeneration of muscle by bone marrow-derived cells was a rare event (Sherwood et al., 2004);
- satellite cells are both true stem cells (Collins et al., 2005); and
- the major muscle-regenerating cells after injury (Collins et al., 2005).

Thus satellite cells were 're-enthroned' as the resident muscle stem cell, as it was described in one editorial. Perhaps the most impressive experiment involved transplanting a single satellite cell (termed 'muscle stem cell' in that study due to the isolation method) into a regenerating muscle. It was estimated that this single cell gave rise to $\approx$20 000–80 000 progeny during several rounds of injury followed by regeneration (Sacco et al., 2008). Thus to conclude, satellite cells are the major resident adult stem cells of skeletal muscle. There are some other stem cells that are capable of making muscle, but their capacity is limited with the exception of vascular-based pericytes and mesoangioblasts.

Before reviewing how satellite cells respond to overload and injury and whether their function is essential for regeneration and hypertrophy, we will also first cover the molecular mechanisms that regulate the identity of satellite cells and the development of skeletal muscle, which is termed **myogenesis**. Myogenesis occurs not only during embryonal development but also when satellite cells respond to injury or growth stimuli.

In the first major molecular myogenesis experiment, 5-azacytidine was shown to convert fibroblasts, which are non-muscle cells, into muscle cells. The muscle-making agent used, 5-azacytidine, reduces DNA methylation, which is an epigenetic regulatory mechanism. This must have 'opened up' some crucial muscle-making genes as the cells swapped their lineage to muscle. The authors then identified several genes that were only expressed after 5-azacytidine treatment. One of these genes, which they had termed **MyoD**, is a transcription factor that was capable and sufficient of turning fibroblasts into muscle cells (myoblasts) (Lassar et al., 1986; Davis et al., 1987). Later the team demonstrated that MyoD could also convert other pigment, nerve, fat and liver cells into myoblasts, showing that MyoD is capable of generally turning non-muscle cells into muscle cells (Weintraub et al., 1989). Soon after other teams identified Myf5, Mrf4 (Myf6) and myogenin as genes that have a similar basic helix-loop-helix (bHLH) structure as MyoD and that can also regulate myogenesis. The genes within this family are known as **myogenic regulatory factors** (MRFs). The *in vivo* function of these MRFs was characterized in several studies by using knockout mouse experiments. So did MyoD knockout prevent the formation of skeletal muscle as one might predict? No, to the surprise of the investigators, MyoD knockout mice developed skeletal muscle. However, soon after it was shown that if MyoD and Myf5 were knocked out together, myogenesis was prevented as there were no muscle cells in the mouse, demonstrating that MyoD and Myf5 have redundant functions (Arnold and Braun, 1996). Another MRF is myogenin. Myogenin is expressed at a later stage than MyoD and Myf5 and regulates the fusion of mononucleated myoblasts into multinucleated myotubes, the precursors of muscle fibres (Edmondson and Olson, 1989; Arnold and Braun, 1996).

How do MRFs function? MyoD functions as a transcription factor which binds to more than 20000 MyoD DNA-binding sites located all over the genome (Cao et al., 2010). By doing so, MyoD probably opens up as a pioneer the ske et al muscle genes by modulating the acetylation of histones which is, like DNA methylation, an epigenetic modification (Cao et al., 2010).

The relevance of MRFs for satellite cells is that they help to turn an existing satellite cells into a muscle fibre. Inactive or quiescent satellite cells generally express low levels of MRF proteins. For satellite cell identity the paired box transcription factor **Pax7** plays a key role. Mice who carry a Pax7 knockout mutation have virtually no satellite cells, showing that Pax7 is key for the specification of satellite cells (Seale et al., 2000). Today, Pax7 is used as a major marker to detect satellite cells by immunohistochemistry.

To put it all together, we will now review how Pax7 and myogenic regulatory factors work in satellite cells that respond to an overload or injury stimulus. In response to injury, satellite cells are activated, proliferate and then either differentiate or self-renew (Zammit et al., 2004):

- **Quiescent satellite cells**. Most satellite cells in an unstimulated adult muscle are quiescent and express Pax7 but not MyoD.

- **Activated satellite cells (also described as satellite cell-derived myoblasts).** When responding to an overload or injury stimulus, satellite cells become activated and break through the basal lamina, migrate on the muscle fibre (Otto et al., 2011) and proliferate to form clusters of activated satellite cells. The hallmark event of satellite cell activation is the expression of MyoD but note that activated satellite cells continue to express Pax7.
- **Differentiating satellite cells.** Activated satellite cells then make a cell fate decision to either self-renew (i.e. return to quiescence) or to differentiate to either merge with a hypertrophying muscle fibre or form new fibres during regeneration after injury (Zammit et al. 2004). Differentiating satellite cells exit the cell cycle and express myogenin, which triggers the fusion process.
- **Self-renewing satellite cells.** Self-renewing satellite cells also exit the cell cycle and lose the expression of MyoD.

After the discovery of Pax7 and the MRFs the next obvious questions were 'What upstream mechanisms regulate the expression of Pax7, MRFs and, therefore, the numbers and fate of satellite cells?' and 'What signals activate satellite cells?' Unfortunately both questions are incompletely answered. A key pathway that governs satellite cell behaviour is the Notch pathway (Conboy and Rando, 2002) but its function is still incompletely understood. Recent reports suggest that high Notch pathway activity is important for satellite cell quiescence (Mourikis et al., 2012). Myostatin has been shown to inhibit satellite cell activation and proliferation (McCroskery et al., 2003) which could be part of the mechanism by which myostatin inhibits muscle growth, especially after birth. In contrast, Smad1, 3 and 5 signalling has been shown to drive the proliferation of activated satellite cells while preventing their differentiation (Ono et al., 2011). Similarly, the Wnt pathway (Otto et al., 2008) and the Hippo pathway member Yap promote the proliferation of satellite cells (Judson et al., 2012) which could be important in expanding satellite cells after a stimulus and before these cells differentiate or self-renew. Several immune and growth factors have been identified as potential regulators of satellite cell fate but the signals that actually activate satellite cells (Bentzinger et al., 2010) after injury or in hypertrophying muscle are still poorly understood.

After this long prelude we will now ask a key question for exercise physiologists in the context of satellite cells, which is whether satellite cells are essential for hypertrophy in response to resistance exercise or other forms of overload. This question has been the subject of an intense debate that has taken place within the field for over 20 years. Closely related to this is also the much discussed **myonuclear domain hypothesis** (Allen et al., 1999), which states that there are mechanisms that maintain a fixed nuclear-to-cytoplasmic volume ratio. If this was true then new nuclei need to be added to hypertrophying fibres whereas atrophying fibres should lose nuclei. The debate whether satellite cells are required for muscle hypertrophy culminated in a point–counterpoint debate in the *Journal of Applied Physiology* (O'Connor and Pavlath, 2007). So is an addition of myonuclei by satellite cells required for hypertrophy and is the myonuclear domain hypothesis valid, since a no answer to the first question leads to a rejection of the myonuclear domain hypothesis? After many years of research a 'no' seems to be the answer to both questions.

First, inducing muscle hypertrophy genetically by a knockout of myostatin (Amthor et al., 2009) did not result in more myonuclei, and the overexpression of constitutively active PKB/Akt (Blaauw et al., 2009) led to an increase in the myonuclear domain. These observations suggest that myonuclei addition from satellite cells is not essential for hypertrophy and thus they contradict the myonuclear domain hypothesis and suggest that satellite cells are not required for PKB/Akt and myostatin knockout-induced hypertrophy.

The gold standard experiment to conclusively address the above two questions is to remove all satellite cells from an adult skeletal muscle and then to test whether the muscle can still hypertrophy if it is subjected to overload. Rosenblatt and Parry (1992) attempted this using γ-irradiation to eliminate satellite cells and synergist ablation to induce hypertrophy (Figure 6.7). They observed that ≈25% of muscle mass were lost after γ-irradiation and overload whereas the same overload without γ-irradiation increased muscle mass by ≈20% (Rosenblatt and Parry, 1992). They took this to mean that functioning satellite cells were essential for muscle hypertrophy in response to overload. However, a major limitation of this experiment was that γ-irradiation has profound side effects, other than just satellite cell loss, that might underlie this detrimental effect. Nearly 20 years later McCarthy et al. (2011) created a transgenic mouse that allowed the investigators to knock out >90% of the satellite cells in adult skeletal muscle. Figure 6.7 shows that removing the satellite cells had no significant impact on the hypertrophic response to synergist ablation (McCarthy et al., 2011) and thereby demonstrates that functioning satellite cells are not required for short-term hypertrophy induced by acute synergist ablation in rodents. However, after eight weeks the hypertrophy is significantly reduced (personal communication by Charlotte A Peterson, 2012). Thus satellite cells are not required for initial hypertrophic responses but are essential for sustaining the hypertrophy in the long term.

Figure 6.7 Satellite cells and muscle hypertrophy. (a) Satellite cell knockout by γ-irradiation prevents synergist ablation-induced hypertrophy of the extensor digitorum muscle (EDL) in rats. Redrawn after Rosenblatt and Parry (1992). (b) In contrast, a severe depletion of Pax7-expressing satellite cells by transgenic knockout does not prevent normal hypertrophy of the synergist ablated plantaris muscle in mice Redrawn after McCarthy et al. (2011).

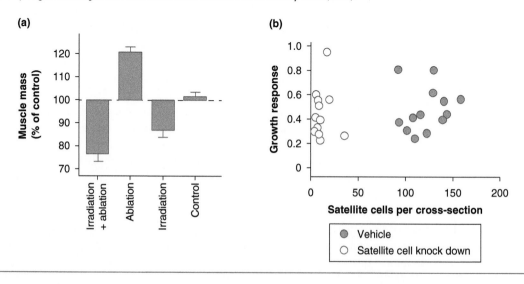

Even though satellite cells play a limited role in short-term muscle hypertrophy, they are important for postnatal muscle growth and essential for muscle regeneration after injury (Lepper et al., 2011) and healthy muscle ageing. Thus both after microinjury induced by eccentric exercise and after muscle injuries in sport, satellite cells will be key players during the regeneration of skeletal muscle.

SUMMARY

Resistance training involves exercising against high loads for short periods of time. The evidence for 'optimal' training regimes is limited, any variation of 1–5 sets of 6–12 repetitions to failure will increase muscle size and strength in most people. The major gross adaptation is muscle hypertrophy, especially of type II fibres. Hyperplasia can occur in some animal models but contributes little to the muscle size increase after weeks or months of resistance training. Resistance exercise acutely increases protein synthesis and breakdown and the protein balance will only be positive if protein or amino acids are additionally ingested. Resistance exercise and protein or amino acids lead to an activation of the mTOR pathway which then increases the rate of protein synthesis by targeting proteins that activate protein synthesis and the production of new ribosomes. The mTOR pathway integrates the input of several positive signals that include hormones, such as IGF-1 and insulin, resistance exercise and essential acids. In addition, mTOR is inhibited by AMPK, which is activated by acute endurance exercise or low glycogen. Myostatin inhibits muscle growth and decreases after both resistance and endurance exercise. Its role in the growth adaptation to resistance exercise is incompletely understood. Satellite cells are the resident adult stem cells of skeletal muscle. They express Pax7 and are normally quiescent in unstimulated adult skeletal muscle. With a growth or injury stimulus, satellite cells become activated, which is marked by MyoD expression. Activated satellite cells then proliferate and either self-renew or differentiate. Satellite cells are not essential for hypertrophy after short-term overload but are essential for the long-term maintenance of a hypertrophied muscle, and for regenerating a skeletal muscle after injury.

REVIEW QUESTIONS

1   What is known about the effects of resistance exercise on protein synthesis?
2   Discuss the evidence for the hypothesis that mTOR is a key mediator of the protein synthesis response to resistance exercise and other forms of skeletal muscle overload.
3   How does mTOR regulate protein synthesis?
4   Discuss the evidence that myostatin is a key regulator of the muscle growth adaptation to resistance exercise.
5   What are satellite cells?
6   Compare and contrast the function of Pax7 and MyoD in relation to satellite cells.
7   What is the 'myonuclear domain hypothesis' and has it been experimentally confirmed?
8   Are satellite cells required for skeletal muscle hypertrophy and regeneration after muscle injury?

FURTHER READING

Egan B and Zierath JR (2013). Exercise metabolism and the molecular regulation of skeletal muscle adaptation. *Cell Metab* 17, 162–184.

REFERENCES

ACSM (2009). American College of Sports Medicine position stand. Progression models in resistance training for healthy adults. *Med Sci Sports Exerc* 41, 687–708.

Allen DL, Roy RR, and Edgerton VR (1999). Myonuclear domains in muscle adaptation and disease. *Muscle Nerve* 22, 1350–1360.

Amthor H, Otto A, Vulin A, Rochat A, Dumonceaux J, Garcia L, et al. (2009). Muscle hypertrophy driven by myostatin blockade does not require stem/precursor-cell activity. *Proc Natl Acad Sci U S A* 106, 7479–7484.

Arnold HH and Braun T (1996). Targeted inactivation of myogenic factor genes reveals their role during mouse myogenesis: a review. *Int J Dev Biol* 40, 345–353.

Baar K and Esser K (1999). Phosphorylation of p70(S6k) correlates with increased skeletal muscle mass following resistance exercise. *Am J Physiol* 276, C120–C127.

Bentzinger CF, von Maltzahn J and Rudnicki MA (2010). Extrinsic regulation of satellite cell specification. *Stem Cell Res Ther* 1, 27.

Blaauw B, Canato M, Agatea L, Toniolo L, Mammucari C, Masiero E, et al. (2009). Inducible activation of Akt increases skeletal muscle mass and force without satellite cell activation. *FASEB J* 23, 3896–3905.

Bodine SC, Stitt TN, Gonzalez M, Kline WO, Stover GL, Bauerlein R, et al. (2001). Akt/mTOR pathway is a crucial regulator of skeletal muscle hypertrophy and can prevent muscle atrophy in vivo. *Nat Cell Biol* 3, 1014–1019.

Bohe J, Low JF, Wolfe RR and Rennie MJ (2001). Latency and duration of stimulation of human muscle protein synthesis during continuous infusion of amino acids. *J Physiol* 532, 575–579.

Cao Y, Yao Z, Sarkar D, Lawrence M, Sanchez GJ, Parker MH, et al. (2010). Genome-wide MyoD binding in skeletal muscle cells: a potential for broad cellular reprogramming. *Dev Cell* 18, 662–674.

Carpinelli RN (2009). Challenging the American College of Sports Medicine 2009 position stand on resistance training. *Medicina Sportiva* 13, 131–137.

Carpinelli RF, Otto RM and Winett RA (2004). A critical analysis of the ACSM position stand on resistancetraining: Insufficient evidence to support recommended training protocols. *J Exerc Physiol* 7, 1–60.

Carroll CC, Carrithers JA and Trappe TA (2004). Contractile protein concentrations in human single muscle fibers. *J Muscle Res Cell Motil* 25, 55–59.

Coleman ME, DeMayo F, Yin KC, Lee HM, Geske R, Montgomery C, et al. (1995). Myogenic vector expression of insulin-like growth factor I stimulates muscle cell differentiation and myofiber hypertrophy in transgenic mice. *J Biol Chem* 270, 12109–12116.

Collins CA, Olsen I, Zammit PS, Heslop L, Petrie A, Partridge TA, et al. (2005). Stem cell function, self-renewal, and behavioral heterogeneity of cells from the adult muscle satellite cell niche. *Cell* 122, 289–301.

Conboy IM and Rando TA (2002). The regulation of Notch signaling controls satellite cell activation and cell fate determination in postnatal myogenesis. *Dev Cell* 3, 397–409.

Davies SP, Reddy H, Caivano M and Cohen P (2000). Specificity and mechanism of action of some commonly used protein kinase inhibitors. *Biochem J* 351, 95–105.

Davis RL, Weintraub H and Lassar AB (1987). Expression of a single transfected cDNA converts fibroblasts to myoblasts. *Cell* 51, 987–1000.

Drummond MJ, Fry CS, Glynn EL, Dreyer HC, Dhanani S, Timmerman KL, et al. (2009). Rapamycin administration in humans blocks the contraction-induced increase in skeletal muscle protein synthesis. *J Physiol* 587, 1535–1546.

Edmondson DG and Olson EN (1989). A gene with homology to the myc similarity region of MyoD1 is expressed during myogenesis and is sufficient to activate the muscle differentiation program. *Genes Dev* 3, 628–640.

Ferrari G, Cusella-De Angelis G, Coletta M, Paolucci E, Stornaiuolo A, Cossu G, et al. (1998). Muscle regeneration by bone marrow-derived myogenic progenitors. *Science* 279, 1528–1530.

Forsberg AM, Nilsson E, Werneman J, Bergstrom J and Hultman E (1991). Muscle composition in relation to age and sex. *Clin Sci (Lond)* 81, 249–256.

Goodman CA, Miu MH, Frey JW, Mabrey DM, Lincoln HC, Ge Y, et al. (2010). A phosphatidylinositol 3-kinase/protein kinase B-independent activation of mammalian target of rapamycin signaling is sufficient to induce skeletal muscle hypertrophy. *Mol Biol Cell* 21, 3258–3268.

Goodman CA, Frey JW, Mabrey DM, Jacobs BL, Lincoln HC, You JS, et al. (2011). The role of skeletal muscle mTOR in the regulation of mechanical load-induced growth. *J Physiol* 589, 5485–5501.

Heron MI and Richmond FJ (1993). In-series fiber architecture in long human muscles. *J Morphol* 216, 35–45.

Holz MK, Ballif BA, Gygi SP and Blenis J (2005). mTOR and S6K1 mediate assembly of the translation preinitiation complex through dynamic protein interchange and ordered phosphorylation events. *Cell* 123, 569–580.

Hornberger TA (2011). Mechanotransduction and the regulation of mTORC1 signaling in skeletal muscle. *Int J Biochem Cell Biol* 43, 1267–1276.

Hubal MJ, Gordish-Dressman H, Thompson PD, Price TB, Hoffman EP, Angelopoulos TJ, et al. (2005). Variability in muscle size and strength gain after unilateral resistance training. *Med Sci Sports Exerc* 37, 964–972.

Inoki K, Zhu T and Guan KL (2003). TSC2 mediates cellular energy response to control cell growth and survival. *Cell* 115, 577–590.

Janssen I, Heymsfield SB, Wang ZM and Ross R (2000). Skeletal muscle mass and distribution in 468 men and women aged 18–88 yr. *J Appl Physiol* 89, 81–88.

Judson RN, Tremblay AM, Knopp P, White RB, Urcia R, De BC, et al. (2012). The Hippo pathway member Yap plays a key role in influencing fate decisions in muscle satellite cells. *J Cell Sci* 125, 6009–6019.

Kadi F, Charifi N, Denis C, Lexell J, Andersen JL, Schjerling P, et al. (2005). The behaviour of satellite cells in response to exercise: what have we learned from human studies? *Pflugers Arch* 451, 319–327.

Katz B (1961). The terminations of the afferent nerve fibre in the muscle spindle of the frog. *Philos Trans R Soc Lond B Biol Sci* 243, 221–240.

Kelley G (1996). Mechanical overload and skeletal muscle fiber hyperplasia: a meta-analysis. *J Appl Physiol* 81, 1584–1588.

Kim E and Guan KL (2009). RAG GTPases in nutrient-mediated TOR signaling pathway. *Cell Cycle* 8, 1014–1018.

Kim PL, Staron RS and Phillips SM (2005). Fasted-state skeletal muscle protein synthesis after resistance exercise is altered with training. *J Physiol* 568, 283–290.

Kim JS, Petrella JK, Cross JM and Bamman MM (2007). Load-mediated downregulation of myostatin mRNA is not sufficient to promote myofiber hypertrophy in humans: a cluster analysis. *J Appl Physiol* 103, 1488–1495.

Kraemer WJ, Adams K, Cafarelli E, Dudley GA, Dooly C, Feigenbaum MS, et al. (2002). American College of Sports Medicine position stand. Progression models in resistance training for healthy adults. *Med Sci Sports Exerc* 34, 364–380.

Lai KM, Gonzalez M, Poueymirou WT, Kline WO, Na E, Zlotchenko E, et al. (2004). Conditional activation of akt in adult skeletal muscle induces rapid hypertrophy. *Mol Cell Biol* 24, 9295–9304.

Lassar AB, Paterson BM and Weintraub H (1986). Transfection of a DNA locus that mediates the conversion of 10T1/2 fibroblasts to myoblasts. *Cell* 47, 649–656.

Lee SJ (2004). Regulation of muscle mass by myostatin. *Annu Rev Cell Dev Biol* 20, 61–86.

Lee SJ (2007). Quadrupling muscle mass in mice by targeting TGF-beta signaling pathways. *PLoS ONE* 2, e789.

Lee SJ (2010). Extracellular regulation of myostatin: a molecular rheostat for muscle mass. *Immunol Endocr Metab Agents Med Chem* 10, 183–194.

Lepper C, Partridge TA and Fan CM (2011). An absolute requirement for Pax7-positive satellite cells in acute injury-induced skeletal muscle regeneration. *Development* 138, 3639–3646.

Lexell J, Taylor CC and Sjostrom M (1988). What is the cause of the ageing atrophy? Total number, size and proportion of different fiber types studied in whole vastus lateralis muscle from 15- to 83-year-old men. *J Neurol Sci* 84, 275–294.

Louis E, Raue U, Yang Y, Jemiolo B and Trappe S (2007). Time course of proteolytic, cytokine, and myostatin gene expression after acute exercise in human skeletal muscle. *J Appl Physiol* 103, 1744–1751.

Mauro A (1961). Satellite cell of skeletal muscle fibers. *J Biophys Biochem Cytol* 9, 493–495.

McCarthy JJ, Mula J, Miyazaki M, Erfani R, Garrison K, Farooqui AB, et al. (2011). Effective fiber hypertrophy in satellite cell-depleted skeletal muscle. *Development* 138, 3657–3666.

McCroskery S, Thomas M, Maxwell L, Sharma M and Kambadur R (2003). Myostatin negatively regulates satellite cell activation and self-renewal. *J Cell Biol* 162, 1135–1147.

McPherron AC, Lawler AM and Lee SJ (1997). Regulation of skeletal muscle mass in mice by a new TGF-beta superfamily member. *Nature* 387, 83–90.

Mendias CL, Marcin JE, Calderon DR and Faulkner JA (2006). Contractile properties of EDL and soleus muscles of myostatin-deficient mice. *J Appl Physiol* 101, 898–905.

Miller BF, Olesen JL, Hansen M, Dossing S, Crameri RM, Welling RJ, et al. (2005). Coordinated collagen and muscle protein synthesis in human patella tendon and quadriceps muscle after exercise. *J Physiol* 567, 1021–1033.

Mosher DS, Quignon P, Bustamante CD, Sutter NB, Mellersh CS, Parker HG, et al. (2007). A mutation in the myostatin gene increases muscle mass and enhances racing performance in heterozygote dogs. *PLoS Genet* 3, e79.

Mourikis P, Sambasivan R, Castel D, Rocheteau P, Bizzarro V and Tajbakhsh S (2012). A critical requirement for notch signaling in maintenance of the quiescent skeletal muscle stem cell state. *Stem Cells* 30, 243–252.

O'Connor RS and Pavlath GK (2007). Point: Counterpoint: Satellite cell addition is/is not obligatory for skeletal muscle hypertrophy. *J Appl Physiol* 103, 1099–1100.

Ono Y, Calhabeu F, Morgan JE, Katagiri T, Amthor H and Zammit PS (2011). BMP signalling permits population expansion by preventing premature myogenic differentiation in muscle satellite cells. *Cell Death Differ* 18, 222–234.

Otto A, Schmidt C, Luke G, Allen S, Valasek P, Muntoni F, et al. (2008). Canonical Wnt signalling induces satellite-cell proliferation during adult skeletal muscle regeneration. *J Cell Sci* 121, 2939–2950.

Otto A, Collins-Hooper H, Patel A, Dash PR and Patel K (2011). Adult skeletal muscle stem cell migration is mediated by a blebbing/amoeboid mechanism. *Rejuvenation Res* 14, 249–260.

Pallafacchina G, Calabria E, Serrano AL, Kalhovde JM and Schiaffino S (2002). A protein kinase B-dependent and rapamycin-sensitive pathway controls skeletal muscle growth but not fiber type specification. *Proc Natl Acad Sci U S A* 99, 9213–9218.

Phillips SM, Tipton KD, Aarsland A, Wolf SE and Wolfe RR (1997). Mixed muscle protein synthesis and breakdown after resistance exercise in humans. *Am J Physiol* 273, E99–107.

Philp A, Hamilton DL and Baar K (2011). Signals mediating skeletal muscle remodeling by resistance exercise: PI3-kinase independent activation of mTORC1. *J Appl Physiol* 110, 561–568.

Rosenblatt JD and Parry DJ (1992). Gamma irradiation prevents compensatory hypertrophy of overloaded mouse extensor digitorum longus muscle. *J Appl Physiol* 73, 2538–2543.

Sacco A, Doyonnas R, Kraft P, Vitorovic S and Blau HM (2008). Self-renewal and expansion of single transplanted muscle stem cells. *Nature* 456, 502–506.

Sale DG (1988). Neural adaptation to resistance training. *Med Sci Sports Exerc* 20, S135–S145.

Scharner J and Zammit PS (2011). The muscle satellite cell at 50: the formative years. *Skelet Muscle* 1, 28.

Schuelke M, Wagner KR, Stolz LE, Hubner C, Riebel T, Komen W, et al. (2004). Myostatin mutation associated with gross muscle hypertrophy in a child. *N Engl J Med* 350, 2682–2688.

Seale P, Sabourin LA, Girgis-Gabardo A, Mansouri A, Gruss P and Rudnicki MA (2000). Pax7 is required for the specification of myogenic satellite cells. *Cell* 102, 777–786.

Sherwood RI, Christensen JL, Conboy IM, Conboy MJ, Rando TA, Weissman IL, et al. (2004). Isolation of adult mouse myogenic progenitors: functional heterogeneity of cells within and engrafting skeletal muscle. *Cell* 119, 543–554.

Spangenburg EE, Le RD, Ward CW and Bodine SC (2008). A functional insulin-like growth factor receptor is not necessary for load-induced skeletal muscle hypertrophy. *J Physiol* 586, 283–291.

Studitsky AN (1964). Free auto- and homografts of muscle tissue in experiments on animals. *Ann N Y Acad Sci* 120, 789–801.

Sutrave P, Kelly AM and Hughes SH (1990). ski can cause selective growth of skeletal muscle in transgenic mice. *Genes Dev* 4, 1462–1472.

Tipton KD, Ferrando AA, Phillips SM, Doyle D, Jr. and Wolfe RR (1999). Postexercise net protein synthesis in human muscle from orally administered amino acids. *Am J Physiol* 276, E628–E634.

Tseng BS, Kasper CE and Edgerton VR (1994). Cytoplasm-to-myonucleus ratios and succinate dehydrogenase activities in adult rat slow and fast muscle fibers. *Cell Tissue Res* 275, 39–49.

Van CM, Duchateau J and Hainaut K (1998). Changes in single motor unit behaviour contribute to the increase in contraction speed after dynamic training in humans. *J Physiol* 513 Pt 1, 295–305.

Weintraub H, Tapscott SJ, Davis RL, Thayer MJ, Adam MA, Lassar AB, et al. (1989). Activation of muscle-specific genes in pigment, nerve, fat, liver, and fibroblast cell lines by forced expression of MyoD. *Proc Natl Acad Sci U S A* 86, 5434–5438.

Welle S, Bhatt K and Pinkert CA (2006). Myofibrillar protein synthesis in myostatin-deficient mice. *Am J Physiol Endocrinol Metab* 290, E409–E415.

Welle S, Bhatt K, Pinkert CA, Tawil R and Thornton CA (2007). Muscle growth after postdevelopmental myostatin gene knockout. *Am J Physiol Endocrinol Metab* 292, E985–E991.

Welle S, Mehta S and Burgess K (2011). Effect of postdevelopmental myostatin depletion on myofibrillar protein metabolism. *Am J Physiol Endocrinol Metab* 300, E993–E1001.

Whittemore LA, Song K, Li X, Aghajanian J, Davies M, Girgenrath S, et al. (2003). Inhibition of myostatin in adult mice increases skeletal muscle mass and strength. *Biochem Biophys Res Commun* 300, 965–971.

Winbanks CE, Weeks KL, Thomson RE, Sepulveda PV, Beyer C, Qian H, et al. (2012). Follistatin-mediated skeletal muscle hypertrophy is regulated by Smad3 and mTOR independently of myostatin. *J Cell Biol* 197, 997–1008.

Zammit PS, Golding JP, Nagata Y, Hudon V, Partridge TA and Beauchamp JR (2004). Muscle satellite cells adopt divergent fates: a mechanism for self-renewal? *J Cell Biol* 166, 347–357.

At the end of the chapter you should be able to:

- Explain why muscle mass and strength are considered to be polygenic traits.

- Discuss the *ACTN3* R577X genotype and other polymorphisms that have been linked to muscle mass, strength and related traits.

- Describe how a rare mutation in the myostatin gene was linked to increased muscle mass.

- List and explain mutations in transgenic mice that affect muscle mass.

- Explain how selective breeding and the analysis of inbred mouse strains have informed us about the genetics of muscle mass and strength.

7 Genetics, muscle mass and strength

Arimantas Lionikas and
Henning Wackerhage

INTRODUCTION

Before reading this chapter you should have read Chapter 2, which introduces sport and exercise genetics, and Chapter 6, which covers the adaptation to resistance training. This adaptation is closely related to the genetics of muscle mass and strength.

Muscle mass and strength are influenced by many limiting factors that depend on DNA sequence variations and on environmental factors such as resistance training. We will start this chapter by exploring the heritability of muscle mass and strength and list heritability obtained in twin and family studies. After that we will discuss DNA sequence variations which are common in the population, termed **polymorphisms**, that have been linked to muscle mass and strength. As a specific example we will discuss the *ACTN3* R577X polymorphism, which is associated with speed and power performance and then summarize the results of other studies where muscle mass and strength-related polymorphisms have been identified. These studies confirm the existence of polymorphisms that affect muscle mass and strength, but these polymorphisms typically only explain a small proportion of the variation of muscle mass and strength. It is currently unclear whether they are only the tip of the iceberg and whether there are many other polymorphisms that affect muscle mass and strength or whether there is a significant contribution of rare DNA sequence variations. One rare human DNA sequence variation with a large effect size, especially on muscle mass, has been reported so far. It is a DNA sequence variation in the first intron of the myostatin gene and a homozygous carrier had approximately a doubled muscle mass in line with similar genotypes in animal species. Finally, we will review transgenic mouse models in which researchers have identified candidate genes where a loss or gain of function affects muscle mass and/or strength. Many of these are linked to mTOR and myostatin–Smad signalling, which both have major effects on muscle mass and strength. In the last part of this chapter we discuss muscle mass and strength in the context of selective breeding for muscle mass and inbred mouse strains.

MUSCLE MASS AND STRENGTH: A POLYGENIC TRAIT

Muscle mass and strength depend, like most other sport and exercise-related traits, on many limiting factors or quantitative traits (summarized in Figure 7.1). The variations of these factors in animal and human populations are explained by both environmental factors, such as training and nutrition, and by DNA sequence variation or genetics. The DNA sequence variations that affect muscle strength can be common alleles with relatively small effect sizes, such as the *ACTN3* R577X genotype (Yang et al., 2003), or rare alleles with sometimes large effect sizes, such as a natural occurring mutation of the myostatin gene (Schuelke et al., 2004). Because muscle mass and strength depend, as we will show in this chapter, on DNA sequence variations in many genes they are considered to be **polygenic traits**. Furthermore the increase of muscle mass and strength in response to a resistance training programme or trainability also varies greatly (Hubal et al., 2005). In contrast to $\dot{V}O_{2max}$ trainability, the heritability of mass and strength trainability has not yet been quantified, but it seems likely that muscle mass and strength trainability are significantly inherited as is the case for endurance trainability (see Chapter 5). The factors that limit strength and muscle mass and their dependency on DNA sequence variation, training and nutrition are illustrated in Figure 7.1.

Heritability of muscle strength and mass

How do we know that muscle mass and strength are inherited? The evidence of the role of genetic factors comes again from twin and family studies, which were introduced in Chapter 2. A popular way of estimating an individual's overall strength is a hand grip test, which is easy to perform and which correlates strongly with arm flexor and leg or knee extensor strength (Silventoinen et al., 2008). In an extensive study Silventoinen et al. (2008) measured hand grip strength in over one million 16- to 25-year-old male subjects and found that it varied between 50 and 999 N. The hand grip strength of 100 of those subjects even exceeded the limit of the dynamometer used. Thus the strongest men were ≈20 times stronger than the weakest. Because the researchers included a large number of monozygous and dizygous twins, they were able to estimate the heritability of grip

Figure 7.1 Muscle strength and mass depend on many limiting factors. Almost all these limiting factors are influenced by DNA sequence variations or quantitative trait loci (QTLs) and environmental factors, of which training and nutrition are the most important ones. n, number; CSA, cross-sectional area (a measure for muscle fibre size); QTL, quantitative trait locus; MPS, muscle protein synthesis; MPB, muscle protein breakdown.

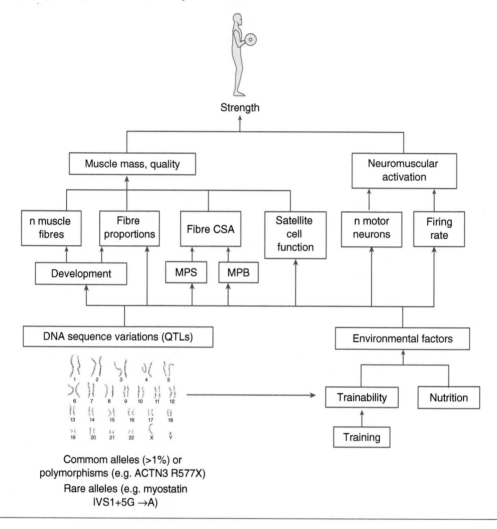

strength. The correlations coefficient r for hand grip strength for the monozygous twins was 0.66 and the one for the dizygous twins was 0.35. If Falconer's formula, which we discussed in Chapter 2, is used, then the heritability h^2 for grip strength can be calculated as $2 \times (r_{MZ} - r_{DZ}) = 0.62$ or 62%. This suggests that grip strength is significantly inherited.

In general, the estimated strength heritabilities vary a lot depending on the type of strength measured and the volunteers investigated. Twin studies generally suffer from low subject numbers and assumptions about the environment and thus the values should only be interpreted as a rough indication. An overview over the heritability of static, dynamic and explosive strength is given in Table 7.1 (Peeters et al., 2009).

Table 7.1 Heritability estimates for static, dynamic and explosive strength		
	TWIN STUDIES	FAMILY/SIB-PAIR STUDIES*
Static strength	14–83% (20 studies)	27–58% (5 studies)
Dynamic strength (isokinetic, concentric and excentric)	29–90% (3 studies)	42–87% (concentric only; 2 studies)
Explosive strength or power (jump tests Wingate test)	34–97% (7 studies)	22–68% (1 study)

From Peeters et al. (2009).
For references also see Peeters et al. (2009). In this table the heritabilities are given as percentages rather than as fractions of '1' as in the original paper.

Table 7.1 shows large variations in the heritability estimates for muscle strength, which is probably mainly due to the limitations of twin and family studies. Nonetheless, the studies demonstrate that strength is, like most other sport and exercise-related traits, significantly inherited.

Muscle strength depends greatly on muscle mass, which cannot be measured directly in humans. However, it is possible to estimate muscle mass by measuring lean body mass. This is body weight minus fat and bone mass, which can be indirectly measured using a DXA scan. In men ≈38% and in women ≈31% of the body mass is skeletal muscle (Janssen et al., 2000) and thus muscle contributes roughly 50% to lean body mass in males and 40% in females, but this varies a lot depending on the body composition. The heritability estimates of lean mass are high, in the order of 60–80% (Hsu et al., 2005; Souren et al., 2007; Bogl et al., 2011). Muscle cross-sectional area estimated from the circumference of the thigh is also highly inherited, with an estimate of 91% (Huygens et al., 2004b). All these findings suggest that muscle mass, which is a major limiting factor for muscle strength, is highly inherited.

The mass of a given muscle depends on the number of muscle fibres and on the average size of the fibres within a muscle. Animal studies show that the number of fibres in the muscle is set during embryogenesis and generally changes little in the adult (Ontell et al., 1993). In both adult animals of one species and humans, fibre numbers for a given muscle can vary greatly between individuals. Lexell and colleagues investigated human cadavers and found that the muscle fibre number of the vastus lateralis varied in men aged 18–22 years between 393 000 and 903 000 fibres (Lexell et al., 1988). Thus, there can be an over 2-fold difference in the number of fibres between individuals in a given muscle, which explains partially why some untrained individuals have larger muscles than others. The heritability of fibre numbers per muscle is unknown, but given the large differences in untrained individuals and the small effect of training and nutrition on fibre numbers in adults, it strongly suggests that DNA sequence variation or heritability, probably together with nutrition in the uterus and during early development, are major causes.

Muscle strength depends on muscle mass, which is significantly inherited as we have just shown and on neuromuscular activation or the ability of the nervous system to maximally innervate existing muscles. Unfortunately, little is known about the heritability of neuromuscular activation as there are no twin or family studies that seek to address this question.

Thus taken together, muscle mass and strength both depend significantly on DNA sequence variations or, in other words, are heritable. Strength is probably around 50% inherited although the heritability estimates vary greatly. Lean body mass and muscle mass are ≈60–90% inherited and the heritability of neuromuscular activation is unknown. Given that muscle mass and strength are significantly inherited we will now review common and rare alleles or DNA sequence variations that are responsible for the heritability of muscle mass and strength.

ASSOCIATION STUDIES AND THE *ACTN3* R577X POLYMORPHISM

Many association studies have been carried out to try to identify common DNA sequence variations or alleles that affect muscle mass, strength and related variables. Perhaps the most discussed polymorphism is the *ACTN3* R577X genotype and for this reason we will now review the research on this polymorphism in depth before summarizing the results of other association studies.

The α-actinins, abbreviated ACTN, are actin-binding proteins located in the Z discs of sarcomeres. There are two ACTN isoforms, of which ACTN2 is expressed in all muscle fibres whereas ACTN3 is only expressed in fast type II muscle fibres. Initially an ACTN3 deficiency was found in patients with myopathies, but then a team led by Kathryn North demonstrated that the absence of ACTN3 is common in individuals with no muscle disease and caused by a so-called *ACTN3* R577X polymorphism (North et al., 1999). Using PCR and DNA sequencing they identified a common C→T single-nucleotide polymorphism (SNP) in exon 16 of the *ACTN3* gene. This one nucleotide difference in the DNA sequence greatly changes the ACTN3 protein because as a consequence amino acid 577, which is normally an arginine (abbreviated as R), is changed to a stop codon, abbreviated as X. Hence the polymorphism is abbreviated as *ACTN3* R577X and refers to the ACTN3 protein.

The consequence of the premature stop codon in position 577 is that a shortened, non-functional version of ACTN3 is produced, which is degraded. Thus the *ACTN3* 577X allele is comparable to a gene knockout.

Because DNA sequencing is expensive and time consuming, the researchers developed a simplified PCR assay, which involves amplifying exon 16 of the *ACTN3* gene using PCR followed by digestion of the PCR product with the restriction enzyme *Dde*I. This enzyme cuts the 577R allele into fragments 205 and 85 base pairs long, whereas the 577X allele is cut into fragments 108, 97 and 86 base pairs long. When the digested PCR product is then electrophoresed on an agarose gel, two bands indicate a homozygous 577R carrier, three bands a homozygous 577X carrier who has no functional *ACTN3*, and a combination of the 205, 97, 86 and 85 bands a heterozygous *ACTN3* R577X carrier (Mills et al., 2001). Using this PCR assay the researchers tested the DNA of individuals from many continents. They found that the frequency of those who were homozygous for the 577X knockout allele was <1% in an African Bantu population while it was ≈18% in Europeans (Yang et al., 2003).

These observations raise some puzzling questions. First, why did the 577X allele accumulate in some populations? Second, why is there no obvious phenotype in individuals who are *ACTN3* knockouts due to being XX carriers? One explanation for the lack of dramatic phenotype is that *ACTN2*, which is expressed in all muscle fibres, might compensate for the loss of *ACTN3* in fast type II fibres. For this reason the 577X allele might be neutral or even slightly beneficial, allowing it to become enriched during human evolution at least in some populations. The research team then reasoned that as *ACTN3* is expressed in fast fibres the *ACTN3* R577X polymorphism might be associated with athletic performance. To investigate this they obtained the genomic DNA from 107 power athletes from sports such as track, swimming, track cycling, judo, speed skating versus 436 controls versus 194 endurance athletes who included long-distance swimmers, cyclists, runners, rowers and cross-country skiers. They even included 50 athletes who had competed in the Olympics. The researchers then performed the *ACTN3* R577X genotyping assay and found that the distribution of the *ACTN3* R577X genotypes of the athletes differed between the different groups, as shown in Figure 7.2.

Figure 7.2 *ACTN3* R577X and performance. (a) Example of result from an *ACTN3* 577R/X test (Yang et al., 2003). The 290 bp PCR product, which has been amplified by PCR, has been cut with *Dde*I, either once into 205 bp and 85 bp fragments (*ACTN3* 577RR) or twice into 108 bp, 97 bp and 86 bp fragments (*ACTN3* 577XX). Note that it is impossible to distinguish the 85 and 86 bp fragments. *Indicates primer-dimers which are an artefact. (b) Association of *ACTN3* R577X genotype and athletic status. RR represents individuals that have the normal arginine in position 577 of the *ACTN3* gene, XX individuals that have a premature stop codon in this position and RX represents heterozygous individuals. Redrawn after Yang et al. (2003). The key finding is that very few power athletes have an *ACTN3* 577XX genotype, suggesting that such a genotype is detrimental for muscle power.

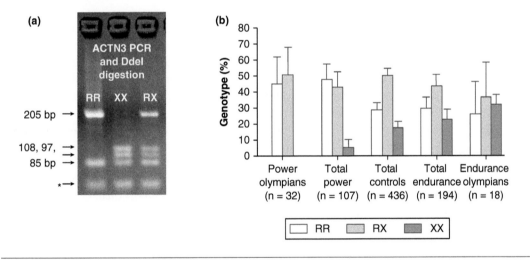

Among 32 power Olympians there was no subject with a XX knockout genotype, although if the distribution was as in the controls one would expect ≈6 individuals with a XX genotype in this cohort. In contrast, the XX genotype was higher in endurance athletes than in the controls, suggesting that it might be beneficial for endurance performance.

So is the *ACTN3* R577X genetic test useful to identify especially future elite power athletes? The test has been marketed as such a talent identification test and is commercially available. The answer is no. The reasons are as follows: First, ≈80% of people will have an RR or RX genotype, which indicates that they carry a genotype that is typical for both power and endurance Olympians, albeit at slightly different frequencies. Only ≈18% in the population will be XX *ACTN3* knockout carriers. These XX carriers might have a tiny advantage in endurance sports and it is likely that they may miss one of the many factors that are required to be a speed or power Olympian. Thus the test is at best an exclusion test for ≈18% of the population. We see no major ethical issues with performing the test unless it is carried out using the DNA of minors or embryos (see Chapter 2).

Follow-up studies have generally confirmed the high frequency of RR carriers in speed and power athletes, especially in Europeans (Alfred et al., 2011). More mechanistic studies showed that *ACTN3* knockout mice displayed a fast-to-slow fibre type shift consistent with decreased power performance (MacArthur et al., 2008). In an association studies in humans it was shown that XX carriers had significantly 5% less type IIx fibres than RR carriers, again consistent with the hypothesis that the XX genotype reduces power performance (Vincent et al., 2007). To conclude, the *ACTN3* R577X genotype is associated with sport performance, and individuals with an XX genotype and the resulting *ACTN3* knockout are unlikely to become elite power athletes. Although the *ACTN3* R577X phenotype is associated with power performance, the effect is small and ≈80% of people will have an RR or RX genotype which is consistent with both elite power and endurance performance.

Many other association studies have been performed in order to test whether specific candidate genetic variants are associated with muscle mass or strength. The results of several of these studies are shown in Table 7.2, which demonstrates that there are common DNA sequence variations or polymorphisms in the human population that explain a fraction of the variability of muscle strength and mass and related traits seen in humans. Thus a key question is whether the so far unexplained heritability of muscle strength and mass is due to undiscovered polymorphisms or whether rare DNA sequence variations play an important role. The answer to this question awaits the large-scale application of whole-genome sequencing to sport and exercise-related questions (Wheeler et al., 2008). The fact that the polymorphisms listed in Table 7.2 only explain a small fraction of the heritability of strength or mass also implies that, for example, a single polymorphism test such as the vitamin D receptor poly-A repeat test will yield little useful information about the muscle strength of the individual. Given that polymorphisms have an additive effect, one possible strategy might be to measure a panel of muscle strength and mass-influencing polymorphisms and to calculate a genetic score, which may give a better indicator of an individual's genetic muscle strength and mass potential.

It has also been attempted to calculate the likelihood of having the perfect genetic score for muscle mass and strength based on the allele frequencies in published studies. The authors found that the likelihood of carrying all or even many of the strength and muscle mass-affecting alleles is extremely small (Hughes et al., 2011).

Genome-wide SNP analyses have also been conducted for muscle strength. An experimental multigene approach was used in several studies by a Belgian team. The team first performed a linkage analysis for 367 male siblings (Huygens et al., 2005) and after that in 283 male siblings a genome-wide SNP-based multipoint linkage analysis (De Mars et al., 2008). In the first study they found linkage for some genes in the myostatin pathway, albeit not myostatin itself, with muscle strength (Huygens et al., 2005). In the second study they found several peaks of suggestive linkage and significant linkage on chromosome 14q24.3 (Figure 7.3).

Table 7.2 Polymorphisms or common DNA sequence variations that have been associated with muscle mass and strength-related traits in humans

GENE	TRAIT AFFECTED AND STUDY DESIGN	NUMBER OF SUBJECTS	REFERENCE
IGF-1 repeat promoter polymorphism	Strength trainability, association study	67	Kostek et al., 2005
IGF2 'Apal' SNP	Hand grip strength, association study	693	Sayer et al., 2002
IGF2 'Apal' SNP	Fat-free mass, strength and sustained power; association study	579	Schrager et al., 2004
IL15RA SNPs	Muscle mass trainability, association study	153	Riechman et al., 2004
Vitamin D receptor poly-A repeat	Muscle strength, association study	175	Grundberg et al., 2004
Vitamin D receptor SNPs	Muscle strength, association study	109	Wang et al., 2006
TNFα promoter SNPs	Muscle mass, association study	1050	Liu et al., 2008
Myostatin K153R	Muscle power, association study	214	Santiago et al., 2011
CNTF SNPs	Muscle strength	494	Roth et al., 2001
ACTN3 R577X	Muscle power trainability, association study	157	Delmonico et al., 2007
Myostatin pathway genes	Strength (QTL mapping)	329	Huygens et al., 2004a
Activin receptor 1B (*ACVR1B*)	Strength (QTL mapping study)	500, 266	Windelinckx et al., 2011

Figure 7.3 Genetics of muscle mass. The top image shows an MRI scan of human legs at mid thigh level. The thigh muscles are seen in grey. Below is the result of a genome-wide linkage scan for QTLs that affect the cross-sectional area of the thigh. An LOD score of 2.2 (dotted line) suggests linkage; an LOD score of 3.3 is considered to be significant linkage. Redrawn after De Mars et al. (2008).

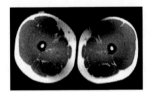

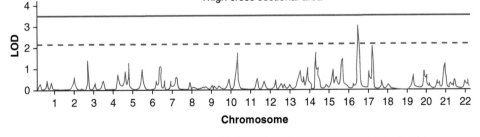

The researchers currently use linkage analysis to fine map the regions identified in their previous studies and so far have identified the activin receptor 1B (*ACVR1B*) as a muscle strength gene (Windelinckx et al., 2011). This gene is related to the myostatin pathway, which we have already discussed in Chapter 6 as a potential regulator of the adaptation to resistance exercise.

A MYOSTATIN KNOCKOUT MUTATION AS AN EXAMPLE OF A RARE DNA SEQUENCE VARIATION WITH A LARGE EFFECT SIZE

We have just reviewed common DNA sequence variations or polymorphisms that are associated with muscle mass and strength. However, such polymorphisms do only explain a fraction of the heritability, for example, of the ≈20-times difference in hand grip strength seen in the human population (Silventoinen et al., 2008). Performing genome-wide association study (GWAS) experiments in large groups that are well phenotyped for muscle mass and strength will help to uncover more polymorphisms that explain the inherited variability of muscle mass and strength.

In addition, there are probably rare DNA sequence variations which are only present in individuals or families that affect muscle mass and strength. As we have stated several times before, the whole-genome sequencing of large cohorts will eventually allow us to determine how important common and rare DNA sequence variations are for muscle mass and strength (Wheeler et al., 2008). In this section we will discuss in detail one rare DNA sequence variation that has a large effect size on muscle mass. This proof-of-principle example is a knockout mutation in the myostatin gene, which has been associated with the a double muscling phenotype in humans. We will discuss this example in detail and then cover transgenic mouse models where the knock-in or knockout of a gene has an effect on muscle strength or mass.

We have already discussed myostatin as a major regulator of muscle mass in Chapter 6. Here, we discuss the case of a boy who was born to a former athlete. Stimulus-induced involuntary twitching was observed after birth, triggering further investigations. During these it was noted that the boy 'appeared extraordinarily muscular, with protruding muscles in his thighs and upper arms' (Schuelke et al., 2004). Ultrasonography showed that the quadriceps muscle mass was 7.2 standard deviations above the mean of age- and sex-matched controls. Anecdotal evidence about unusually strong family members emerged and the research team decided to test whether the mutation was due to a knockout mutation in the myostatin gene, as such mutations could lead to double muscling in mice and other species (McPherron et al., 1997; McPherron and Lee, 1997). Thus a mutation of the myostatin gene was a candidate cause for the phenotype and the researchers started to test this hypothesis.

The myostatin gene comprises three exons and the researchers designed primers to amplify all exons and the introns between, followed by DNA sequencing of the PCR products. The researchers found no unusual DNA sequence in the exons but noted a DNA sequence variation in the first intron where a G was mutated to an A, abbreviated as **IVS1+5 G→A**. This indicates a DNA sequence variation in intron 1 (IVS stands for intervening sequence which means the same as 'intron'), where after 5 base pairs a guanine (G) is changed to an alanine (A).

The researchers now had to demonstrate that the IVS1+5 G→A was a rare DNA sequence variation, otherwise other carriers should also have the unusual double muscling phenotype. First, they developed a simplified genotyping assay in order to test for the IVS1+5 G→A variation in a larger cohort. They developed a restriction fragment length polymorphism (RFLP) PCR assay which involves, similar to that used for the *ACTN3* R577X genotype, a PCR reaction to amplify a 166 bp PCR product followed by digestion of the PCR product with a restriction enzyme *Acc*I into fragments of 135 base pairs and 31 base pairs, while the mutated IVS1+5 G→A fragment of the boy was left uncut. After running the digested PCR product on an agarose gel the rare, uncut IVS1+5 A allele appears as a 166 bp band, whereas the common IVS1+5 G allele appears as a 135 bp band (Figure 7.4; note that the common 31 bp band is not

Figure 7.4 Myostatin toddler. (a) Gross leg phenotype of a toddler showing a pronounced skeletal muscle hypertrophy. (b) The researchers hypothesized DNA variations in the myostatin gene as a cause and identified an IVS1+5 G→A mutation in the myostatin gene in the patient using PCR and Sanger sequencing. Schematic result of a IVS1+5 G→A PCR assay. After PCR amplification of the part of the myostatin gene with the DNA sequence variation, the restriction enzyme *Acc*I is used to cut the wildtype PCR product into 135 bp and 31 bp pieces, while the mutated DNA remains an uncut and is 166 bp. This schematic gel shows that the patient is homozygous for the mutation, the mother heterozygous and the control is homozygous for the normal DNA sequence. (c) Schematic drawing showing the analysis of the serum of the patient and of rat and human control serum with a JA16 anti-myostatin antibody which detects several forms of myostatin. The key band is the myostatin propeptide, which is absent in the toddler. Redrawn after Schuelke et al. (2004).

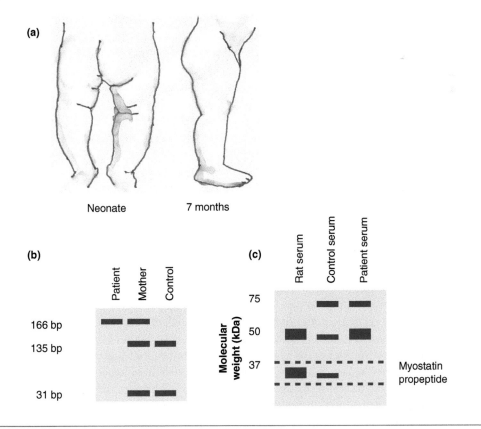

shown). Using this assay the authors demonstrated that the boy was homozygous for this mutation, while the mother was heterozygous. They also showed that all controls were homozygous for the common IVS1+5 G allele, which is in line with the hypothesis that the IVS1+5 A allele was a rare DNA sequence variation.

At this stage it still was unknown whether the IVS1+5 A allele had an effect on the serum concentration of myostatin. To study this the authors obtained serum from animals and patients who are known to produce normal myostatin, performed an immunoprecipitation to concentrate myostatin and then performed a Western blot to detect the different forms of myostatin to which the antibody binds. The Western blot demonstrated that the myostatin propeptide is absent in the boy while it is present in other individuals. This was strong, albeit not perfect, evidence that the IVS1+5 G→A mutation resulted in no myostatin propeptide as a likely cause for the double muscling phenotype.

This research identified the first rare DNA sequence variation with a large effect on muscle mass and strength and reemphasizes the important role of myostatin. This fits in well with the literature showing that a knockout of the myostatin gene results in skeletal muscle hypertrophy in mice (McPherron et al., 1997). Similarly a naturally occurring knockout mutation of the myostatin gene has been associated with a large muscle mass in Piedmontese and Belgian Blue cattle breeds (McPherron and Lee, 1997). In two species the myostatin gene has been linked to athletic performance, suggesting that not only muscle mass depends on myostatin genotypes. First it has been shown that 'bully' whippet race dogs are heterozygous for a naturally occurring myostatin knockout mutation (Mosher et al., 2007). Second, researchers performed a GWAS in thoroughbred race horses and found that myostatin was a good predictor of race distance (Binns et al., 2010). All the studies in the myostatin field demonstrate that myostatin knockout mutations can occur naturally and that they can lead to a large increase in muscle mass. However, in some cases a myostatin knockout mutation only increases muscle mass, whereas strength does not rise proportionally (Amthor et al., 2007).

Taken together, this research suggests the existence of rare DNA sequence variations in genes such as the myostatin gene with a sometimes large effect size on muscle mass and strength. Whole-genome sequencing will accelerate the discovery of such rare DNA sequence variations to inform researchers, whereas rare DNA sequence variations contribute significantly to the variation of muscle strength and mass seen in humans and other species (Wheeler et al., 2008).

TRANSGENIC MICE WITH A MUSCLE STRENGTH AND/OR PHENOTYPE

The identification of polymorphisms or rare DNA sequence variations depends on previous knowledge of genes that affect muscle mass and strength. Genes where this is demonstrated are termed **candidate genes**. Such candidate genes are commonly identified via gain- or loss-of-function experiments in cultured muscle cells *in vitro* or in transgenic mice or other species *in vivo*. We have listed key examples of transgenic mouse models where muscle mass or size changes result from a gene knock-in or knockout in Table 7.3.

Most of the transgenic mice listed in Table 7.3 have either mutations in the muscle growth-inhibiting myostatin–Smad pathway (myostatin knockout, FLRG knock-in, c-ski knock-in) or the muscle growth-promoting PKB/Akt–mTOR pathway (PLA$_2$

Table 7.3 Examples of genetically altered mice with a skeletal muscle mass and/or strength-related phenotype

PROMOTER AND TRANSGENE	PHENOTYPE	REFERENCE
Global myostatin knockout from birth or in adult muscle	Doubling of muscle mass	McPherron et al., 1997
Global myostatin knockout and follistatin knock-in	Quadrupling of muscle mass	Lee, 2007
MSV-driven c-ski knock-in in muscle fibres	Fast-fibre hypertrophy	Sutrave et al., 1990
Global knockout of PLA$_2$	Small skeletal muscle hypertrophy and cardiac hypertrophy (Akt–mTOR signalling-related)	Haq et al., 2003
Avian α-actin-driven human IGF-1 knock-in in muscle	Muscle fibre hypertrophy	Coleman et al., 1995
Human skeletal actin promoter driven overexpression of constitutively active PKB/Akt in skeletal muscle	≈1.5–2-fold muscle fibre hypertrophy	Lai et al., 2004
p70 S6k knockout	Skeletal muscle atrophy	Ohanna et al., 2005
MCK-driven MMP9 knock-in in skeletal muscle fibres	Skeletal muscle hypertrophy and slow to fast fibre type switch	Dahiya et al., 2011
Global *Pax7*[a] knockout	No/few satellite cells, reduced skeletal muscle size	Seale et al., 2000
Global *ACTN3*[a] knockout	Reduced force generation and fast fibre diameter in skeletal muscle	MacArthur et al., 2008

MSV, Moloney murine sarcoma virus; PLA$_2$, cytosolic phospholipase A$_2$; CreERt2, oestrogen receptor-driven expression of Cre recombinase (this construct can be used to induce a knockout by giving tamoxifen to the animal); IGF-1, insulin-like growth factor 1; PKB/Akt, protein kinase B; *Pax7*, paired box gene 7; *ACTN3*, actinin 3; MCK, muscle creatine kinase promoter; MMP9, metalloproteinase-9.

knockout, IGF-1 knock-in, PKB/Akt knock-in). This confirms what we have said about these signalling pathways in resistance training (Chapter 6). We have shown two of the phenotypes in Figure 7.5 in order to give a better idea about the phenotype.

So what have we learned? First, both the PKB/Akt–mTOR pathway (Lai et al., 2004) and the myostatin–Smad pathway (McPherron et al., 1997; Whittemore et al., 2003) can induce skeletal muscle hypertrophy when genes are knocked in or out both at birth or in the adult. Both pathways also affect fibre numbers when the genetic modification is present from birth and fibre size at all stages.

A second, common observation is that muscle hypertrophy induced by targeting genes belonging to the PKB/Akt–mTOR (Lai et al., 2004) and myostatin–Smad (McPherron and Lee, 2002) pathways can both reduce fat, resulting in mice that are muscular and lean. Myostatin knockout mice have an increased myofibrillar protein synthesis rate per muscle (Welle et al., 2011) and activating mutations in the PKB/Akt–mTOR pathway increase protein synthesis as this is the major function of this pathway (Proud, 2004). Thus in these mice, protein synthesis will rise above protein breakdown until a steady state with a presumably high protein turnover is reached. The energetic cost of protein synthesis or translation is the equivalent of 4 ATP for one

Figure 7.5 Transgenic mice with a muscle hypertrophy phenotype. Overexpression or knock-in of constitutively active PKB/Akt results in muscle fibre hypertrophy. Redrawn after Lai et al. (2004). A combined myostatin knockout and FLRG (F66) knock-in increases muscle mass by ≈4-fold, the most extreme example of a transgenic muscle hypertrophy mouse so far. Redrawn after Lee (2007).

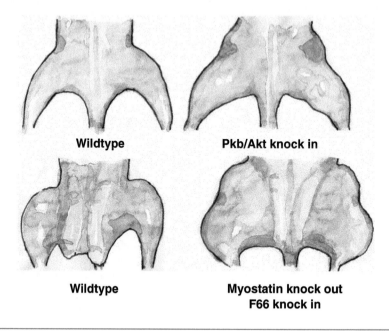

peptide bond between amino acids. Using such information the energy cost of protein synthesis has been estimated to be 3.6 kJ per g of newly synthesized protein (Hall, 2010). Thus to put it simply, an activation of the PKB/Akt–mTOR pathway or inhibition of the myostatin–Smad pathway will increase whole-body protein synthesis and breakdown, which in turn increases basal energy turnover. In line with this, the metabolic rate of myostatin knockout mice is significantly higher than that of wildtype mice (McPherron and Lee, 2002). The high energetic costs of maintaining a high muscle mass also seems one plausible reason why DNA sequence variations that increase muscle mass did not enrich in the population as they will increase metabolic rate and thus limit survival time during periods when nutrients are limited.

SELECTIVE BREEDING AND INBRED MOUSE STRAINS

At the end of this chapter we will discuss the results of selective breeding and inbred mouse strain studies as a non-biased strategy to identify genes that affect muscle mass and/or strength. Geneticists have performed selective breeding experiments (a) to identify the cumulative effect of selecting, in the ideal case, all DNA sequence variations within a population that affect muscle size and (b) to identify these DNA sequence variations. Selection studies for body weight have also led to an accumulation of genetic variants or alleles that increase or decrease muscle mass, as muscle mass is related to body mass. For example, the gastrocnemius weight in male mice of the so-called DUH strain with high body mass is ≈247 mg, while the gastrocnemius weight of mice selected for small body weight reaches only ≈66 mg (Lionikas and Bünger, unpublished). Among the selected alleles there might be some that affect the growth of all cells, such as genetic variations in the growth hormone system, and alleles that affect muscle mass specifically, such as those in the myostatin–Smad pathway.

Mice have also been helpful for understanding of whether genetic factors affect number and size of muscle-comprising fibres. The soleus muscle of mice is a particularly useful muscle for such studies for several reasons. First, it is a small muscle where the number of fibres within a cross-section can be counted reliably in reasonable time. Second, mouse solei consist almost only of slow-twitch type I and fast-twitch type IIa fibres, which can be easily distinguished using ATPase-based fibre typing or immuno-histochemistry using readily available antibodies. Unlike other appendicular muscles, it does not express type IIb fibres, which are present in many other rodent but not human muscles (Smerdu et al., 1994). Such analyses have revealed that the soleus muscles of different inbred strains had from ≈250 fibres in *m. spretus* (Totsuka et al., 2003) to between ≈500 and ≈800 fibres in commonly used laboratory strains such as C57BL/6, C3H, DBA/2 (Nimmo et al., 1985) and further up to ≈1200 fibres in the DUH strain (Lionikas and Bünger, unpublished). This implies that DNA sequence variations determine much of the variation in the number of fibres between these mouse strains.

Importantly, however, a larger muscle does not always mean more fibres (Figure 7.6). The LG/J and SM/J strains, whose names originated from the fact that they were selected for large and small body weight, respectively, differ 2-fold in the size of the soleus and other muscles (Lionikas et al., 2010). Yet the number of fibres in the muscle is not different between these two strains. Most of the difference in muscle mass is explained by the cross-sectional area of the fibres (Carroll and Lionikas, unpublished). The latter example illustrates that difference in the phenotype can occur by selection of alleles that affect either fibre size or muscle fibre number or both.

SUMMARY

Muscle mass and strength is limited by the number of muscle fibres, the size of muscle fibres and neuromuscular activation. These limiting factors depend on DNA sequence variations, which can be common or rare, and on environmental factors such as resistance training and nutrition. Lean body mass, muscle mass and strength all vary greatly (hand grip strength varies more than 20-fold in humans) and are significantly inherited. The heritability estimates vary considerably and it is impossible to give a reliable overall estimate. Muscle mass and/or strength depends on both common DNA sequence variations or polymorphisms such as the *ACTN3* R577X genotype and on rare DNA sequence variations such as the myostatin IVS1+5 G→A allele. It is currently unclear how much of the muscle mass and strength variation in human populations is due polymorphisms and rare DNA sequence variations. Transgenic mouse models and inbred mouse strains show that the variation of muscle mass and strength depends on DNA sequence variations that affect the number of muscle fibres within a given muscle and on the size of muscle fibres. Also

Figure 7.6 Selective breeding for low and high muscle mass. Differences in soleus muscle weight, fibre numbers and fibre size in two mice belonging to mouse strains selected for low and high body weight. In this example, the Berlin low strain has 630 fibres in the soleus and the type I and IIa fibres are on average 700 and 750 µm² in cross-sectional area, respectively. In contrast, the Dümmersdorf high example has 1223 fibres in the soleus and the type I and IIa fibres are 2000 and 2600 µm², respectively. This extreme example suggests that there are genetic variations that affect fibre numbers, fibre size and fibre proportions in a given muscle.

Berlin low (2.4 mg)   **Dümmersdorf high (22.4 mg)**

Mouse	Low	High	
Fibre number:	630	1223	(Arimantas Lionikas)
Type I (µm²):	700	2000	
Type IIa (µm²):	750	2600	

mouse studies demonstrate that a combination of gain or loss-of-function mutations can increase muscle mass at least 4-fold compared to wildtype mice. Key candidate genes are found within the PKB/Akt–mTOR and myostatin–Smad signal transduction pathways, which have also been implicated in the adaptation to resistance exercise.

REVIEW QUESTIONS

1 Draw a diagram to illustrate factors that limit muscle mass and strength.
2 What is known about the heritability of human muscle size and strength?
3 Describe the discovery of the *ACTN3* R577X polymorphism. Can an *ACTN3* R577X genetic test alone be used to identify with good likelihood someone who has the potential to become a world class sprinter?
4 Describe the experimental strategy that researchers have used to identify a mutation in the myostatin gene as a rare DNA sequence variation responsible for a doubling of muscle mass in a toddler.
5 Explain and compare two transgenic mouse models where a transgene in either the PKB/Akt–mTOR pathway or the myostatin–Smad pathway has increased muscle size. What is the maximal muscle size increase that has been achieved in a transgenic mouse model when compared to the wildtype?

FURTHER READING

Bouchard C and Hoffman EP (2011). Genetic and molecular aspects of sports performance, Vol. 18 of Encyclopedia of sports medicine. John Wiley & Sons, Chichester.

Peeters MW, Thomis MA, Beunen GP and Malina RM (2009). Genetics and sports: an overview of the pre-molecular biology era. Med Sport Sci 54, 28–42.

Pescatello LS and Roth SM (2011). Exercise genomics Springer, New York.

Schuelke M, Wagner KR, Stolz LE, Hubner C, Riebel T, Komen W, et al. (2004). Myostatin mutation associated with gross muscle hypertrophy in a child. N Engl J Med 350, 2682–2688.

Yang N, MacArthur DG, Gulbin JP, Hahn AG, Beggs AH, et al. (2003). ACTN3 genotype is associated with human elite athletic performance. Am J Hum Genet 73, 627–631.

REFERENCES

Alfred T, Ben-Shlomo Y, Cooper R, Hardy R, Cooper C, Deary IJ, et al. (2011). ACTN3 genotype, athletic status, and life course physical capability: meta-analysis of the published literature and findings from nine studies. *Hum Mutat* 32, 1008–1018.

Amthor H, Macharia R, Navarrete R, Schuelke M, Brown SC, Otto A, et al. (2007). Lack of myostatin results in excessive muscle growth but impaired force generation. *Proc Natl Acad Sci U S A* 104, 1835–1840.

Binns MM, Boehler DA and Lambert DH (2010). Identification of the myostatin locus (MSTN) as having a major effect on optimum racing distance in the Thoroughbred horse in the USA. *Anim Genet* 41 Suppl 2, 154–158.

Bogl LH, Latvala A, Kaprio J, Sovijarvi O, Rissanen A and Pietilainen KH (2011). An investigation into the relationship between soft tissue body composition and bone mineral density in a young adult twin sample. *J Bone Miner Res* 26, 79–87.

Coleman ME, DeMayo F, Yin KC, Lee HM, Geske R, Montgomery C, et al. (1995). Myogenic vector expression of insulin-like growth factor I stimulates muscle cell differentiation and myofiber hypertrophy in transgenic mice. *J Biol Chem* 270, 12109–12116.

Dahiya S, Bhatnagar S, Hindi SM, Jiang C, Paul PK, Kuang S and Kumar A (2011). Elevated levels of active matrix metalloproteinase-9 cause hypertrophy in skeletal muscle of normal and dystrophin-deficient mdx mice. *Hum Mol Genet* 20, 4345–4359.

De Mars G, Windelinckx A, Huygens W, Peeters MW, Beunen GP, Aerssens J, et al. (2008). Genome-wide linkage scan for maximum and length-dependent knee muscle strength in young men: significant evidence for linkage at chromosome 14q24.3. *J Med Genet* 45, 275–283.

Delmonico MJ, Kostek MC, Doldo NA, Hand BD, Walsh S, Conway JM, et al. (2007). Alpha-actinin-3 (ACTN3) R577X polymorphism influences knee extensor peak power response to strength training in older men and women. *J Gerontol A Biol Sci Med Sci* 62, 206–212.

Grundberg E, Brandstrom H, Ribom EL, Ljunggren O, Mallmin H and Kindmark A (2004). Genetic variation in the human vitamin D receptor is associated with muscle strength, fat mass and body weight in Swedish women. *Eur J Endocrinol* 150, 323–328.

Hall KD (2010). Mathematical modelling of energy expenditure during tissue deposition. *Br J Nutr* 104, 4–7.

Haq S, Kilter H, Michael A, Tao J, O'Leary E, Sun XM, et al. (2003). Deletion of cytosolic phospholipase A2 promotes striated muscle growth. *Nat Med* 9, 944–951.

Hsu FC, Lenchik L, Nicklas BJ, Lohman K, Register TC, Mychaleckyj J, et al. (2005). Heritability of body composition measured by DXA in the diabetes heart study. *Obes Res* 13, 312–319.

Hubal MJ, Gordish-Dressman H, Thompson PD, Price TB, Hoffman EP, Angelopoulos TJ, et al. (2005). Variability in muscle size and strength gain after unilateral resistance training. *Med Sci Sports Exerc* 37, 964–972.

Hughes DC, Day SH, Ahmetov II and Williams AG (2011). Genetics of muscle strength and power: Polygenic profile similarity limits skeletal muscle performance. *J Sports Sci* 29, 1425–1434.

Huygens W, Thomis MA, Peeters MW, Aerssens J, Janssen R, Vlietinck RF, et al. (2004a). Linkage of myostatin pathway genes with knee strength in humans. *Physiol Genomics* 17, 264–270.

Huygens W, Thomis MA, Peeters MW, Vlietinck RF and Beunen GP (2004b). Determinants and upper-limit heritabilities of skeletal muscle mass and strength. *Can J Appl Physiol* 29, 186–200.

Huygens W, Thomis MA, Peeters MW, Aerssens J, Vlietinck R and Beunen GP (2005). Quantitative trait loci for human muscle strength: linkage analysis of myostatin pathway genes. *Physiol Genomics* 22, 390–397.

Janssen I, Heymsfield SB, Wang ZM and Ross R (2000). Skeletal muscle mass and distribution in 468 men and women aged 18–88 yr. *J Appl Physiol* 89, 81–88.

Kostek MC, Delmonico MJ, Reichel JB, Roth SM, Douglass L, Ferrell RE, et al. (2005). Muscle strength response to strength training is influenced by insulin-like growth factor 1 genotype in older adults. *J Appl Physiol* 98, 2147–2154.

Lai KM, Gonzalez M, Poueymirou WT, Kline WO, Na E, Zlotchenko E, et al. (2004). Conditional activation of akt in adult skeletal muscle induces rapid hypertrophy. *Mol Cell Biol* 24, 9295–9304.

Lee SJ (2007). Quadrupling muscle mass in mice by targeting TGF-beta signaling pathways. *PLoS ONE* 2, e789.

Lexell J, Taylor CC and Sjostrom M (1988). What is the cause of the ageing atrophy? Total number, size and proportion of different fiber types studied in whole vastus lateralis muscle from 15- to 83-year-old men. *J Neurol Sci* 84, 275–294.

Lionikas A, Cheng R, Lim JE, Palmer AA and Blizard DA (2010). Fine-mapping of muscle weight QTL in LG/J and SM/J intercrosses. *Physiol Genomics* 42A, 33–38.

Liu D, Metter EJ, Ferrucci L and Roth SM (2008). TNF promoter polymorphisms associated with muscle phenotypes in humans. *J Appl Physiol* 105, 859–867.

MacArthur DG, Seto JT, Chan S, Quinlan KG, Raftery JM, Turner N, et al. (2008). An Actn3 knockout mouse provides mechanistic insights into the association between alpha-actinin-3 deficiency and human athletic performance. *Hum Mol Genet* 17, 1076–1086.

McPherron AC and Lee SJ (1997). Double muscling in cattle due to mutations in the myostatin gene. *Proc Natl Acad Sci U S A* 94, 12457–12461.

McPherron AC and Lee SJ (2002). Suppression of body fat accumulation in myostatin-deficient mice. *J Clin Invest* 109, 595–601.

McPherron AC, Lawler AM and Lee SJ (1997). Regulation of skeletal muscle mass in mice by a new TGF-beta superfamily member. *Nature* 387, 83–90.

Mills M, Yang N, Weinberger R, Vander Woude DL, Beggs AH, Easteal S and North K (2001). Differential expression of the actin-binding proteins, alpha-actinin-2 and -3, in different species: implications for the evolution of functional redundancy. *Hum Mol Genet* 10, 1335–1346.

Mosher DS, Quignon P, Bustamante CD, Sutter NB, Mellersh CS, Parker HG, et al. (2007). A mutation in the myostatin gene increases muscle mass and enhances racing performance in heterozygote dogs. *PLoS Genet* 3, e79.

Nimmo MA, Wilson RH and Snow DH (1985). The inheritance of skeletal muscle fibre composition in mice. *Comp Biochem Physiol A Comp Physiol* 81, 109–115.

North KN, Yang N, Wattanasirichaigoon D, Mills M, Easteal S and Beggs AH (1999). A common nonsense mutation results in alpha-actinin-3 deficiency in the general population. *Nat Genet* 21, 353–354.

Ohanna M, Sobering AK, Lapointe T, Lorenzo L, Praud C, Petroulakis E, et al. (2005). Atrophy of S6K1(-/-) skeletal muscle cells reveals distinct mTOR effectors for cell cycle and size control. *Nat Cell Biol* 7, 286–294.

Ontell MP, Sopper MM, Lyons G, Buckingham M and Ontell M (1993). Modulation of contractile protein gene expression in fetal murine crural muscles: emergence of muscle diversity. *Dev Dyn* 198, 203–213.

Peeters MW, Thomis MA, Beunen GP and Malina RM (2009). Genetics and sports: an overview of the pre-molecular biology era. *Med Sport Sci* 54, 28–42.

Proud CG (2004). mTOR-mediated regulation of translation factors by amino acids. *Biochem Biophys Res Commun* 313, 429–436.

Riechman SE, Balasekaran G, Roth SM and Ferrell RE (2004). Association of interleukin-15 protein and interleukin-15 receptor genetic variation with resistance exercise training responses. *J Appl Physiol* 97, 2214–2219.

Roth SM, Schrager MA, Ferrell RE, Riechman SE, Metter EJ, Lynch NA, et al. (2001). CNTF genotype is associated with muscular strength and quality in humans across the adult age span. *J Appl Physiol* 90, 1205–1210.

Santiago C, Ruiz JR, Rodriguez-Romo G, Fiuza-Luces C, Yvert T, Gonzalez-Freire M, et al. (2011). The K153R polymorphism in the myostatin gene and muscle power phenotypes in young, non-athletic men. *PLoS ONE* 6, e16323.

Sayer AA, Syddall H, O'Dell SD, Chen XH, Briggs PJ, Briggs R, et al. (2002). Polymorphism of the IGF2 gene, birth weight and grip strength in adult men. *Age Ageing* 31, 468–470.

Schrager MA, Roth SM, Ferrell RE, Metter EJ, Russek-Cohen E, Lynch NA, et al. (2004). Insulin-like growth factor-2 genotype, fat-free mass, and muscle performance across the adult life span. *J Appl Physiol* 97, 2176–2183.

Schuelke M, Wagner KR, Stolz LE, Hubner C, Riebel T, Komen W, et al. (2004). Myostatin mutation associated with gross muscle hypertrophy in a child. *N Engl J Med* 350, 2682–2688.

Seale P, Sabourin LA, Girgis-Gabardo A, Mansouri A, Gruss P and Rudnicki MA (2000). Pax7 is required for the specification of myogenic satellite cells. *Cell* 102, 777–786.

Silventoinen K, Magnusson PK, Tynelius P, Kaprio J and Rasmussen F (2008). Heritability of body size and muscle strength in young adulthood: a study of one million Swedish men. *Genet Epidemiol* 32, 341–349.

Smerdu V, Karsch-Mizrachi I, Campione M, Leinwand L and Schiaffino S (1994). Type IIx myosin heavy chain transcripts are expressed in type IIb fibers of human skeletal muscle. *Am J Physiol* 267, C1723–C1728.

Souren NY, Paulussen AD, Loos RJ, Gielen M, Beunen G, Fagard R, et al. (2007). Anthropometry, carbohydrate and lipid metabolism in the East Flanders Prospective Twin Survey: heritabilities. *Diabetologia* 50, 2107–2116.

Sutrave P, Kelly AM and Hughes SH (1990). ski can cause selective growth of skeletal muscle in transgenic mice. *Genes Dev* 4, 1462–1472.

Totsuka Y, Nagao Y, Horii T, Yonekawa H, Imai H, Hatta H, et al. (2003). Physical performance and soleus muscle fiber composition in wild-derived and laboratory inbred mouse strains. *J Appl Physiol* 95, 720–727.

Vincent B, De Bock K, Ramaekers M, Van den Eede E, Van Leemputte M, Hespel P, et al. (2007). ACTN3 (R577X) genotype is associated with fiber type distribution. *Physiol Genomics* 32, 58–63.

Wang P, Ma LH, Wang HY, Zhang W, Tian Q, Cao DN, Zheng GX and Sun YL (2006). Association between polymorphisms of vitamin D receptor gene ApaI, BsmI and TaqI and muscular strength in young Chinese women. *Int J Sports Med* 27, 182–186.

Welle S, Mehta S and Burgess K (2011). Effect of postdevelopmental myostatin depletion on myofibrillar protein metabolism. *Am J Physiol Endocrinol Metab* 300, E993–E1001.

Wheeler DA, Srinivasan M, Egholm M, Shen Y, Chen L, McGuire A, et al. (2008). The complete genome of an individual by massively parallel DNA sequencing. *Nature* 452, 872–876.

Whittemore LA, Song K, Li X, Aghajanian J, Davies M, Girgenrath S, et al. (2003). Inhibition of myostatin in adult mice increases skeletal muscle mass and strength. *Biochem Biophys Res Commun* 300, 965–971.

Windelinckx A, De Mars G, Huygens W, Peeters MW, Vincent B, Wijmenga C, et al. (2011). Comprehensive fine mapping of chr12q12–14 and follow-up replication identify activin receptor 1B (ACVR1B) as a muscle strength gene. *Eur J Hum Genet* 19, 208–215.

Yang N, MacArthur DG, Gulbin JP, Hahn AG, Beggs AH, Easteal S, and North K (2003). ACTN3 genotype is associated with human elite athletic performance. *Am J Hum Genet* 73, 627–631.

8 Molecular sport nutrition

D Lee Hamilton, Stuart Galloway, Oliver Witard and Henning Wackerhage

LEARNING OBJECTIVES

At the end of the chapter you should be able to:

- Explain how glucose is sensed and how this leads to the production of insulin and neural responses in the hypothalamus.

- Explain how insulin stimulates glucose uptake and glycogen synthesis.

- Discuss the mechanisms by which appetite is regulated in the brain.

- Explain how amino acids are sensed and how this leads to a transient increase of protein synthesis.

- Use molecular mechanisms to justify nutritional interventions in athletes.

INTRODUCTION

When it comes to nutrients, things are constantly on the move: Waves of proteins, carbohydrates, fats, micronutrients and water flood our digestive tracts after a meal, leading to substantial changes in circulating levels of amino acids, carbohydrates, lipids and micronutrients. These fluxes in nutrients alter homeostasis. However, our cells and organ systems have evolved a number of integrated mechanisms that allow for the partitioning and storage of glucose as glycogen, the use of amino acids to synthesize new proteins and the processing of lipids either into triglyceride stores or cell membranes. For this system to function, nutrients and storage molecules must be sensed both in the blood and within tissues. Some of these nutrient-sensing mechanisms overlap with the mechanisms by which we adapt to exercise. Therefore, there is the potential to exploit nutrient provision in terms of meal timing and meal composition to augment exercise adaptations and optimize performance gains. In the first section of this chapter we will discuss some examples of how nutrients are sensed and how this sensing directs nutrients to cells that are nutrient depleted or have an elevated nutrient demand.

In the preceding chapters we have discussed how endurance exercise and resistance exercise activate intramuscular signalling cascades; namely AMPK–PGC-1 and mTOR signalling, respectively. In addition to exercise, these pathways also can be modulated by nutrients as well as ergogenic aids and drugs. In the second part of this chapter we will therefore discuss how manipulating nutrient intake and meal composition can exploit these pathways in a manner that synergizes with the adaptive responses to training that may ultimately lead to improved performance.

MOLECULAR NUTRIENT SENSING

The body responds to the ebb and flow of nutrients by switching between nutrient usage for energy metabolism and growth or storage. This dynamic response to changes in nutrient availability prevents the detrimental effects of excessive nutrient concentrations as well as the consequences of running empty. So what systems sense nutrients and regulate nutrient consumption and storage accordingly? We start by explaining the function of key nutrient sensors. In our bodies two types of systems sense nutrients:

- **Nutrient sensors in the brain, digestive tract and endocrine organs**, such as taste receptors in the tongue or the glucose-sensing pancreas, which sense nutrients and release hormones to regulate the systemic, whole-body disposal of nutrients.
- **Cellular or local nutrient sensors**, such as the glucose-sensing system in the pancreas or the amino acid-sensing Rag proteins.

In this section we will discuss key examples of sensors from both classes. For each class, we first explain the nutrient-sensing mechanism and then discuss the importance of the sensing and signalling response for the body, especially in relation to exercise.

ENDOCRINE NUTRIENT SENSORS

As mentioned previously, there are a number of endocrine nutrient sensors that secrete a variety of hormones or neurotransmitters in a nutrient-dependent fashion. These hormones/neurotransmitters are the effectors that link nutrient status to systemic metabolic responses to maintain homeostasis. Key endocrine nutrient sensors are the endocrine pancreas, which includes, among others, the β-cells and α-cells. In the hypothalamus of the brain there are glucose-sensing neurons termed **glucose excited (GE)** and **glucose inhibited (GI)** neurons, which are important in maintaining glucose homeostasis. At a number of points in the digestive tract there are nutrient sensors that either play an important role in the reward pathways associated with food, such as the taste receptors, or regulate the expression of hormones that signal satiation (fullness) or hunger. Finally, adipose tissue, which is traditionally thought of as just a storage depot, also acts as an endocrine organ by secreting a number of hormones to signal nutrient storage status. We will briefly overview the mechanisms of action of some of the nutrient sensors and how their effectors modify metabolism to maintain homeostasis.

The endocrine pancreas

To achieve glucose homeostasis, blood glucose concentrations must be tightly controlled around 4–6 mmol/L, as both high and low concentrations of blood glucose damage the body. Blood glucose concentrations that are consistently higher than 7 mmol/L, termed **hyperglycaemia**, result in the detrimental glycosylation of proteins that is seen in poorly controlled diabetes. This damages nerves (neuropathy), eyes (retinopathy) and/ or kidneys (nephropathy). In contrast, blood glucose concentrations below 4 mmol/L (symptoms usually develop at ≈2.8–3 mmol/L), termed **hypoglycaemia**, cause a series of symptoms depending upon the severity, magnitude and/or length of hypoglycaemia. The symptoms of hypoglycaemia are linked to the effects of the counter-regulatory hormones adrenaline and glucagon, which defend against low blood glucose concentrations, and also to the effects of low blood glucose on the brain, which are termed **neuroglycopenic symptoms**. Mild symptoms of hypoglycaemia include feelings of tiredness, hunger and dizziness. If hypoglycaemia persists, other symptoms such as shaking, sweating and irritability may develop. Persistent and deep hypoglycaemia will eventually lead to unconsciousness and finally brain damage, because the brain needs glucose to survive.

Blood glucose homeostasis is regulated by the integrated responses of a number of glucose storage organs. These organs are regulated by the antagonistic actions of the glucose storage-stimulating hormone **insulin** and the glucose-releasing hormone **glucagon**. Insulin and glucagon are secreted from β-cells and α-cells in the islets of Langerhans of the pancreas, respectively. For athletes, both of these hormones are important for training, recovery and competition. Insulin is important because, without it, little glycogen would be synthesized after a meal. Insulin also reduces muscle protein breakdown. In contrast, glucagon is important for regulating hepatic glucose output during exercise. Glucagon acts on the liver to stimulate the export of glucose from the liver to support the energy demands of muscle contraction. Without an exercise-induced glucagon response, as can occur in some type 1 diabetes patients, it is difficult to maintain blood glucose homeostasis during exercise and a lack of glucagon can limit performance and cause hypoglycaemia unless glucose is ingested.

How is glucose sensed and how is the insulin gene transcribed and insulin secreted when glucose is high?

The chain of events that links high blood glucose concentrations to the secretion of insulin by the β-cells is known as **stimulus-secretion coupling** (Andrali et al., 2008; Henquin, 2009).

The cassette of GLUT2 and β-cell-specific glucokinase forms part of the glucose-sensing mechanism that allows β-cells to be 'energized' in response to postprandial levels of glucose. The sequence of events leading to insulin secretion is described below (see Figure 8.1):

1   **Glucose entry into β-cells**: Glucose enters β-cells primarily via GLUT2 transporters. Unlike GLUT4 transporters in skeletal muscle and adipose tissue, GLUT2 transporters in muscle do not require insulin to function. GLUT2 is known as a

Figure 8.1 Glucose-stimulated insulin release by β-cells in the pancreas. (1) Glucose enters β-cells via GLUT2. (2) Glucose stimulates the expression of the insulin gene. (3) Glucose is phosphorylated by glucokinase (not shown) and metabolised in the mitochondria to increase the $[ATP]/[ADP]\cdot[P_i]$ ratio or energize the β-cell. (4) In energised β-cells, the K_{ATP} potassium channels are closed. As a consequence, the membrane is depolarises leading to increased calcium flux and insulin secretion as Ca^{2+}-activated motors pull insulin-containing vesicles to the membrane.

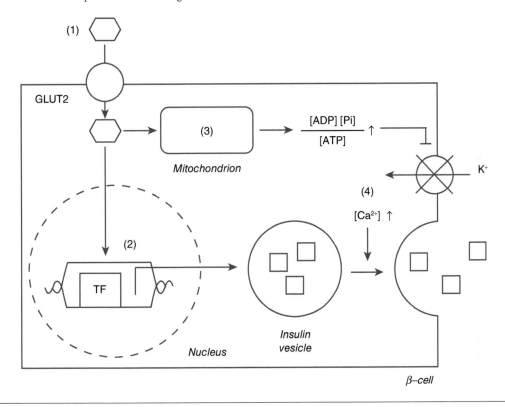

low-affinity but high-capacity glucose transporter. High concentrations of glucose are required to achieve high rates of transport as it has a K_m of 15–20 mM. The low affinity for glucose for this transporter means that at fasted glucose levels, relatively little glucose is transported into the β-cell.

2   **Glucose-induced insulin gene transcription**: A high glucose concentration inside β-cells leads to the activation of insulin expression-promoting transcription factors, which in turn activates the transcription of the insulin gene. Once transcribed and translated, insulin is stored in vesicles.

3   **Glucose-induced metabolic activation and cell depolarization**: In addition to the low-affinity GLUT transporter, β-cells also express a low-affinity isoform of glucokinase, an enzyme that phosphorylates glucose to glucose-6-phosphate, which requires high concentrations of glucose to work efficiently. The combination of GLUT2 and glucokinase in the β-cell means that, at fasted glucose levels, relatively little glucose is transported and phosphorylated and as a result glycolytic flux is

repressed. However, after a meal the glucose levels are closer to optimal working concentration for GLUT2. This results in a large influx of glucose, which is then phosphorylated by glucokinase and energizes the β-cell.

4   **High or stimulatory concentrations of glucose** within the β-cell then fuel glycolysis and the Krebs cycle. This in turn elevates the $[ATP]/[ADP] \cdot [P_i]$ ratio, which closes the ATP-dependent K_{ATP} channel. Closure of K_{ATP} channels subsequently depolarizes the β-cell or, in other words, an action potential occurs.

5   **Depolarization-induced Ca^{2+} entry and insulin secretion**: β-cell depolarization then triggers the entry of Ca^{2+} into the β-cell and a further Ca^{2+} release from the endoplasmic reticulum that is located inside the β-cell. The high Ca^{2+} concentration then activates tiny motors that pull insulin-containing vesicles to the membrane, where insulin is released into the blood.

To summarize, glucose activates the transcription of the insulin gene and energizes the β-cell, which leads to depolarization-induced increases in intracellular $[Ca^{2+}]$, causing insulin secretion into the blood. As a result, a rise in insulin concentration follows a rise in glucose concentration after a meal. Insulin then stimulates the uptake of glucose into tissues which lowers blood glucose concentration. Finally, a decline in blood glucose concentration reduces the secretion of insulin into the blood.

Some amino acids also trigger the release of insulin

Hyperaminoacidemia refers to an increase in blood amino acid concentrations in response to ingestion of a high protein meal. This change in blood chemistry also increases insulin secretion in a manner similar to hyperglycaemia. As with glucose, the branched chain amino acids, in particular leucine, stimulate β-cells to secrete more insulin (Newsholme et al., 2010). This effect would lead to hypoglycaemia in the absence of exogenous glucose provision. Therefore, amino acids also cause the secretion of glucagon (Rocha et al., 1972) to counter the effect of insulin on blood glucose.

Insulin increases glucose uptake and glycogen synthesis in insulin-responsive tissues

There are several insulin-sensitive tissues within the body. These tissues include the liver, adipose tissues and skeletal muscle. However, skeletal muscle accounts for up to 80% of glucose disposal under insulin-stimulated conditions. Therefore, the ability of skeletal muscle to dispose of glucose after a meal is an important determinant of insulin sensitivity and whole-body metabolism (Petersen et al., 2007). In a sport and exercise context, insulin-stimulated glucose disposal occurs after exercise and a meal, when the subsequent rise of insulin leads to the glucose uptake and glycogen recovery, especially in the exercised muscles.

In response to insulin, skeletal muscle increases the uptake and storage of glucose as glycogen. The mechanisms that link a high insulin concentration in the blood to glucose uptake and glycogen synthesis are introduced in Figure 8.2.

Figure 8.2 Insulin-stimulated glucose uptake. Simplified diagram showing how insulin stimulates glucose uptake and glycogen synthesis in organs such as skeletal muscle. (1) Insulin binding to the insulin receptor leads to the activation of PKB/Akt, which (2) phosphorylates and inhibits the action of AS160. (3) The relieved inhibition on the Rab GTPases then induces GLUT4 translocation to increase glucose uptake. (4) PKB phosphorylation and inhibition of GSK3 leads to the dephosphorylation of glycogen synthase (GS) and increased glycogen synthesis.

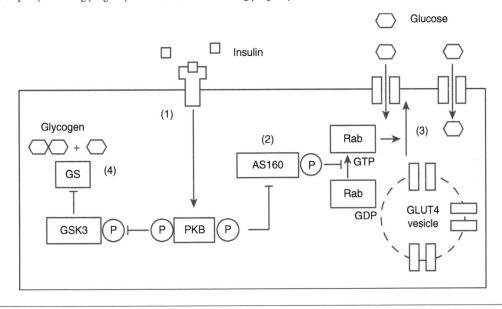

Figure 8.2 is, although already detailed, a simplification of the events that link insulin to glucose uptake and glycogen synthesis. In more detail, the events are as follows:

1 First, insulin binds to the **insulin receptor**. The insulin receptor is a tetramer made up of two extracellular α-subunits and two intracellular β-subunits. Insulin binds to the extracellular α-subunits of the insulin receptor and causes a conformational change in the receptor, which brings the intracellular β-subunits closer together. The β-subunits are then phosphorylated at tyrosine residues which act as a docking signal for insulin receptor substrate (IRS) proteins.

2 IRS proteins then tyrosine phosphorylate and act as docking signals for SH2 (Src homology 2) domain-containing proteins. A key SH2 domain-containing protein is the lipid kinase phosphoinositide 3-kinase (PI3K), which, after docking, phosphorylates PIP2 (phosphotidylinositol-4,5-bisphosphate) to generate PIP3 (phosphotidylinositol-3,4,5-trisphosphate).

3 PIP3 then acts as a docking signal for proteins containing PH (pleckstrin homology) domains. Two key PH domain-containing proteins are the kinases PKB/Akt (protein kinase B) and PDK1 (phosphotidylinositol-dependent protein kinase). An increase in the concentration of PIP3 in the membrane brings PDK1 and PKB close together and unveils a phosphorylation site on PKB/Akt. PDK1 then phosphorylates PKB/Akt on Thr308, which increases its activity and leads to its phosphorylation on PKB/Akt Ser473, leading to a fully active kinase.

4   PKB/Akt then leaves the membrane and translocates from the membrane to the nucleus, where it regulates a number of key factors that control glucose metabolism. PKB/Akt phosphorylates the protein AS160 (Akt substrate of 160 kDa). The phosphorylation of AS160 is a critical step allowing, via Rab proteins, the vesicles that contain GLUT4 to translocate to the membrane. This increases glucose uptake.

5   PKB/Akt also phosphorylates and inhibits GSK3 (glycogen synthase kinase 3). GSK3, as its name suggests, phosphorylates and inhibits the action of glycogen synthase. The action of PKB/Akt relieves an inhibition on glycogen synthase and as a consequence glycogen is synthesized at a greater rate.

How does insulin act on other insulin-responsive tissues?

Fat and heart cells also express the translocating glucose transporter GLUT4. Therefore, insulin-stimulated glucose uptake in these tissues occurs via very similar means to skeletal muscle. The liver, on the other hand, expresses the insulin receptor but not GLUT4. However, the liver is able to increase its glucose uptake in an insulin-dependent fashion. This increased glucose uptake is driven, in part, by the activation of the liver-specific glucokinase (Massa et al., 2011). Glucokinase phosphorylates glucose with a high capacity and maintains a low intracellular concentration of free glucose. Consequently, a concentration gradient is achieved that favours the uptake of glucose by facilitated diffusion through GLUT2 and GLUT1. To summarize, glucose-stimulated secretion of insulin links the glucose status of the circulation to the metabolism of glucose in insulin-responsive tissues. Insulin effectively coordinates a whole-body response and thus favours the uptake and storage of glucose in liver, adipose and skeletal muscle. Consequentially, in response to a meal or a glucose challenge, glucose can be effectively cleared from the blood.

Practical application: carbohydrate loading

We have just discussed the mechanisms that lead to the production of insulin and the mechanisms by which insulin stimulates glucose uptake and glycogen synthesis. These mechanisms are active during **carbohydrate loading** or **carboloading**, which is a term used to describe a method of increasing skeletal muscle glycogen concentration above normal resting concentrations prior to endurance competitions. This strategy is based on early muscle biopsy studies that demonstrate that time to fatigue is longer if the pre-exercise glycogen concentration is high (Bergstrom and Hultman, 1967). Other studies have shown that glycogen depletion by exercise followed by nutrient intake increases the concentration of glycogen above resting or pre-exercise levels in skeletal muscle (Bergstrom and Hultman, 1966) and liver (Nilsson and Hultman, 1973). This has been termed **glycogen supercompensation** and it seems to be an evolutionarily conserved response from rodents to humans, suggesting that it was an important aspect of survival. As we have discussed, we now have a good idea about how insulin will stimulate glucose uptake and glycogen synthesis but a key challenge for molecular exercise physiologists is to identify the mechanisms that allows glycogen to supercompensate during a carbo-loading regime. The data from the aforementioned studies suggest that carbohydrate loading strategies can extend the time to fatigue on steady-state exercise protocols by up to 20%. However, no sporting competitions depend upon judging who takes the longest to fatigue; performance is generally judged on who can cover the set distance the quickest. In protocols where a set distance is covered as quickly as possible carbohydrate loading can improve performance by 2–3% (Hawley et al., 1997a). Carbohydrate loading will not benefit every athlete. For instance, there is no benefit in high-intensity bouts of less than 5 min. Although there may be some small benefits for moderate-intensity bouts of less than 90 min, there does appear to be a clear benefit to be gained from carbohydrate loading if the events are longer than 90 min (Hawley et al., 1997).

Provision of carbohydrate drinks during exercise has also been shown to almost universally improve exercise capacity over long durations (>90 min). This effect was originally thought to be due to a sparing of endogenous muscle glycogen stores. In performance-based studies it is interesting to note that there is no clear relationship between carbohydrate feeding 'sparing' of muscle glycogen during exercise. Therefore, particularly in tasks lasting 60–90 min, improvements in performance must occur through other mechanisms. Why then does carbohydrate feeding improve performance when no clear relationship exists with sparing of muscle glycogen utilization? Researchers interested in this phenomenon have investigated other aspects of carbohydrate provision that could influence muscle function and performance. One such area that has received considerable attention is oral glucose sensing. Many studies have now been conducted examining oral rinsing with carbohydrate solutions in fed and fasted states compared with a placebo solution. These studies have observed increases in endurance cycling and running performance when carbohydrate is rinsed compared to when an artificially sweetened matched placebo is rinsed in the mouth (Chambers et al., 2009b; Rollo and Williams, 2011). This oral sensing effect of carbohydrate has been shown to improve performance in both a fed and fasted state in untrained participants (Fares and Kayser, 2011). However, the effect is largest when the mouth rinsing is used during exercise in a fasted state in trained cyclists (Lane et al., 2013). This effect is often not evident in trained individuals who are in a fed state (Beelen et al., 2009).

To get to the bottom of how this oral sensing works there have been a number of brain imaging studies conducted. It appears that the presence of carbohydrate in the mouth increases activity in brain regions associated with reward and motor control (Chambers et al., 2009). This impact on motor function has been explored by Gant and colleagues (2010), who observed that the presence of non-sweet carbohydrate in the mouth has the capacity to alter the excitability of the corticomotor pathway and change peripheral motor output. This mechanism could explain why a higher work rate can be sustained when mouth-rinsing is used. Importantly, this response is not associated with peripheral endocrine responses, as blood glucose concentration remains unchanged. These data suggest that the sensing of energy intake in the mouth can impact immediately upon muscle function. If carbohydrate is also ingested, then secondary peripheral changes in blood glucose and endocrine responses will occur. These responses will ensure that sufficient substrate is supplied to the muscle to sustain a higher work rate.

Based on all this research, several carbohydrate ingestion strategies have emerged. The carbohydrate feeding strategies outlined below are especially useful for runs of 30 km or longer (Burke et al., 2007) and similar endurance events:

- **Classic carbohydrate loading (also known as Saltin diet in some countries)**: A glycogen depletion phase of 3–4 days of hard training with low carbohydrate intake followed by 3–4 days of high carbohydrate intake (8–12 g/kg per day) and little exercise.
- **Modified carbohydrate loading**: Two or three days of high carbohydrate intake (8–12 g/kg per day or 70–85% of total energy) and an exercise taper. This strategy can double muscle glycogen content (Hawley et al., 1997).
- **Carbohydrate ingestion during exercise**: Irrespective of the carbohydrate loading strategy, carbohydrate should be consumed during exercise at a rate of 30–60 g/h to sustain prolonged exercise bouts (>90 min duration).
- **Carbohydrate mouth rinse** to boost performance in tasks lasting 60–90 min.

How does exercise-related glycogen depletion lead to glycogen supercompensation? This is still very much a mystery. We have covered how insulin regulates glucose transport in skeletal muscle, with the final step being the regulation of the amount of GLUT4 at the muscle membrane. Exercise has an insulin-sensitizing effect on skeletal muscle in that the concentration of insulin required to achieve 50% of maximum GLUT4 translocation and glucose uptake is substantially reduced following exercise. Therefore, post-exercise feeding will allow for greater amounts of glucose to be taken up by the muscle than would have been taken up pre-exercise. In spite of how well characterized the insulin-signalling pathway is, we are still unaware of how exercise mediates the post-exercise insulin sensitization, but it may be related to the energy/fuel sensor AMPK (Holloszy, 2005). Interestingly, the effect of carbohydration on AMPK may have implications for exercise adaptations, but we will discuss this later (Hawley et al., 2011).

How is glucagon secretion regulated and how does it act?

The hormone glucagon regulates glucose homeostasis during fasting and exercise but has opposite effects to insulin. In contrast to the insulin secretion by β-cells, the mechanisms for stimulus–secretion coupling in the glucagon-producing α-cells are poorly understood. Some evidence suggests that nutrient-sensing neurons in the vagus nerve might be involved (Tanaka et al., 1990). Like β-cells, α-cells also express K_{ATP} channels and therefore have the capacity to link nutrient status to membrane excitability. Surprisingly α-cells secrete glucagon in response to exogenously added glucose *in vitro* (Bansal and Wang, 2008), but in a living organism the opposite happens due to alternative mechanisms. It appears that α-cells respond secondary to the nutrient-sensing mechanism of the β-cells. Glucagon secretion is repressed by insulin and, interestingly, the zinc associated with insulin secretion. When insulin secretion is repressed by low blood glucose concentrations (below ≈4.4 mM), the reduced insulin and zinc concentration in the interstitial fluid bathing the α-cells leads to a de-repression of glucagon secretion, and glucagon secretion from the α-cells is increased (Bansal and Wang, 2008).

The primary target organ for glucagon action is the liver. Under physiological conditions, the primary action of glucagon is to stimulate hepatic glucose output.

Glucagon stimulates both **gluconeogenesis**, which is the production of glucose from precursor carbon compounds such as lactate, pyruvate or alanine, and the breakdown of glycogen, which is termed **glycogenolysis** (Jiang and Zhang, 2003). Glucagon mediates these actions by acting on the G-protein coupled **glucagon receptor**. The sequence of events that lead to increased hepatic glucose output are described below:

1   Circulating glucagon attaches to the extracellular loops of the glucagon receptor. This binding event changes the conformation status of the glucagon receptor which activates the G-proteins coupled to the receptor.
2   The release of the G-protein $G_s\alpha$ from the receptor activates adenylate cyclase, leading to increased levels of cyclic AMP (cAMP).
3   cAMP activates the kinase protein kinase A (PKA), which phosphorylates and activates phosphorylase kinase.
4   Phosphorylase kinase phosphorylates and activates phosphorylase, which breaks down glycogen to generate glucose.
5   PKA also phosphorylates and activates the transcription factor CREB, which leads to the increased expression of Pgc1. Pgc1 then increases the expression of enzymes that facilitate gluconeogenesis, which is the production of glucose from lactate, pyruvate or alanine.

To summarize, the binding of secreted glucagon to the glucagon receptor activates PKA. The activation of PKA enhances glycogenolysis, inhibits glycolysis and potentiates gluconeogenesis and ultimately results in increased hepatic glucose output. The action of glucagon is essential for preventing too low blood glucose concentrations (hypoglycaemia) during exercise and in response to fasting.

NUTRIENT SENSING BY THE HYPOTHALAMUS IN THE BRAIN

The brain and, more specifically, the part of the brain called the **hypothalamus**, is another organ that regulates blood glucose levels. The hypothalamus is located at the base of the brain and is the principle control centre for regulating energy balance. Figure 8.3 gives an overview of the parts of the hypothalamus and their connections. The hypothalamus receives peripheral signals from neural, nutrient and endocrine cues that signal satiation and control glucose homeostasis and the metabolism of the whole body (Simpson et al., 2009). Several nuclei are present in the hypothalamus that are important for appetite control. These nuclei include the paraventricular nucleus (PVN), ventromedial nucleus (VMN), dorsomedial nucleus (DMN), lateral hypothalamic area (LHA) and the arcuate nucleus (ARC). Our discussion of appetite will focus on the ARC. However, first we will address the hypothalamic control of glucose homeostasis.

One of the most critical brain centres for the regulation of glucose homeostasis is the ventromedial hypothalamus (VMH) located in the hypothalamus. The VMH contains two types of glucose-sensitive neurons, **glucose excited (GE)** and **glucose**

Figure 8.3 Anatomy and wiring of the hypothalamus in the brain. The arcuate nucleus (ARC) contains neurons expressing the neuropeptides NPY/AgRP and CART/POMC. These neurons have inputs into the other nuclei and regulate reward and the feeling of fullness. NPY/AgRP-producing neurons induce hunger and CART/POMC-producing neurons suppress hunger.

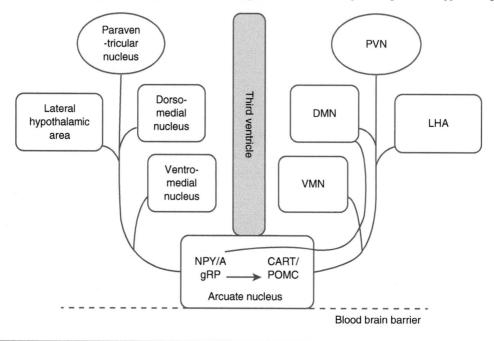

inhibited (GI). In a nutshell, GE neurons increase their firing rate in response to high levels of glucose and GI neurons suppress their firing rate in response to high levels of glucose. These neurons project into regions of the brain that regulate sympathetic drive and neural input onto the endocrine pancreas (Beall et al., 2012). GE neurons act to suppress the induction of hormones which increase blood glucose and GI neurons act to enhance the induction of hormones which increase blood glucose. In this way, brain glucose concentration is sensed and relayed via an accelerator/brake control mechanism into an endocrine response to maintain glucose levels in an optimal range.

How do glucose-excited (GE) and glucose-inhibited (GI) neurons sense glucose?

The presence of some components of the β-cell glucose-sensing machinery, such as glucokinase and the potassium channel K_{ATP}, has led to the hypothesis that GE neurons sense glucose in a manner similar to that of β-cells (Mountjoy and Rutter, 2007). A number of reports have suggested this to be the case (Mountjoy and Rutter, 2007; Beall et al., 2012). However, what is the sequence of events that leads to the suppression of GE neurons in response to low blood glucose concentrations. The events are shown in Figure 8.4 and explained in the text below.

Figure 8.4 Glucose-excited and glucose-inhibited neurons. Schematic diagram explaining the effect of reduced glucose in (a) glucose-excited (GE) and (b) glucose-inhibited (GI) neurons. The mechanisms are explained in detail in the text.

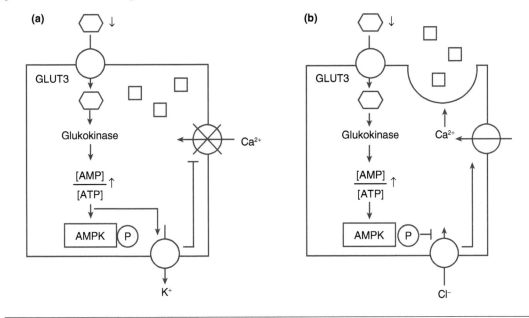

The response of GE neurons is as follows:

1 A fall in extracellular glucose concentrations leads to reduced intracellular glucose concentrations and an increased [AMP] to [ATP] ratio.
2 The increased [AMP] to [ATP] ratio leads to the activation of both AMPK and K_{ATP} channels.
3 AMPK, via an unknown mechanism, is required for the suppression of GE activity and the more active K_{ATP} channels lead to the hyperpolarization of the membrane potential.
4 The hyperpolarized membrane leads to reduced activity of voltage-gated calcium channels (VDCC), leading to reduced intracellular calcium.
5 The reduced intracellular calcium prevents neurotransmitter release and the cell firing is inhibited.

Like the α-cells of the endocrine pancreas, GI neurons also are slightly less well characterized than GE neurons. However, it is believed that AMPK also plays a critical role in the activation of these neurons in response to hypoglycaemia, albeit by different mechanisms. The proposed events in GI neurons after a lowering of the glucose concentration are as follows (Figure 8.4b):

1 A decrease in glucose concentration leads to a reduced energy charge of the cell.
2 The altered AMP: ATP ratio leads to an increase in AMPK activity.
3 Activated AMPK somehow leads to an inhibition of the chloride channel CFTR, decreasing chloride flux.

4   The decreasing chloride flux activates VDCC, increasing intracellular calcium.

5   The increasing intracellular calcium concentration leads to neurotransmitter release.

As glucose concentration decreases, so too does the energy charge of both GE and GI neurons and the altered firing rates in both cell types is dependent upon the activation of AMPK. The resultant changes in firing rates, which are inhibited in GE neurons and increased in GI neurons lead, via interactions with other hypothalamic regions, to a change in neurotransmitter and hormone concentrations which counter hypoglycaemia. The above described defence mechanism is termed the **counter-regulatory response**.

What are the endocrine cues regulated by hypoglycaemia?

The primary mechanism by which hypoglycaemia is reversed is an increased sympathetic drive leading to increased **adrenaline** (epinephrine), **noradrenaline** (norepinephrine) and also an increased secretion of **glucagon**, as discussed earlier. These responses are dependent upon the action of glucose-sensing regions in the hypothalamus and lead to an increased glucose output from the liver. In response to exercise, the increased sympathetic drive leads to a glucagon response (Luyckx and Lefebvre, 1974) and this is considered to be important for maintaining normal blood glucose concentrations during exercise (Ramnanan et al., 2011). In healthy individuals, these counter-regulatory mechanisms are so finely tuned that hypoglycaemia is an extremely rare event. However patients with type 1 diabetes mellitus have impaired counter-regulatory responses to both exercise and hypoglycaemia (Cryer, 2009). For instance the glucagon response to hypoglycaemia is lost in a large proportion of patients with type 1 diabetes mellitus (Cryer, 2005) and the glucagon response to exercise also is substantially blunted (Galassetti et al., 2006). Therefore, exercise for patients with type 1 diabetes mellitus needs to be planned with this in mind. Hypoglycaemic episodes are one of the major limiting factors in type 1 diabetes mellitus therapy (Cryer, 2008). In addition, fear of hypoglycaemia is cited as one of the greatest barriers for patients with type 1 diabetes mellitus adhering to an exercise programme (Brazeau et al., 2008). How does type 1 diabetes mellitus lead to this blunting effect? One key explanation for the defect in glucagon regulation is due to the loss of insulin that is thought to impair the ability of α-cells to respond to hypoglycaemia. However, a key aspect of standard **type 1 diabetes mellitus** care is the use of intensive insulin therapy. Insulin therapy presents the patient with a major side effect which is hypoglycaemia. The greater the number of hypoglycaemic episodes, the more impaired hypoglycaemia sensing and counter-regulation becomes. There also appears to be a so-called **cross-tolerance effect** in that exercise impairs hypoglycaemia sensing and hypoglycaemia impairs counter-regulatory responses to exercise (Cryer, 2009).

One explanation presented is that the defect lies with the glucose-sensing cells of the hypothalamus. AMPK is critical for relaying the signal of an energy deficit. However, as mentioned previously, AMPK activation in muscle generates an adaptive signal which improves the ability of muscle to deal with subsequent energy stresses in the context of exercise. It is postulated that the recurrent activation of AMPK in the glucose-sensing neurons leads to a similar adaptive process that allows the cells to be in a greater state of energy stress before AMPK becomes activated to generate the hypoglycaemic signal (Ashford et al., 2012). To date the only treatment for impaired hypoglycaemia sensing is to avoid hypoglycaemic events.

Recommendations to prevent hypoglycaemia in athletes and in patients with type 1 diabetes mellitus

Hypoglycaemia exposure causes the aforementioned cross-tolerance effect, leading to the impaired counter-regulatory responses not only to hypoglycaemia but also to exercise. The cross-tolerance effect also works the other way in that exercise impairs counter-regulatory responses to hypoglycaemia. As a result, glycaemic management for athletes or regular exercisers with type 1 diabetes mellitus is a very complicated issue. Patients with type 1 diabetes mellitus on intensive insulin therapy will typically experience hypoglycaemia every other day, leading to significant impairments in their ability to maintain euglycaemia during exercise (Gallen et al., 2011). There is also the added complication of post-exercise insulin sensitization, which can lead to hypoglycaemia late in the day (6–15 h after exercise has finished) if insulin or carbohydrate delivery is not adjusted accordingly. This effect is compounded by the fact that antecedent exercise impairs hypoglycaemic detection and counter-regulatory responses. Therefore exercisers with type 1 diabetes mellitus must be extra vigilant about their glycaemic control during and after exercise.

Glycaemic management strategies will differ depending upon individual circumstances. Factors that require consideration include the type and mode of insulin delivery, the type of exercise to be performed and the training status of the individual. Long-duration, moderate-intensity endurance exercise tends to lead to hypoglycaemia during exercise with a risk of post-exercise hypoglycaemia, whereas high-intensity or anaerobic-type exercise, including weight lifting, has a tendency to induce post-exercise hyperglycaemia (Gallen et al., 2011). Therefore, strategies to achieve glycaemic control should be highly individualized. There are a number of general recommendations (Gallen et al., 2011) that can be followed to reduce the risk of these complications:

- **Consume enough carbohydrates** in the daily diet to match the demands of the training load. For example light exercise (3–5 g/kg per day), moderate-intensity exercise for 1 h/day (5–7 g/kg per day), and moderate- to high-intensity exercise for 1–3 h/day (7–10 g/kg per day) and moderate- to high-intensity exercise for 4–5 h/day (≥10–12 g/kg per day).
- **Supplement carbohydrates** during exercise depending upon the type and duration of exercise. For example, for events of 1–3 h consume 30–60 g/h of effort, for events greater than 3 h consume 30–60 g/h of effort, for events longer than 3 h consume 30–60 g/h or potentially 70 g or more/h of effort.
- **Consider timing of insulin delivery relative to exercise** such that if exercise is to be performed within 90–120 min of an insulin bolus then the meal insulin bolus should be adjusted and the subject may require less carbohydrates during exercise. However, commencing exercise within 2 h of an insulin bolus may require the athlete to supplement more carbohydrates during the exercise bout.

- **Assess pre-exercise blood glucose concentrations**. If blood glucose concentration is less than 7 mmol/L the subject should consume 15–30 g of carbohydrate prior to any effort. On the other hand, if blood glucose concentration is above 10 mmol/L then carbohydrate feeding should be delayed.

- **Account for hypoglycaemia history** because hypoglycaemic events in the days prior to exercise will increase the glucose requirement during exercise (Davis et al., 2000). In some instances, it may be recommended to avoid exercise in the days following severe hypoglycaemia.

- **Exercise type** will have an influence on the glucose supplementation strategies required during and after exercise. Mixed sports such as football that require intermittent sprinting, sprint intervals and weight lifting have a tendency to induce hyperglycaemia. On the other hand, steady-state endurance exercise tends to cause hypoglycaemia.

- **Be aware of the potential for post-exercise hypoglycaemia**. Regardless of the exercise type the post-exercise insulin sensitization will pose a real hypoglycaemia threat for the next 6–15 h. If the exercise occurred late in the day then night time hypoglycaemia should also be closely monitored for. Finally, if unaccustomed damaging exercise is performed, such as a large proportion of downhill running, then the subject may suffer from post-exercise insulin resistance and they should take this into account on an individual basis by altering insulin delivery or carbohydrate provision.

Although type 1 diabetes mellitus significantly complicates exercise, it does not have to be performance-limiting in the realm of sport. There are/have been a number of high-performing athletes across a wide range of sports with type 1 diabetes mellitus, such as the Olympic rower Sir Steven Redgrave, professional road cyclist Javier Mejias, Olympic swimmer Gary Hall, Jr and NBA basketball player Gary Forbes to name but a few.

Control of appetite and metabolism by the hypothalamus

A major factor in appetite control is the physical distension of the stomach by food and liquid. Stomach distension activates stretch receptors that send afferent signals to the brain leading to changes in the activation of a number of brain regions which indicate fullness and reward (Wang et al., 2008). Equally important are nutrient-sensitive regulators that relay information on nutrient status to the hypothalamus. These other regulators fine-tune feeding responses that regulate meal size and frequency to accurately match energy intake to energy expenditure. In this section we will focus our attention on the **arcuate nucleus** in the hypothalamus for the following reasons (Figure 8.3):

- control of food intake by antagonistic neuronal pathways;
- proximity to the incomplete blood–brain barrier at the median eminence. This location allows for molecules in the plasma to enter this area of the brain with greater ease. As a consequence the arcuate nucleus can 'taste' the nutrient and endocrine status of the blood and control appetite and metabolism.

There are two primary interconnecting hypothalamic circuits that are antagonistic in action. In the arcuate nucleus, neurons that co-express neuropeptide Y (NPY) and agouti-related peptide (AgRP) increase appetite (the technical term is 'orexigenic'), promote food intake and reduce systemic energy expenditure. On the other hand, the neuronal pathways which co-express cocaine- and amphetamine-regulated transcript (CART) and pro-opiomelanocortin (POMC) neurons suppress appetite (the technical term is 'anorexigenic') and thus food intake when active. These two circuits project into and interact with other hypothalamic nuclei as shown in Figure 8.3.

NUTRIENT AND HORMONAL CONTROL OF APPETITE CENTRES

Nutrient-sensitive hormones such as leptin, ghrelin, insulin and several other hormones also act on the hypothalamus. These hormones bind to membrane receptors and modulate neuronal pathways such as POMC/CART and NPY/AgRP pathways to ultimately regulate appetite (Blundell, 2006).

These peripheral signals are actively released from several tissues, both in a circadian rhythm and in a manner dependent upon energy/feeding status. For instance, the gastrointestinal tract releases a number of hormones, including peptide YY (PYY), ghrelin, glucagon-like peptide-1 (GLP-1), oxyntomodulin (OXM), and cholecystokinin (CCK). The pancreas produces insulin and pancreatic polypeptide (PP) and the adipose tissue expresses adiponectin and leptin. These hormones or signals can be classified into episodic signals and tonic signals (Blundell, 2006). Episodic signals are those secreted periodically in a manner dependent upon the feeding state. The gastrointestinal tract is the main source of these signals, which include ghrelin, GLP-1, PYY and PP. Tonic signals are defined as those mediators that are released at a relatively constant rate throughout the circadian rhythm in a manner proportionate to body fat mass. Tonic signalling is mainly mediated by leptin and insulin release and they are thought to be collectively responsible for the chronic control of feeding and maintenance of body weight control. For energy homeostasis to be achieved, episodic and tonic signals must be integrated to effectively regulate individual meal size and feeding frequency to achieve energy balance.

Episodic hormones or signals

Episodic hormones or signals are indicators of acute nutritional status and influence food intake by determining the size and duration of a meal. They tend to be expressed in the digestive tract and their expression or release from the digestive tract occurs in a nutrient-sensitive manner. Most signal via the hypothalamus by regulating the expression of neuropeptides that affect appetite.

Ghrelin

Ghrelin is one of the main players in initiating a meal and is the only gastrointestinal hormone characterized to increase appetite (Cummings and Overduin, 2007). Ghrelin expression is increased in the fasted state and potently stimulates appetite. It is released from the fundus of the stomach and mediates its actions through the growth hormone secretagogue receptor (GHSR). Ghrelin release is inhibited by the presence of fat and amino acids in the stomach (Al et al., 2010) and its release is increased when the stomach is empty.

PYY

Apart from ghrelin, most acute regulators of food intake tend to be appetite lowering. These acute regulators include PYY, which is released after a meal from the L-cells in the small intestine (Batterham and Bloom, 2003). PPY secretion is rapidly stimulated within 15 min of meal termination and peaks at 90 min post meal (Batterham and Bloom, 2003) by the presence of short-chain fatty acids, fibre and bile salts (Plaisancie et al., 1996). In response to feeding, the binding of PYY to its receptor in the arcuate nucleus inhibits the expression of NPY and AgRP, and via that induces the feeling of fullness.

Pancreatic polypeptide (PP)

PP, alongside PYY_{3-36}, is a member of the PP-fold family (Suzuki et al., 2010). PP is released after a meal into the circulation from the islets of Langerhans. During the digestive process, levels of PP increase and the post-prandial levels of the hormone positively correlate to the caloric content of a meal (Simpson et al., 2009). PP acts in the hypothalamus to also signal fullness.

Proglucagon-derived peptides

Proglucagon-derived peptides, such as GLP-1 and OXM, are now recognized as major controllers of appetite. GLP-1 is co-secreted along with PYY from the L-cells of the small intestine in response to the presence of food intake in the proximal small intestine (Strader and Woods, 2005). It reduces food intake, gastrointestinal motility and delays gastric emptying, all of which contribute to lower appetite (Cummings and Overduin, 2007). GLP-1 also stimulates insulin secretion (D'Alessio et al., 1994). GLP-1 binding to its receptor stimulates adenyl cyclase activity and cAMP production to suppress further food intake (Holst, 2007).

Oxyntomodulin (OXM) shares several similarities with GLP-1 and is released from the L-cells of the small intestine. OXM inhibition of food intake is greater than that of GLP-1 despite disproportionate binding to the GLP-1R, suggesting potential secondary mechanisms of OXM action (Simpson et al., 2009).

CCK

CCK has long been regarded to be a satiety signal following its discovery in the late 1950s (Jorpes et al., 1959). Food in the digestive tract stimulates the secretion of CCK from I-cells. In particular, the presence of fat and protein in the partly digested food that enters the gut from the stomach, termed **chime**, are potent stimulators of CCK secretion (Lewis and Williams, 1990). CCK acts primarily by influencing gastrointestinal functions, including gut motility, gastric emptying and gallbladder contraction. The interaction between CCK and CCK receptors stimulates brainstem reflexes in the hypothalamus, signalling satiation and terminates the meal (Moran, 2000). CCK appears to act in a short-term fashion and its effects are limited to the period of feeding. The satiety effects CCK promotes can be prolonged through interactions with tonic signals of energy status. This interaction between CCK and signals such as leptin and insulin suggests that insulin and leptin influence body fat mass in cooperation with the short-term regulation of energy status exerted by CCK.

Tonic hormones or signals

Tonic signals are those signals that are regulated in a manner dependent upon the circadian rhythm and are modified by nutritional status. They act through the hypothalamus, but in contrast to episodic signals they exert a long-term or chronic influence on appetite.

Leptin

Leptin is a product of the *ob* gene that is expressed primarily in adipose tissue. Leptin is considered to be a tonic signal of satiety (Zhang et al., 1994). In the absence of leptin, obesity caused by overeating and metabolic dysfunction is inevitable. Instead of being regulated by acute feeding, the release of leptin into the circulation is a function of an individual's fat mass and it seems to act as an indicator of fuel storage. As leptin is an appetite-lowering factor, it may be logical to presume increases in plasma leptin in obese individuals would facilitate suppression of appetite and food intake. However, obesity is associated with severe leptin resistance. Several mechanisms are proposed to produce the leptin resistance associated with obesity. These include a reduced receptor function in the hypothalamus, impaired transport across the blood–brain barrier and the influence of inhibitory factors such as SOCS3 (the inhibitor suppressor of cytokine signalling 3), and the leptin binding factor c-reactive protein (CRP) (Bjorbaek et al., 1998; Banks et al., 2004; Chen et al., 2006).

The physiological roles of leptin are broad and diverse. These roles include processes such as regulating the proliferation and synthesis of T lymphocyte cells, skeletal muscle metabolism and the regulation of puberty (Friedman and Halaas, 1998). However, the central action of leptin in the hypothalamus reduces food intake, increases energy expenditure and facilitates fat and glucose metabolism. To summarize, leptin integrates information about the size of fat depots and relays that signal to the brain to change feeding behaviour and metabolism.

Insulin

As previously discussed, insulin is secreted by pancreatic β-cells in a manner dependent on blood glucose levels. Interestingly, insulin can cross the blood–brain barrier and act upon the hypothalamus to reduce food intake (Schwartz, 2000). The increased insulin release associated with weight gain could be hypothesized to suppress appetite but, instead, the onset of weight gain often diminishes insulin sensitivity both centrally and peripherally, blocking the appetite-lowering effect of insulin (Spanswick et al., 2000).

NUTRIENT SENSING BY THE HYPOTHALAMUS

In addition to the episodic and tonic hormones or signals that we have just covered, appetite and metabolism is also affected by nutrients that are directly sensed by the hypothalamus. We touched on how glucose is sensed by glucose-sensing neurons and how these play a key role in regulating glucose homeostasis. There is some evidence that fats can also be sensed by the hypothalamus (Jordan et al., 2010), although the evidence for this in humans is limired. In contrast, protein has a clear satiating effect which can be exploited by high-protein diets. High concentrations of amino acids have a strong influence on the hypothalamus. Leucine activates hypothalamic mTORC1 and this suppresses the expression of AgRP/NPY and thereby may contribute reduced appetite (Andre and Cota, 2012).

EXPLOITING NUTRIENT COMPOSITION TO AID IN THE CONTROL OF APPETITE AND TO CONTROL WEIGHT

Are there any practical lessons that can be learned from the complicated regulation of appetite by the hypothalamus in response to the many signals that we have just discussed? Can this information be used, for example, to reduce body weight? The answer is 'yes', as appetite can be controlled by the composition and frequency of meals to improve satiety. The first strategy is to exploit the role of stomach distension in controlling appetite (Kissileff et al., 2003). This can be achieved by consuming sufficient fibre to bulk out the meal and also by consuming plenty of fluids with a meal so as to distend the stomach (Kristensen and Jensen, 2011). The second step is to limit ghrelin secretion and thereby appetite, by consuming protein and fat in each meal. The presence of fat and amino acids have been shown to suppress ghrelin release from the stomach (Al et al., 2010). The third step is to increase the amount of appetite-suppressing hormones such as insulin, PYY, GLP-1 and CCK. This can be achieved by consuming a diet high in protein. Dietary protein increases the secretion of insulin, CCK and amino acids in the hypothalamus and suppresses the expression of NPY, thereby exerting anorectic effects at several levels of appetite regulation. In addition, CCK has a greater appetite suppressant effect when the stomach is distended (Kissileff et al., 2003). Finally, eating more frequently can enhance satiation and reduce food consumption at subsequent meals, possibly by limiting the amount of circulating ghrelin (Allirot et al., 2013). Leaving the molecular mechanisms aside, these recommendations can be summarized as follows:

- Eat sufficient fibre ($\approx$25 g/day for women, $\approx$38 g/day for men but no more than 50–60 g/day) and fluid at each meal. In particular, focus on consuming fibre with a high viscosity (Vuksan et al., 2009) such as can be found in nuts, seeds and fruit and vegetables. Additional satiation may be achieved by consuming fibre-supplemented drinks (Lyly et al., 2010).
- Consume sufficient protein, at least 1.2 g/kg (Westerterp-Plantenga et al., 2012) and consume protein and some fat with each meal.

- Increase the frequency of meal consumption. Eating a meal in four sessions separated by 1 h each can reduce caloric consumption at subsequent meals (Allirot et al., 2013). Therefore eating small frequent meals up to every hour can help with calorie control.

Can nutrient composition be manipulated to maintain muscle mass during weight loss interventions?

Athletes such as boxers, lightweight rowers and weight-lifters compete in weight-categorized sports. However, losing weight to compete may impair performance by inducing the loss of muscle mass. For instance, reducing the habitual energy intake for two weeks to 60% of the normal intake results in the loss of ≈3 kg of total body mass, of which ≈50% consists of muscle mass (Mettler et al., 2010). One possible explanation for the loss of muscle mass during low-energy dieting is a reduced response of muscle protein synthesis. Indeed 10 days of moderate energy restriction in physically active individuals leads to ≈20% decrease in muscle protein synthesis (Pasiakos et al., 2010). Lower protein intake reduces the concentrations of essential amino acids, which reduces in a feed forward mechanism the expression of the transporters that transport them into skeletal muscle (Peyrollier et al., 2000; Drummond et al., 2012). In addition, a reduction of insulin during dieting reduces the presence of amino acid transporters in the plasma membrane in a fashion similar to GLUT4 translocation (Hyde et al., 2002). These two changes reduce the intracellular amino acid concentration and, via that mechanism, protein synthesis. With these molecular mechanisms in mind, we can again develop a strategy to limit the skeletal muscle mass loss during weight loss interventions.

By increasing the dietary protein intake of an energy-restricted diet, it seems possible to maintain the muscle mass of athletes (Walberg et al., 1988; Mettler et al., 2010). A low-energy diet with a traditional macronutrient composition (15% protein, 50% carbohydrates, 35% fat) produced a significant amount of muscle loss, but a high-protein diet of equivalent energy (35% protein, 50% carbohydrates, 15% fat) preserved muscle mass while still promoting weight loss. Given the information that muscle protein synthesis rates are reduced during weight loss (Pasiakos et al., 2010), that the response of muscle protein synthesis to protein ingestion is not different between sexes and as 20–25 g of protein are sufficient to maximally stimulate muscle protein synthesis (Moore et al., 2009), we can recommend that post-workout drinks containing 20–25 g of protein may provide an important component of a nutritional strategy designed for 'toning up' in both men and women. Protein in the diet therefore not only serves to assist with satiation by the regulation of satiety hormones and the hypothalamus, but it also serves to preserve muscle mass by stimulating muscle protein synthesis, possibly by upregulating the expression of amino acid transporters.

PROTEIN NUTRITION AND MUSCLE PROTEIN SYNTHESIS

Resistance exercise and the ingestion of proteins, which are digested to amino acids, both increase muscle protein synthesis via the mTORC1 pathway. Insulin, which also rises after a meal, affects mainly breakdown in adult humans at least. In contrast, when high [ADP] and [AMP] or low glycogen activate AMPK then this inhibits protein synthesis due to the inhibitory effect of AMPK on mTOR signalling as we have already discussed in Chapter 6. In this section we will review how amino acids stimulate protein synthesis. The activation of protein synthesis by amino acids is demonstrated in Figure 8.5, which demonstrates a transient increase of human muscle protein synthesis when the concentration of essential amino acids is increased (Bohe et al., 2001). Moreover, rapamycin, a mTOR inhibitor, prevents the amino acid-induced increase in human muscle protein synthesis (Dickinson et al., 2011).

Figure 8.5 Amino acids stimulate muscle protein synthesis via mTOR. (a) Effect of continuous amino acid infusion on skeletal muscle protein synthesis. Redrawn after Bohe et al. (2001). (b) Rapamycin, an mTOR inhibitor, completely prevents an increase of protein synthesis in response to essential amino acid feeding in human skeletal muscle Redrawn after Dickinson et al. (2011).

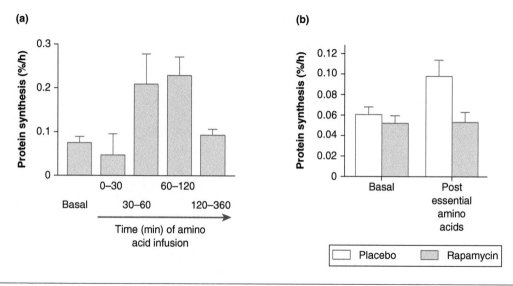

For molecular exercise physiologists the key question is: 'How is amino acid sensing transduced via mTOR into a signal for protein synthesis?'

The elusive amino acid sensor

Much is now known about the complicated regulation of the many elements in the central mTOR pathway, but the sensing of resistance exercise (see Chapter 6) and amino acids is not quite as well understood despite major research efforts. Our current knowledge on amino acid sensing is as follows (see Figure 8.6):

- **Amino acid import**: Amino acids enter cells via amino acid transporters. A key amino acid transporter is a so-called antiporter which is dependent on glutamine (Nicklin et al., 2009).

Figure 8.6 Mechanisms of amino acid-stimulated protein synthesis. Essential amino acid (EAA) update and mTORC1 activation. (1) The glutamine transporter SLC1A5 generates a glutamine gradient (inside to outside). (2) The glutamine gradient is used to drive the uptake of EAAs via the antiporter SLC7A5. The enhanced EAAs increase the association of Rags in the GTP-loaded state. (3) The more Rags existing in the GTP-loaded state, the more active mTORC1 is and the higher the rate of protein synthesis.

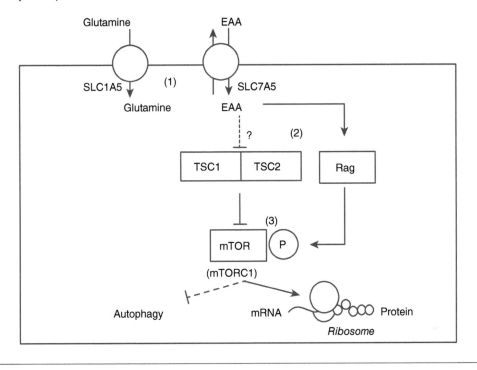

- **Amino acid sensing**: The actual amino acid sensing involves Rag proteins which activate the mTORC1 complex via unknown mechanisms (Kim and Guan, 2009). The mTOR pathway from TSC1/2 and above is not required for amino acid sensing and instead mediates the activation of protein synthesis by IGF-1.

We will now discuss the actual amino acid-sensing mechanism, which is incompletely understood, in more detail (Kim and Guan, 2009). It is known that PKB–TSC2 signalling is not required, while molecules further down the mTOR pathway, starting with Rheb, are required. In humans, this is best demonstrated by the fact that rapamycin treatment completely blocks the normal increase of protein synthesis in response to amino acid stimulation (Dickinson et al., 2011) (see Figure 8.5). At the core of the amino acid-sensing process are Rag GTPase proteins. The GTP-loading status of Rags alters the localization and activity of components of the mTOR complex (Sancak et al., 2010). In general, amino acids somehow increase the proportion of Rags in the GTP-bound state, which aids in controlling mTORC1 localization and activity. In addition to the Rags, other proteins are also involved in the amino acid-sensing process. hVPS34, for instance, is a nutrient-sensitive lipid kinase whose lipid products are thought to assist in the co-localization of the various mTOR complex components. MAP4K3 is another factor implicated in mTORC1 activation by EAAs. MAP4K3 is a kinase whose activity is amino acid sensitive and is required for mTORC1 activation, however the mechanism of action is still unclear (Dodd and Tee, 2012). Finally, a role for the leucine tRNA synthetase has been suggested in amino acid sensing (Jewell et al., 2013), but again the complete mechanism is unclear. Taken together, researchers active in this area see the nuts and bolts of the machine but do not understand how the whole amino acid sensor works and whether there is one or whether there are several amino acid sensors.

How sensitive are the sensing mechanisms? At the cellular level at least it appears that the amino acid-sensing mechanisms are highly sensitive to changes in intracellular amino acid concentrations. For instance only a 17% increase in intracellular leucine is required to maximally stimulate mTORC1 signalling (Christie et al., 2002). In addition to this only 30% of maximal mTORC1 signalling is required to saturate the protein synthesis response (Crozier et al., 2005). So, like most signalling mechanisms, there is a substantial amplification of the signal such that minor changes in intracellular amino acid concentrations lead to substantial changes in protein synthesis.

How much dietary protein is sufficient to maximally stimulate the production of new muscle protein after a workout?

The provision of amino acids from ingested dietary protein stimulates an increased response of muscle protein synthesis during recovery following a workout. However, the increased post-exercise rate of muscle protein synthesis with protein ingestion is not infinite. There seems to be a maximal rate of muscle protein synthesis that can be attained by the ribosomes or protein synthetic machinery of the muscle (Bohe et al., 2003; Cuthbertson et al., 2005; Moore et al., 2009). That is, at high doses of ingested protein, rather than being utilized for muscle protein synthesis, excess amino acids are either used for non-anabolic processes such as substrate oxidation for energy production or are simply excreted. Therefore, in the context of optimizing gains in muscle mass during exercise recovery, the dose of protein contained in a post-workout drink or meal is important.

A number of recent studies have examined the response of muscle protein synthesis to various sources and doses of protein (Boirie et al., 1997; Tang et al., 2009a, 2009b; Pennings et al., 2011; Yang et al., 2012). A whey source of protein, compared with soy or casein protein, has been shown to stimulate a superior response of muscle protein synthesis after resistance exercise in young and old adult males (Tipton et al., 1999a, 1999b). This beneficial response is likely explained by the characteristic faster release of amino acids from the ingested protein into the blood and to the higher leucine content of whey protein (Moore et al., 2009). Indeed, of all 20 amino acids, leucine is thought to be the main stimulator of muscle protein synthesis (Kimball and Jefferson, 2005).

So, what is the optimal daytime pattern of protein intake for building muscle mass? Four 20 g protein feedings over the course of the day results in a greater post-exercise response of muscle protein synthesis compared with eight 10 g protein feedings or two 40 g protein feedings (Areta et al., 2013). Hence, these data provide an indication that the repeated ingestion of moderate amounts of dietary protein at regular intervals throughout the day will benefit athletes wishing to enhance muscle mass.

To summarize, it is now generally accepted that the primary determinant of muscle protein balance is the dynamic regulation of protein synthesis. Muscle protein breakdown remains relatively stable in most situations however protein synthesis is dynamically regulated in response to feeding and the impact of resistance exercise is not only to acutely stimulate increased protein synthesis but to sensitize the muscle to nutrition for potentially up to 72 h (Figure 8.7). This means that a single bout of resistance exercise increases the amount of protein laid down in response to each meal for potentially up to 72 h after completion of the work out (Miller et al., 2005; Churchward-Venne et al., 2012). Reminiscent of the effect of exercise on insulin sensitivity, little is known about how resistance exercise sensitizes muscle to nutrition but a clue may be in the resistance exercise-dependent regulation of amino acid transport expression (Drummond et al., 2012). Resistance exercise increases the expression of a number of amino acid transporters which, if translated into a greater transport capacity, could explain the increased sensitivity to feeding. Greater transport expression would lead to greater amino acid uptake into the exercised muscles and greater uptake of amino acids would translate to enhanced protein synthesis via the activation of mTORC1.

Figure 8.7 Simplified schematic depicting the combined effect of resistance training and meals on skeletal muscle protein balance. (a) In non-exercising subjects meals cause a transient shift to a positive protein balance (above the middle line) followed by a negative protein balance during subsequent fasting periods. Overall, the protein balance is zero, which means that skeletal muscle is neither gained nor lost. (b) After a bout of resistance exercise in those that have a trainability for hypertrophy, muscle protein synthesis and net protein balance are shifted upwards for up to 72 h (Miller et al., 2005). When combined with regular meals (dotted line) the average protein balance is positive, which results over time in skeletal muscle hypertrophy.

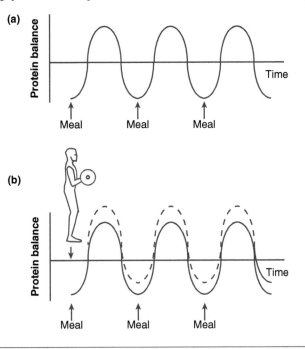

Recent evidence suggests that the most effective strategy for increasing muscle protein synthesis by feeding is to follow a bout of resistance exercise with four boluses of 20 g of protein every 3 h (Areta et al., 2013a). Additional work suggests that high-quality protein from dairy or poultry sources is more effective at stimulating muscle protein synthesis than plant-based sources. Based on these data we can therefore make the following nutritional recommendations to maximize the growth effects of resistance exercise:

- High-quality protein from dairy or eggs is superior for stimulating protein synthesis so consume a casein, whey or egg white source of protein.
- Consume a minimum of 20 g of protein with each meal and immediately post training.
- Ideally consume protein at least every 3 h, especially after a session of resistance exercise.
- Also consume protein immediately post work out.

FAT METABOLISM

Fats or triglycerides comprise three fatty acids that are bound via ester bounds to the three OH groups of a glycerol molecule which acts as a backbone. The majority of fats are stored in the adipose tissue but some are stored within skeletal muscle in the form of lipid droplets. Fats are stored as triglycerides, but when needed the fatty acids are cut off by the enzyme **hormone-sensitive lipase (HSL)** to generate free fatty acids (FFA) and glycerol. Even in a lean athlete the energy stored in adipose tissue and elsewhere is enough for many hours and even days of continuous activity (Yeo et al., 2011). The benefit of storing energy as fat is due to its lack of osmotic potential. Storing 1 g of fat is associated with no water and yields 37 kJ ($\approx$9 kcal) of energy. In contrast, 1 g of stored glycogen is associated with $\approx$4 g of water and yields 16 kJ ($\approx$4 kcal). In addition, the ability to utilize fat as a substrate spares muscle glycogen, and so being able to utilize fat as a substrate at a given intensity improves performance in long-endurance events in particular.

Fat adaptation

FFAs can act as ligands for transcription factors such as those belonging to the peroxisome proliferator-activated receptor (PPAR) family, which regulate fatty acid metabolism. When FFAs bind to PPARs this induces the binding of PPARs to retinoic acid X receptors (RXR) (Mangelsdorf and Evans, 1995) which in turn affects gene expression. Activated PPARs increase the expression of a number of fatty acid metabolism genes that improve lipid handling. This then raises the question as to whether or not fat supplementation can be used as a tool to further drive endurance exercise adaptations. This has been trialled with so-called **fat adaptation diets**. Fat adaptation diets have been proven to improve the rates of fat oxidation during steady-state exercise even in the presence of exogenous and repleted endogenous glucose stores (Yeo et al., 2011). But what is the mechanism behind this response? The improvement is thought to occur through a range of potential mechanisms (Yeo et al., 2011). However two key regulators could be responsible: AMPK activation and PPAR activation. Fat adaptation diets have been shown to activate AMPK (Yeo et al, 2008). As discussed in Chapter 4, AMPK is a key adaptive signal to drive endurance training adaptations such as mitochondrial biogenesis. It also controls the protein acetyl-CoA carboxylase (ACC), which facilitates fatty acid entry into mitochondria (Winder et al., 1997, 2000). Therefore the enhanced AMPK activation in response to a high-fat diet in trained athletes could contribute to the enhanced fatty acid metabolism. The second mechanism is the activation of PPARs. PPARs are transcription factors that are ligand-activated nuclear receptors (i.e. they act in the nucleus to increase the expression of target genes in a ligand-dependent manner). As mentioned earlier, various lipids and FFAs act as ligands to activate PPARs. Although no evidence has yet been presented that PPARs are activated by fat adaption diets, based on the gene expression changes, many of which are PPAR targets, it is very likely that PPARs are activated by the diet. For instance, a number of fatty acid transport and metabolism genes are increased in expression at the level of both the mRNA and the protein (Yeo et al., 2011).

SUMMARY

In this chapter we have identified the molecular mechanisms that allow cells to sense glucose, amino acids and free fatty acids. This sensed input either leads to a direct adaptation or causes endocrine cells or neurons to release hormones and neurotransmitters such as insulin or ghrelin. These hormones and neurotransmitters then regulate nutrient uptake, metabolism and also neural responses such as appetite. Much of this is relevant in the context of exercise, which is associated with nutrient usage, and for athletes, which exploit these mechanisms to manipulate their nutrition to carboload, build muscle mass or lose weight while preserving muscle mass. Also, for athletes with type 1 diabetes it helps to know the molecular mechanisms in order to improve their nutrition to prevent low blood glucose, termed hypoglycaemia, during training sessions and in competitions.

REVIEW QUESTIONS

1 Via what hormones and signalling mechanisms does glucose regulate its own circulating concentration during fasting and feeding?
2 Name three feeding-induced hormones or signals which control appetite and describe their action.
3 Describe why exercise is a challenge for patients with type 1 diabetes mellitus and give exercise recommendations for these patients.
4 How can diet composition be altered on a weight loss diet to preserve muscle mass?
5 Describe the mechanism by which low glycogen training leads to increased fat metabolism.
6 How are amino acids sensed?

FURTHER READING

Atherton PJ and Smith K (2012). Muscle protein synthesis in response to nutrition and exercise. *J Physiol* 590, 1049–1057.
Bilan PJ, Samokhvalov V, Koshkina A, Schertzer JD, Samaan MC and Klip A (2009). Direct and macrophage-mediated actions of fatty acids causing insulin resistance in muscle cells. *Arch Physiol Biochem* 115, 176–190.

Gallen IW, Hume C and Lumb A (2011). Fuelling the athlete with type 1 diabetes. *Diabetes Obes Metab* 13, 130–136.

Cummings DE and Overduin J (2007). Gastrointestinal regulation of food intake. *J Clin Invest* 117, 13–23.

Hawley JA, Schabort EJ, Noakes TD and Dennis SC (1997). Carbohydrate-loading and exercise performance. An update. *Sports Med* 24, 73–81.

Philp A, Hargreaves M and Baar K (2012). More than a store: regulatory roles for glycogen in skeletal muscle adaptation to exercise. *Am J Physiol Endocrinol Metab* 302, E1343–E1351.

Watt MJ and Hoy AJ (2012). Lipid metabolism in skeletal muscle: generation of adaptive and maladaptive intracellular signals for cellular function. *Am J Physiol Endocrinol Metab* 302, E1315–E1328.

REFERENCES

Al MO, Pardo M, Roca-Rivada A, Castelao C, Casanueva FF and Seoane LM (2010b). Macronutrients act directly on the stomach to regulate gastric ghrelin release. *J Endocrinol Invest* 33, 599–602.

Allirot X, Saulais L, Seyssel K, Graeppi-Dulac J, Roth H, Charrie A, et al. (2013b). An isocaloric increase of eating episodes in the morning contributes to decrease energy intake at lunch in lean men. *Physiol Behav* 110–111, 169–178.

Andrali SS, Sampley ML, Vanderford NL and Ozcan S (2008). Glucose regulation of insulin gene expression in pancreatic beta-cells. *Biochem J* 415, 1–10.

Andre C and Cota D (2012). Coupling nutrient sensing to metabolic homoeostasis: the role of the mammalian target of rapamycin complex 1 pathway. *Proc Nutr Soc* 71, 502–510.

Areta JL, Burke LM, Ross ML, Camera DM, West DW, Broad EM, et al. (2013b). Timing and distribution of protein ingestion during prolonged recovery from resistance exercise alters myofibrillar protein synthesis. *J Physiol* 591, 2319–2331.

Ashford M, Beall C and McCrimmon R (2012). Hypoglycaemia: exercise for the brain? *J Neuroendocrinol* 24, 1365–1366.

Banks WA, Coon AB, Robinson SM, Moinuddin A, Shultz JM, Nakaoke R, et al. (2004). Triglycerides induce leptin resistance at the blood–brain barrier. *Diabetes* 53, 1253–1260.

Bansal P and Wang Q (2008a). Insulin as a physiological modulator of glucagon secretion. *Am J Physiol Endocrinol Metab* 295, E751–E761.

Batterham RL and Bloom SR (2003a). The gut hormone peptide YY regulates appetite. *Ann N Y Acad Sci* 994, 162–168.

Beall C, Ashford ML and McCrimmon RJ (2012a). The physiology and pathophysiology of the neural control of the counterregulatory response. *Am J Physiol Regul Integr Comp Physiol* 302, R215–R223.

Beelen M, Berghuis J, Bonaparte B, Ballak SB, Jeukendrup AE and van Loon LJ (2009). Carbohydrate mouth rinsing in the fed state: lack of enhancement of time-trial performance. *Int J Sport Nutr Exerc Metab* 19, 400–409.

Bergstrom J and Hultman E (1966). Muscle glycogen synthesis after exercise: an enhancing factor localized to the muscle cells in man. *Nature* 210, 309–310.

Bergstrom J and Hultman E (1967). A study of the glycogen metabolism during exercise in man. *Scand J Clin Lab Invest* 19, 218–228.

Bjorbaek C, Elmquist JK, Frantz JD, Shoelson SE and Flier JS (1998). Identification of SOCS-3 as a potential mediator of central leptin resistance. *Mol Cell* 1, 619–625.

Blundell JE (2006). Perspective on the central control of appetite. *Obesity (Silver Spring)* 14 Suppl 4, 160S–163S.

Bohe J, Low JF, Wolfe RR and Rennie MJ (2001). Latency and duration of stimulation of human muscle protein synthesis during continuous infusion of amino acids. *J Physiol* 532, 575–579.

Bohe J, Low A, Wolfe RR and Rennie MJ (2003). Human muscle protein synthesis is modulated by extracellular, not intramuscular amino acid availability: a dose-response study. *J Physiol* 552, 315–324.

Boirie Y, Dangin M, Gachon P, Vasson MP, Maubois JL and Beaufrere B (1997). Slow and fast dietary proteins differently modulate postprandial protein accretion. *Proc Natl Acad Sci U S A* 94, 14930–14935.

Brazeau AS, Rabasa-Lhoret R, Strychar I and Mircescu H (2008). Barriers to physical activity among patients with type 1 diabetes. *Diabetes Care* 31, 2108–2109.

Burke LM, Millet G and Tarnopolsky MA (2007). Nutrition for distance events. *J Sports Sci* 25 Suppl 1, S29–S38.

Chambers ES, Bridge MW and Jones DA (2009). Carbohydrate sensing in the human mouth: effects on exercise performance and brain activity. *J Physiol* 587, 1779–1794.

Chen K, Li F, Li J, Cai H, Strom S, Bisello A, et al. (2006). Induction of leptin resistance through direct interaction of C-reactive protein with leptin. *Nat Med* 12, 425–432.

Christie GR, Hajduch E, Hundal HS, Proud CG and Taylor PM (2002). Intracellular sensing of amino acids in Xenopus laevis oocytes stimulates p70 S6 kinase in a target of rapamycin-dependent manner. *J Biol Chem* 277, 9952–9957.

Churchward-Venne TA, Burd NA and Phillips SM (2012). Nutritional regulation of muscle protein synthesis with resistance exercise: strategies to enhance anabolism. *Nutr Metab (Lond)* 9, 40.

Crozier SJ, Kimball SR, Emmert SW, Anthony JC and Jefferson LS (2005). Oral leucine administration stimulates protein synthesis in rat skeletal muscle. *J Nutr* 135, 376–382.

Cryer PE (2005). Mechanisms of hypoglycemia-associated autonomic failure and its component syndromes in diabetes. *Diabetes* 54, 3592–3601.

Cryer PE (2008). Hypoglycemia: still the limiting factor in the glycemic management of diabetes. *Endocr Pract* 14, 750–756.

Cryer PE (2009). Exercise-related hypoglycemia-associated autonomic failure in diabetes. *Diabetes* 58, 1951–1952.

Cummings DE and Overduin J (2007). Gastrointestinal regulation of food intake. *J Clin Invest* 117, 13–23.

Cuthbertson D, Smith K, Babraj J, Leese G, Waddell T, Atherton P, et al. (2005). Anabolic signaling deficits underlie amino acid resistance of wasting, aging muscle. *FASEB J* 19, 422–424.

D'Alessio DA, Kahn SE, Leusner CR and Ensinck JW (1994). Glucagon-like peptide 1 enhances glucose tolerance both by stimulation of insulin release and by increasing insulin-independent glucose disposal. *J Clin Invest* 93, 2263–2266.

Dickinson JM, Fry CS, Drummond MJ, Gundermann DM, Walker DK, Glynn EL, et al. (2011). Mammalian target of rapamycin complex 1 activation is required for the stimulation of human skeletal muscle protein synthesis by essential amino acids. *J Nutr* 141, 856–862.

Dodd KM and Tee AR (2012). Leucine and mTORC1: a complex relationship. *Am J Physiol Endocrinol Metab* 302, E1329–E1342.

Drummond MJ, Dickinson JM, Fry CS, Walker DK, Gundermann DM, Reidy PT, et al. (2012). Bed rest impairs skeletal muscle amino acid transporter expression, mTORC1 signaling, and protein synthesis in response to essential amino acids in older adults. *Am J Physiol Endocrinol Metab* 302, E1113–E1122.

Fares EJ and Kayser B (2011). Carbohydrate mouth rinse effects on exercise capacity in pre- and postprandial states. *J Nutr Metab* 2011, 385962.

Friedman JM and Halaas JL (1998). Leptin and the regulation of body weight in mammals. *Nature* 395, 763–770.

Galassetti P, Tate D, Neill RA, Richardson A, Leu SY and Davis SN (2006). Effect of differing antecedent hypoglycemia on counterregulatory responses to exercise in type 1 diabetes. *Am J Physiol Endocrinol Metab* 290, E1109–E1117.

Gallen IW, Hume C and Lumb A (2011). Fuelling the athlete with type 1 diabetes. *Diabetes Obes Metab* 13, 130–136.

Gant N, Stinear CM and Byblow WD (2010). Carbohydrate in the mouth immediately facilitates motor output. *Brain Res* 1350, 151–158.

Hawley JA, Schabort EJ, Noakes TD and Dennis SC (1997). Carbohydrate-loading and exercise performance. An update. *Sports Med* 24, 73–81.

Hawley JA, Burke LM, Phillips SM and Spriet LL (2011). Nutritional modulation of training-induced skeletal muscle adaptations. *J Appl Physiol* 110, 834–845.

Henquin JC (2009). Regulation of insulin secretion: a matter of phase control and amplitude modulation. *Diabetologia* 52, 739–751.

Holloszy JO (2005). Exercise-induced increase in muscle insulin sensitivity. *J Appl Physiol* 99, 338–343.

Holst JJ (2007). The physiology of glucagon-like peptide 1. *Physiol Rev* 87, 1409–1439.

Hyde R, Peyrollier K and Hundal HS (2002). Insulin promotes the cell surface recruitment of the SAT2/ATA2 system A amino acid transporter from an endosomal compartment in skeletal muscle cells. *J Biol Chem* 277, 13628–13634.

Jewell JL, Russell RC and Guan KL (2013). Amino acid signalling upstream of mTOR. *Nat Rev Mol Cell Biol* 14, 133–139.

Jiang G and Zhang BB (2003). Glucagon and regulation of glucose metabolism. *Am J Physiol Endocrinol Metab* 284, E671-E678.

Jordan SD, Konner AC and Bruning JC (2010). Sensing the fuels: glucose and lipid signaling in the CNS controlling energy homeostasis. *Cell Mol Life Sci* 67, 3255–3273.

Jorpes E, Mutt V and Olbe L (1959). On the biological assay of cholecystokinin and its dosage in cholecystography. *Acta Physiol Scand* 47, 109–114.

Kim E and Guan KL (2009). RAG GTPases in nutrient-mediated TOR signaling pathway. *Cell Cycle* 8, 1014–1018.

Kimball SR and Jefferson LS (2005). Role of amino acids in the translational control of protein synthesis in mammals. *Semin Cell Dev Biol* 16, 21–27.

Kissileff HR, Carretta JC, Geliebter A and Pi-Sunyer FX (2003b). Cholecystokinin and stomach distension combine to reduce food intake in humans. *Am J Physiol Regul Integr Comp Physiol* 285, R992–R998.

Kristensen M and Jensen MG (2011). Dietary fibres in the regulation of appetite and food intake. Importance of viscosity. *Appetite* 56, 65–70.

Lane SC, Bird SR, Burke LM and Hawley JA (2013). Effect of a carbohydrate mouth rinse on simulated cycling time-trial performance commenced in a fed or fasted state. *Appl Physiol Nutr Metab* 38, 134–139.

Lewis LD and Williams JA (1990). Regulation of cholecystokinin secretion by food, hormones, and neural pathways in the rat. *Am J Physiol* 258, G512–G518.

Luyckx AS and Lefebvre PJ (1974). Mechanisms involved in the exercise-induced increase in glucagon secretion in rats. *Diabetes* 23, 81–93.

Lyly M, Ohls N, Lahteenmaki L, Salmenkallio-Marttila M, Liukkonen KH, Karhunen L, et al. (2010). The effect of fibre amount, energy level and viscosity of beverages containing oat fibre supplement on perceived satiety. *Food Nutr Res* 54.

Mangelsdorf DJ and Evans RM (1995). The RXR heterodimers and orphan receptors. *Cell* 83, 841–850.

Massa ML, Gagliardino JJ and Francini F (2011). Liver glucokinase: An overview on the regulatory mechanisms of its activity. *IUBMB Life* 63, 1–6.

Mettler S, Mitchell N and Tipton KD (2010). Increased protein intake reduces lean body mass loss during weight loss in athletes. *Med Sci Sports Exerc* 42, 326–337.

Miller BF, Olesen JL, Hansen M, Dossing S, Crameri RM, Welling RJ, et al. (2005). Coordinated collagen and muscle protein synthesis in human patella tendon and quadriceps muscle after exercise. *J Physiol* 567, 1021–1033.

Moore DR, Robinson MJ, Fry JL, Tang JE, Glover EI, Wilkinson SB, et al. (2009). Ingested protein dose response of muscle and albumin protein synthesis after resistance exercise in young men. *Am J Clin Nutr* 89, 161–168.

Moran TH (2000). Cholecystokinin and satiety: current perspectives. *Nutrition* 16, 858–865.

Mountjoy PD and Rutter GA (2007). Glucose sensing by hypothalamic neurones and pancreatic islet cells: AMPle evidence for common mechanisms? *Exp Physiol* 92, 311–319.

Newsholme P, Gaudel C and McClenaghan NH (2010). Nutrient regulation of insulin secretion and beta-cell functional integrity. *Adv Exp Med Biol* 654, 91–114.

Nicklin P, Bergman P, Zhang B, Triantafellow E, Wang H, Nyfeler B, et al. (2009). Bidirectional transport of amino acids regulates mTOR and autophagy. *Cell* 136, 521–534.

Nilsson LH and Hultman E (1973). Liver glycogen in man–the effect of total starvation or a carbohydrate-poor diet followed by carbohydrate refeeding. *Scand J Clin Lab Invest* 32, 325–330.

Pasiakos SM, Vislocky LM, Carbone JW, Altieri N, Konopelski K, Freake HC, et al. (2010b). Acute energy deprivation affects skeletal muscle protein synthesis and associated intracellular signaling proteins in physically active adults. *J Nutr* 140, 745–751.

Pennings B, Boirie Y, Senden JM, Gijsen AP, Kuipers H and van Loon LJ (2011). Whey protein stimulates postprandial muscle protein accretion more effectively than do casein and casein hydrolysate in older men. *Am J Clin Nutr* 93, 997–1005.

Petersen KF, Dufour S, Savage DB, Bilz S, Solomon G, Yonemitsu S, et al. (2007). The role of skeletal muscle insulin resistance in the pathogenesis of the metabolic syndrome. *Proc Natl Acad Sci U S A* 104, 12587–12594.

Peyrollier K, Hajduch E, Blair AS, Hyde R and Hundal HS (2000). L-Leucine availability regulates phosphatidylinositol 3-kinase, p70 S6 kinase and glycogen synthase kinase-3 activity in L6 muscle cells: evidence for the involvement of the mammalian target of rapamycin (mTOR) pathway in the L-leucine-induced up-regulation of system A amino acid transport. *Biochem J* 350 Pt 2, 361–368.

Plaisancie P, Dumoulin V, Chayvialle JA and Cuber JC (1996). Luminal peptide YY-releasing factors in the isolated vascularly perfused rat colon. *J Endocrinol* 151, 421–429.

Ramnanan CJ, Edgerton DS, Kraft G and Cherrington AD (2011). Physiologic action of glucagon on liver glucose metabolism. *Diabetes Obes Metab* 13 Suppl 1, 118–125.

Rocha DM, Faloona GR and Unger RH (1972). Glucagon-stimulating activity of 20 amino acids in dogs. *J Clin Invest* 51, 2346–2351.

Rollo I and Williams C (2011). Effect of mouth-rinsing carbohydrate solutions on endurance performance. *Sports Med* 41, 449–461.

Sancak Y, Bar-Peled L, Zoncu R, Markhard AL, Nada S and Sabatini DM (2010). Ragulator-Rag complex targets mTORC1 to the lysosomal surface and is necessary for its activation by amino acids. *Cell* 141, 290–303.

Schwartz MW (2000). Biomedicine. Staying slim with insulin in mind. *Science* 289, 2066–2067.

Simpson KA, Martin NM and Bloom SR (2009). Hypothalamic regulation of food intake and clinical therapeutic applications. *Arq Bras Endocrinol Metabol* 53, 120–128.

Spanswick D, Smith MA, Mirshamsi S, Routh VH and Ashford ML (2000). Insulin activates ATP-sensitive K+ channels in hypothalamic neurons of lean, but not obese rats. *Nat Neurosci* 3, 757–758.

Strader AD and Woods SC (2005). Gastrointestinal hormones and food intake. *Gastroenterology* 128, 175–191.

Suzuki K, Simpson KA, Minnion JS, Shillito JC and Bloom SR (2010). The role of gut hormones and the hypothalamus in appetite regulation. *Endocr J* 57, 359–372.

Tanaka K, Inoue S, Nagase H and Takamura Y (1990). Modulation of arginine-induced insulin and glucagon secretion by the hepatic vagus nerve in the rat: effects of celiac vagotomy and administration of atropine. *Endocrinology* 127, 2017–2023.

Tang JE, Moore DR, Kujbida GW, Tarnopolsky MA and Phillips SM (2009a). Ingestion of whey hydrolysate, casein, or soy protein isolate: effects on mixed muscle protein synthesis at rest and following resistance exercise in young men. *J Appl Physiol* 107, 987–992.

Tipton KD, Ferrando AA, Phillips SM, Doyle D, Jr. and Wolfe RR (1999a). Postexercise net protein synthesis in human muscle from orally administered amino acids. *Am J Physiol* 276, E628–E634.

Vuksan V, Panahi S, Lyon M, Rogovik AL, Jenkins AL and Leiter LA (2009). Viscosity of fiber preloads affects food intake in adolescents. *Nutr Metab Cardiovasc Dis* 19, 498–503.

Walberg JL, Leidy MK, Sturgill DJ, Hinkle DE, Ritchey SJ and Sebolt DR (1988). Macronutrient content of a hypoenergy diet affects nitrogen retention and muscle function in weight lifters. *Int J Sports Med* 9, 261–266.

Wang GJ, Tomasi D, Backus W, Wang R, Telang F, Geliebter A, et al. (2008). Gastric distention activates satiety circuitry in the human brain. *Neuroimage* 39, 1824–1831.

Westerterp-Plantenga MS, Lemmens SG and Westerterp KR (2012). Dietary protein – its role in satiety, energetics, weight loss and health. *Br J Nutr* 108 Suppl 2, S105–S112.

Winder WW, Wilson HA, Hardie DG, Rasmussen BB, Hutber CA, Call GB, et al. (1997). Phosphorylation of rat muscle acetyl-CoA carboxylase by AMP-activated protein kinase and protein kinase A. *J Appl Physiol* 82, 219–225.

Winder WW, Holmes BF, Rubink DS, Jensen EB, Chen M and Holloszy JO (2000). Activation of AMP-activated protein kinase increases mitochondrial enzymes in skeletal muscle. *J Appl Physiol* 88, 2219–2226.

Yang Y, Breen L, Burd NA, Hector AJ, Churchward-Venne TA, Josse AR, et al. (2012). Resistance exercise enhances myofibrillar protein synthesis with graded intakes of whey protein in older men. *Br J Nutr* 1–9.

Yeo WK, Carey AL, Burke L, Spriet LL and Hawley JA (2011). Fat adaptation in well-trained athletes: effects on cell metabolism. *Appl Physiol Nutr Metab* 36, 12–22.

Zhang Y, Proenca R, Maffei M, Barone M, Leopold L and Friedman JM (1994). Positional cloning of the mouse obese gene and its human homologue. *Nature* 372, 425–432.

At the end of the chapter you should be able to:

- Explain insulin resistance and β-cell dysfunction and discuss how these two factors lead to the development of type 2 diabetes mellitus.

- Discuss how genetic variation and epigenetics contribute to the risk of developing type 2 diabetes genetics.

- Explain how human evolution might have shaped our genome in a way so that a sedentary lifestyle and obesity increase the risk of developing type 2 diabetes mellitus.

- Explain the molecular mechanisms by which acute exercise increases the uptake of glucose into skeletal muscle and the possible role of exercise as an insulin-mimetic.

- Explain the molecular mechanisms by which exercise improves and maintains insulin sensitivity for several days after exercise.

9 Human evolution, type 2 diabetes mellitus and exercise

Kian-Peng Goh, Angela Koh and Henning Wackerhage

INTRODUCTION

Type 2 diabetes is a disease where high concentrations of blood glucose occur due to the combination of **insulin resistance** and β-**cell dysfunction**. In this chapter, we will first explain type 2 diabetes and then discuss how environmental factors, genetic predisposition and possibly epigenetic mechanisms lead to the development of type 2 diabetes. After introducing the disease we will review human evolution and argue that there is a now a divergence between our ancient genomes that have been selected for high physical activity and feast famine cycles and our current lifestyles, which are marked by low physical activity and high food intake. The consequences are chronic diseases such as type 2 diabetes mellitus, which we discuss in this chapter. Finally, we will review the mechanisms by which exercise increases glucose uptake acutely and increases insulin-stimulated glucose uptake in the days after exercise.

For this chapter you should have read the sections on insulin release by the β-cells of the pancreas and insulin-mediated glucose uptake in Chapter 8 and the signalling pathways that regulate mitochondrial biogenesis in Chapter 4.

TYPE 2 DIABETES MELLITUS AND OTHER CHRONIC DISEASES

In 2001, 1.36 million (17.3% of total deaths) individuals in high-income countries died of **ischaemic heart disease**, especially heart attacks, 1.03 million (13%) of lung, colon, rectum, breast and stomach **cancer**, 0.78 million (9.9%) of **cerebrovascular disease** or stroke and 200 000 (2.6%) of **diabetes mellitus** (Lopez et al., 2006). These data hide the true effect of diabetes somewhat because diabetes is a risk factor for both ischaemic heart disease and cerebrovascular disease and thus many of these deaths can also be attributed to diabetes mellitus.

The diagnosis of diabetes mellitus is based on plasma glucose levels, either in the fasting state or 2 h after a 75 g oral glucose tolerance test (OGTT). Since 1997, the diagnostic cut-off points for fasting plasma glucose and 2 h values are 7.0 mmol/L and 11.1 mmol/L respectively (ADA, 2013). There is an intermediate group of patients whose blood glucose levels do not meet the diagnostic criteria for diabetes, yet are higher than normal. They form an entity known as prediabetes and can be subdivided into two groups (ADA, 2013):

- those with impaired fasting glucose (fasting plasma glucose of 5.6–6.9 mmol/L)
- those with impaired glucose tolerance (2 h glucose in OGTT of 7.8–11.0 mmol/L).

The importance of recognizing these individuals lies in their higher risk of developing diabetes mellitus in the future, as well as the effectiveness of lifestyle measures in lowering this risk, which we shall discuss later.

From an epidemiological perspective, it is estimated that between 221 and 347 million individuals out of a global population of ≈7 billion individuals were diagnosed with diabetes mellitus in 2010 and this trend is still increasing (Zimmet et al., 2001; Danaei et al., 2011). The increasing prevalence in diabetes mellitus consistently follows the increase in obesity rates, suggesting that body weight and lifestyle factors are major contributors (Finucane et al., 2011). In populations experiencing increased affluence, obesity and type 2 diabetes mellitus have also been found to occur earlier in life. For example, impaired glucose tolerance was detected in 25% of obese children (4–10 years) and type 2 diabetes in 4% of 11–18 year olds (Sinha et al., 2002). Hence, type 2 diabetes is no longer the mature-onset disease it once used to be.

Diabetes mellitus can be broadly classified into two major classes: type 1 and type 2 (ADA, 2013):

- **Type 1** diabetes mellitus (T1DM) accounts for 5–10% of patients with diabetes and is an immune-mediated disease in which the body's own immune system destroys the insulin-secreting β-cells of the pancreas. The consequence of β-cell destruction is insulin deficiency. This means that the normal spike in insulin levels after eating is absent. Hence, insulin-mediated glucose uptake by skeletal muscle, liver and fat becomes impaired and hyperglycaemia ensues. As a lack of insulin (insulinopaenia) is the main defect in T1DM, patients are usually treated with insulin injections.

- **Type 2** diabetes mellitus (T2DM) occurs due to combined insulin resistance and β-cell dysfunction. It is the commonest form of diabetes, accounting for 80–90% of those with diabetes. Insulin resistance refers to a decreased glucose uptake, especially by skeletal muscle, fat and the liver at a given insulin concentration. On the other hand, β-cell dysfunction occurs when the β-cells of the pancreas release less insulin at a given blood glucose concentration.

Poorly controlled diabetes can lead to potentially life-threatening emergencies such as diabetic ketoacidosis or a hyperosmolar, hyperglycaemic state. However, most of the deaths in diabetic patients result from the long-term complications of sustained hyperglycaemia. These complications are often touted as the true burden of the disease and can affect any organ of the body. The major complications include the following:

- **Diabetic retinopathy** is damage of the retina of the eye. If untreated it can progress to blindness.
- **Diabetic nephropathy** is kidney disease which can develop into renal failure requiring dialysis.
- **Diabetic neuropathy** is damage to the peripheral or autonomic nerves.
- **Macrovascular complications** are the result of atherosclerosis (blockage of the blood vessels as a result of cholesterol accumulation and deposition) of major vessels which can lead to myocardial infarctions or strokes.

ENVIRONMENTAL CAUSES AND PATHOPHYSIOLOGY OF TYPE 2 DIABETES MELLITUS

To explain how type 2 diabetes mellitus develops, we will first introduce insulin resistance, which is one of the hallmarks of the disease. Insulin resistance is defined as a state of decreased insulin-mediated glucose uptake by the body in response to physiological insulin concentrations (Levy-Marchal, 2010). In other words, at given insulin concentrations less glucose is transported from the blood into the organs. The consequence is an elevated blood glucose concentration which is known as hyperglycaemia.

At first, insulin resistance is no major problem because the body compensates by secreting up to 4–5 times more insulin than usual (Kahn et al., 2006). This is known as compensatory hyperinsulinaemia and ensures that the blood glucose levels are near normal despite insulin resistance. If left untreated then the insulin-releasing β-cells in the pancreas eventually become impaired and secrete less insulin than is necessary to maintain normal blood glucose levels. Mild hyperglycaemia is the consequence, as insufficient glucose is taken up by the organs. As the disease progresses, β-cells deteriorate further until hyperglycaemia progresses to the point where the blood glucose level crosses the diagnostic threshold and diabetes is finally declared. These stages likely represent a continuum rather than distinct entities; a typical progression of the disease is illustrated in simplified form in Table 9.1.

	IMPAIRED GLUCOSE TOLERANCE/ FASTING GLUCOSE		TYPE 2 DIABETES MELLITUS		
	STAGE I	STAGE II	STAGE III	STAGE IV	STAGE V
HbA1c (%)	<5.5	5.5–6.1	6.2–7.5	7.6–10	>10
Fasting glucose (mmol/L)	<6.1	6.1–6.9	7.0–8.9	8.9–13.3	>13.3
Insulin resistance	Moderate	Moderate	Moderate	Moderate-severe	Severe
Insulin release	↑↑↑	↑↑	↑	↓ or ↔	↓↓↓

From Reasner and DeFronzo (2001).

Note that the 'stages' shown are fluid and not universally accepted.

HbA1c (%) is glycated haemoglobin, which is a measure of average glucose concentrations over the last three months. High HbA1c percentages indicate poor glycaemic control.

PATHOGENESIS OF INSULIN RESISTANCE

We will now review the molecular defects that trigger **insulin resistance** and **β-cell dysfunction,** or the pathogenesis of type 2 diabetes mellitus.

Most glucose is taken up by skeletal muscle and then by the liver. In insulin-resistant subjects, insulin-stimulated glucose uptake and glycogen synthesis, especially by skeletal muscle, are reduced when compared to that in healthy subjects (Petersen and Shulman, 2006; Abdul-Ghani and DeFronzo, 2010). Many mechanisms have been proposed to link a sedentary lifestyle and high food intake to insulin resistance. Here we will briefly discuss the most important of these hypotheses.

Insulin resistance due to fatty acids and fat-related metabolites hypothesis

One key factor appears to be the fat content of organs that take up glucose, as high fat concentrations in liver and skeletal muscle cause insulin resistance. This phenomenon is perhaps best demonstrated in 'fatless' or lipodystrophic mice, which have no white fat. In these mice, ingested fat cannot be deposited in fat tissue and thus ends up in skeletal muscle and the liver instead. This in turn triggers severe insulin resistance in muscle and liver (Kim et al., 2000). If white fat is transplanted into 'fatless' mice, then ingested fat is deposited in the transplanted fat tissue instead of muscle and liver and as a result insulin resistance normalizes (Figure 9.1).

The lipodystrophic mouse model demonstrates the link between fat content and insulin resistance, but what is the underlying mechanism? There are several theories. One possibility is that free fatty acids (FFAs), which are generated from fat, and glucose compete for uptake and oxidation (Randle et al., 1963). Thus if there are many FFAs then glucose oxidation is low. Support for Randle's hypothesis comes from so–called euglycaemic–hyperinsulinaemic and hyperglycaemic–hyperinsulinaemic clamp studies. These studies show that fat infusion reduces glucose uptake at a given insulin concentration, which

Figure 9.1 Fat and insulin sensitivity. Relationship between the (a) fat concentration in skeletal muscle and (b) insulin-stimulated glucose transport/uptake in skeletal muscle glucose transport into skeletal muscle in three types of mice. Fatless (A-ZIP/F-1 or lipodystrophic) mice have no adipose tissue and as a consequence more fat is deposited in skeletal muscle. This leads to insulin resistance. If fat tissue is transplanted into fatless mice then less fat is deposited into skeletal muscle and insulin-stimulated glucose uptake into skeletal muscle is normalized. This demonstrates that the fat content of organs such as skeletal muscle and the liver (not shown here) negatively affects insulin-stimulated glucose uptake. Redrawn after Kim et al. (2000).

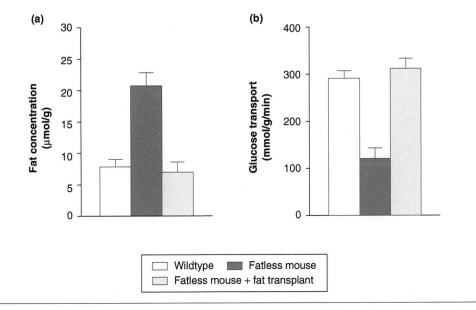

again identifies a high fat concentration as a cause of insulin resistance (Ferrannini et al., 1983), probably because of the competition between fat and glucose oxidation (Randle et al., 1963). However, there might be another reason. Some FFAs are broken down into **DAG** (diacylglycerol), **fatty acyl-CoA** and **ceramides** (Kahn et al., 2006). These factors, especially ceramides, inhibit the insulin-signalling cascade and further reduce insulin-stimulated glucose uptake (Kahn et al., 2006; Petersen and Shulman, 2006). However, while there seems to be a link and plausible mechanisms between a high fat content of skeletal muscle and liver, there is one observation which does not fit this pattern. Highly trained endurance athletes have a commonly increased muscle fat concentration, but in contrast to sedentary subjects, endurance athletes do not generally suffer from insulin resistance (Amati et al., 2011). The reason for this paradox is incompletely understood but it may be related to the difference in the types of fat and metabolites in an athlete's muscle when compared to the muscle of a sedentary subject (Amati et al., 2011).

Insulin resistance due to adipokines and pro-inflammatory mediators hypothesis

Fat tissue communicates with the other tissues of the body by releasing fat tissue hormones called **adipokines**. Some examples of adipokines include leptin, adiponectin, retinol-binding protein-4 (RBP4) and resistin. Some adipokines affect insulin sensitivity in other organs such as skeletal muscle and liver by activating AMPK (Kahn et al., 2006; Muoio and Newgard, 2008) and thus a defective adipokine release in subjects with high fat mass may contribute to insulin resistance. In addition, high levels of FFA can trigger the release of pro-inflammatory mediators such as TNFα, IL-6 and IL-1β from immune cells and the liver. These mediators can also contribute to insulin resistance (Kahn et al., 2006; Muoio and Newgard, 2008).

Insulin resistance due to mitochondrial dysfunction or deficiency hypothesis

Mitochondria generate the ATP that fuels muscle contractions and the basic energy demands of virtually all cells. In Chapter 4 we have reviewed how exercise increases the expression and activity of PGC-1α, which is the master regulator of mitochondrial biogenesis. Interestingly, the expression of PGC-1α has been reported to be lower in individuals with prediabetes and type 2 diabetes mellitus when compared to non-diabetic controls (Mootha et al., 2003; Patti et al., 2003). Moreover, a ≈30% reduction in the concentration of mitochondria was found in individuals with insulin resistance and type 2 diabetes mellitus. It has been suggested that fewer and/or less-functional mitochondria reduce the ability of the cell to oxidize glucose and other nutrients and that this contributes to insulin resistance (Lowell and Shulman, 2005). This is an intriguing hypothesis but it is disputed, because at rest in a healthy person both fat and carbohydrate oxidation are far below their maximal exercise rates in skeletal muscle and thus even a much decreased number of mitochondria should have no effect on glucose oxidation and hence insulin sensitivity (Holloszy, 2009). Thus it is unclear whether mitochondrial dysfunction contributes to type 2 diabetes mellitus.

Insulin resistance due to endoplasmic reticulum (ER) stress hypothesis

The endoplasmic reticulum (ER) is an organelle where proteins, lipids and other molecules are synthesized and metabolized. In the ER proteins like insulin are folded and secreted. ER stress can be induced by obesity and high concentrations of FFAs; it triggers the unfolded protein response which is believed to worsen insulin resistance and β-cell dysfunction (Cnop et al., 2011).

Insulin resistance due to O-GlcNAc modification hypothesis

The so-called **hexosamine biosynthetic pathway** converts glucose into a signalling monosaccharide called O-linked β-N-acetylglucosamine (O-GlcNAc). O-GlcNAc can be added to or removed from other proteins by the action of enzymes, a process similar to phosphorylation and dephosphorylation. One protein that is affected by O-GlcNAc modification is the insulin-receptor substrate (IRS). Normally activated by insulin, IRS conveys information from the insulin receptor to glucose uptake and glycogen synthesis through an impressive signalling cascade. O-GlcNAc modification of IRS and other key proteins in the insulin-signalling pathway may downregulate this pathway and result in insulin resistance (Hart et al., 2011).

To summarize, there are multiple mechanisms likely to contribute to the development of insulin resistance. It seems that excess fat in skeletal muscle and liver is a key driver of insulin resistance via several mechanisms. A reduced number of mitochondria, ER stress and O-GlcNAc modification are all potential contributing mechanisms.

PATHOGENESIS OF β-CELL DYSFUNCTION

The second defect of type 2 diabetes mellitus is β-**cell dysfunction**. This indicates that β-cells, which are located within the islets of Langerhans of the pancreas, secrete insufficient amounts of insulin to maintain a normal glucose concentration in the blood. This β-cell dysfunction is an important factor in the development of type 2 diabetes mellitus and can develop early on during the progression of an individual towards type 2 diabetes (Reasner and DeFronzo, 2001; Petrie et al., 2011). What causes β-cell dysfunction? First, several studies suggest that β-cell mass is reduced by up to 63% in patients with type 2 diabetes mellitus compared to controls (Kahn et al., 2009). However, a reduction of β-cell mass is probably not the sole explanation because a significant proportion of individuals with up to 50% of the pancreas removed failed to develop diabetes mellitus (Kahn et al., 2009). Thus while a reduced β-cell mass might contribute to the development of type 2 diabetes mellitus, it is probably not the only factor responsible. Another reason for β-cell dysfunction is glucolipotoxicity leading to ER stress, which may also contribute to insulin resistance as we have discussed above. The ER is a particularly good candidate for the decline of β-cell function because in the ER insulin and other secreted proteins are synthesized and folded (Cnop et al., 2011).

GENETIC CAUSES OF TYPE 2 DIABETES MELLITUS

Type 2 diabetes mellitus is also partially inherited. Generally, a person's genetic predisposition determines the threshold at which diabetes-related misbehaviour, such as a sedentary lifestyle or overeating, will trigger type 2 diabetes mellitus. The importance of genetics for the development of type 2 diabetes mellitus is perhaps best demonstrated by the effect of a family history on developing type 2 diabetes mellitus. The lifetime risk is 38% if one parent is affected and almost 60% if both parents are affected (Stumvoll et al., 2005). As we have already said, in most cases type 2 diabetes mellitus is a polygenic disease that is triggered by lifestyle changes. However, there are some rare instances of families where the mutation of a single gene is responsible. One example is the PKBβ/Akt2 gene, which mediates insulin-induced glycogen synthesis and GLUT4-dependent glucose uptake. In this case, a glycine-to-alanine substitution in position 274 of the protein results in severe insulin resistance and type 2 diabetes mellitus (George et al., 2004). Such single-gene mutations also play a role in maturity-onset diabetes of the young (MODY), also known as monogenic diabetes. It is an autosomal dominant form of diabetes caused by a mutation in a gene responsible for normal insulin secretion. Eleven genes have been implicated so far, giving rise to the subtypes MODY 1–11. However, such single-gene mutations are rare.

More common DNA sequence variations or polymorphisms have been identified with genome-wide association studies (GWAS). When GWAS were first introduced, there was an expectation that many type 2 diabetes mellitus susceptibility DNA sequence variations would be identified. The first landmark GWAS by the Wellcome Trust Case Control Consortium identified the following as candidate genes for type 2 diabetes mellitus: *PPARG* (peroxisomal proliferative activated receptor gamma), *KCNJ11* (the inwardly rectifying Kir6.2 component of the pancreatic beta-cell KATP channel), *FTO* (fat-mass and obesity-associated), *CDKAL1* (CDK5 regulatory subunit-associated protein 1-like 1), *HHEX* (homeobox, haematopoietically expressed), *IDE* (insulin-degrading enzyme) and *TCF7L2* (transcription factor 7-like 2) (The Wellcome Trust Case Control Consortium, 2007). To date 38 single-nucleotide polymorphisms (SNPs) have been linked to type 2 diabetes mellitus. This is in addition to nearly two dozen others that are associated with glycaemic traits (Billings and Florez, 2010). Interestingly, the majority of loci identified so far point to defects related to β-cells with the exception of the *FTO* gene variants, which mediate their effects on disease susceptibility through increased adiposity. More importantly, the currently identified loci confer a modest effect size of only up to ≈10% in terms of disease susceptibility (Billings and Florez, 2010). Hence, a large portion of the heritability remains unexplained, which has been termed the '**missing heritability**'. It is unclear whether this is because the heritability that is explained by these variants has been underestimated or whether rare alleles explain much of the heritability of type 2 diabetes mellitus.

FETAL NUTRITION, EPIGENETICS AND TYPE 2 DIABETES MELLITUS

At the end of the first part of this chapter we will discuss links between the nutrition of a fetus in the womb and the risk of developing type 2 diabetes mellitus and other diseases later in life. Such relationship was first demonstrated by comparing the relationship between birth weight, which depends on fetal nutrition, and disease later in life (Barker, 1997). The results of this study showed that a lower birth weight was significantly correlated with an increased risk of developing insulin resistance, type 2 diabetes mellitus and other related diseases in adulthood (Table 9.2).

Table 9.2 Prevalence of type 2 diabetes and impaired glucose tolerance in men aged 59–70 years		
BIRTHWEIGHT (KG)	INSULIN RESISTANCE	IMPAIRED GLUCOSE TOLERANCE AND/ OR TYPE 2 DIABETES MELLITUS
≤2.5	30%	40%
Up to 2.95	19%	34%
Up to 3.41	17%	31%
Up to 3.86	12%	22%
Up to 4.31	6%	13%
>4.31	6%	14%

From Barker (1997).

Further support for a link between fetal nutrition and disease risk comes from three retrospective famine studies on pregnant women and children during the Dutch Winter Famine (1944–1945), Leningrad Siege (1941–1944) and Chinese Famine (1959–1961). This has led to the hypothesis that undernutrition is sensed by the fetus or child and leads to long-term adaptive changes which prepare the individual for a life of malnourishment to ensure the best chance of survival (Gluckman et al., 2009). This is also known as the 'Barker hypothesis', 'fetal origins hypothesis' or, most recently, the 'thrifty phenotype hypothesis'.

So what could be the molecular mechanism that explains this fetal or early life programming that persists for the rest of the person's life? It is not completely understood, but epigenetic regulation or epigenetics is a likely cause. Epigenetics refers to mechanisms that cause 'long-term changes in gene expression that occur in the absence of changes to the DNA sequence' (Dolinoy, 2008). Some of these epigenetic changes include DNA methylation and histone modification, which open up or close down stretches of DNA and genes. Studies in rats on either a protein-restricted or normal diet showed that the liver of the protein-restricted offspring had a lower methylation but increased expression of the PPARα gene and higher expression of the key β-oxidation enzyme acyl-CoA oxidase regulated by PPARα (reviewed by Gluckman et al., 2009). This is a proof-of-principle experiment, and more mechanistic evidence is needed to develop this hypothesis.

To summarize the first part of this chapter, type 2 diabetes mellitus is a high concentration of blood glucose due to the combination of **insulin resistance** and **β-cell dysfunction**. Insulin resistance is caused by genetic predisposition and factors such as high concentrations of fat and fat-related metabolites, possibly a reduced number and function of mitochondria, the signalling of adipokines, pro-inflammatory mediators, ER stress and O-GlcNAc modification. In contrast, a loss of β-cells and ER stress are discussed as possible environmental causes of β-cell dysfunction. Type 2 diabetes mellitus is partially inherited as a family history of type 2 diabetes increases an individual's risk of developing type 2 diabetes mellitus. Specific mutations have been associated with maturity-onset diabetes of the young (MODY) and GWAS studies have identified some genes where DNA sequence variations explain a small fraction of the heritability of type 2 diabetes. Finally, low birth weight increases the risk of developing type 2 diabetes mellitus and it is believed that this may be due to epigenetic regulation.

EVOLUTION OF HUMAN EXERCISE CAPACITY, METABOLISM AND TYPE 2 DIABETES MELLITUS

In the first part of this chapter we defined type 2 diabetes mellitus and discussed environmental, genetic and epigenetic causes of the disease on a molecular level, where possible. In the second part of this chapter we will first make the case that humans are, as a result of evolution, an exercise-dependent species and that therefore the lack of physical activity in combination with overeating causes chronic diseases such as type 2 diabetes mellitus. We will then go on to explain the molecular mechanisms by which exercise improves insulin resistance and type 2 diabetes mellitus.

Evolution describes the change in the inherited characteristics of living beings over time and how it gives rise to diversity. It occurs because advantageous DNA sequence variations are naturally selected. Evolution as a theory is well supported not only by indirect observations but also directly by experiments. For example, when yeast is cultured for ≈200 generations in a low-nutrient medium, DNA sequence variations that increase survival in a low-nutrient medium are selected, proving experimentally a key principle of evolution (Gresham et al., 2008).

How did humans evolve? Fossil finds suggest that hominins or humans evolved in the arid area east of the East African rift from ≈7 million years ago, while the great apes evolved in the rainforest to the west. The East African rift is a barrier that catches rain clouds coming from the west and, as a consequence, a dry open savannah lies to the east (Bramble and Lieberman, 2004; Cerling et al., 2011). Consistent with this theory, tooth-mark analysis of early hominins suggests a predominantly grass-based diet. This is in contrast to the mainly fruit-eating great apes (Ungar and Sponheimer, 2011).

There is evidence that the evolution of hominins in the savannah included changes to parts of the anatomy and physiology that are related to locomotion and exercise capacity. One of the earliest and most important changes was bipedalism, found in Australopithecus fossils and subsequent hominins from ≈4 million years ago (Ward, 2002). This was a fundamental evolutionary development, as the ability to walk on two legs led to an increase in the achievable travel distance compared to that of the non-human primates or great apes (Isbell and Young, 1996). Hominins then progressed further to evolve as endurance runners later on from ≈2 million years ago (Bramble and Lieberman, 2004). This is supported by other specific developments around that time. Some of these include long Achilles tendons for energy storage, less body hair and more sweat glands for improved thermoregulation and an anatomy that better stabilizes the trunk and head during running (Bramble and Lieberman, 2004). The evidence for this comes from fossil finds and anatomical studies.

As a consequence of the selection of endurance running traits, human endurance capacity today compares well to other species with a high endurance capacity such as horses. This is demonstrated by the fact that few animals will be able to run, for example, a marathon in just over 2 h, which is the current best human time. This evolution of a high endurance running capacity presumably enabled our ancestors to cover larger distances for foraging and to increase their protein intake by searching for cadavers and hunting in the open savannah environment east of the East African rift system.

Limited evidence also exists showing that our muscle physiology and biochemistry have evolved for endurance running. For example the great apes, which have evolved in the rainforest to the west of the East African rift, have more muscle power than humans and can jump higher (Scholz et al., 2006). Histological differences are also consistent with this finding, as chimpanzees have more fast type II fibres in some of their leg muscles than the average human (Myatt et al., 2011). Together, these observations support the hypothesis that hominins have evolved as a low-muscle-power, high-endurance species.

The out-of-Africa model proposes that modern *Homo sapiens*, the only hominin species to date, appeared ≈200 000 years ago and subsequently migrated through Eurasia ≈70 000 years ago to the rest of the world (Soares et al., 2012). Presumably their high intelligence and perhaps their endurance capacity enabled humans to migrate, settle, adapt and further evolve in very diverse environments ranging from polar regions to deserts. Humans also acquired the knowledge and skill to shield themselves sufficiently in order to better survive large environmental fluctuations such as ice ages (the last ice age peaked ≈20 000 years ago), diseases, natural disasters, conflicts and predators. Even though intelligence and technologies such as clothing, weapons and later medicine reduced natural selection, it was not completely eliminated.

The adaptation to specific environments is evident, for example, by the variations in skin colour and the selection of a lactase persistence genotype ≈10 000–5000 years ago after the domestication of cattle. The lactase persistence DNA sequence variations are one of the few examples where we can trace the effect of evolution on our DNA sequences. Normally the milk sugar lactose is poorly digested after childhood, but carriers of lactase persistence alleles retain a high activity of the lactose-digesting enzyme lactase later in life. This enables carriers to digest milk well throughout life (Bersaglieri et al., 2004). Today, carriers of the lactase persistence alleles are frequent in most of Europe but are much less common in Africa and Asia.

To summarize, hominins evolved from an area east of the East African rift. The ability of hominins to cover larger distances increased first with bipedalism ≈4 million years ago and second with the selection for endurance running capacity from ≈2 million years ago. *Homo sapiens* or modern-day humans then migrated out of Africa ≈70 000 years ago and evolved in many diverse environments. Little is known about the specific DNA sequence changes that were selected during hominin evolution and that code for our ability to perform endurance exercise.

EVOLUTION AND TYPE 2 DIABETES MELLITUS

We will now try to link human evolution and the resultant increase of our endurance exercise capacity to type 2 diabetes mellitus and other chronic diseases. The key argument is that industrialization and modern medicine, along with developments in transportation, housing, technology and mass food production have changed our lifestyles profoundly so that we now lead lives for which our genomes have not been selected. Moreover, selective pressures such as predators, diseases and famines are eliminated or can be controlled with modern medicine and because of that individuals who carry disease-predisposing alleles are much more likely to survive and reproduce. There are two main consequences:

- **Genome–lifestyle divergence**: Humans have most likely evolved as a species adapted for a physically active lifestyle and periods of starvation. As a consequence of modernization, physical activity has declined, together with marked changes in food availability (Chakravarthy and Booth, 2004). This has led to genome–lifestyle divergence; we could be said to be leading space age lives with a Stone Age genome. This divergence has been suggested to be a key reason for the rise in chronic diseases, including type 2 diabetes mellitus (Chakravarthy and Booth, 2004).

- **Neutral, disease-predisposing alleles**: As a result of industrialization and modern medicine, previous negative pressures such as famine, disease (in particular infections and malnutrition), and predators have greatly decreased. Consequently, alleles that were detrimental in the past, such as disease susceptibility, poor famine survival, poor predator evasion alleles, are now likely to be neutral because diseases are treated and because famines and predators are no longer a threat. The result is a 'genetic drift' and reduced clearance of disease-predisposing alleles from the overall human genome (Speakman, 2008).

The concept of genome–lifestyle divergence is indirectly supported by the observation that less physical activity (Blair et al., 1996; Archer and Blair, 2011) and a nutrition high in overall calories, saturated fat, salt and sugar (Hill et al., 2009) are major risk factors for the development of many chronic diseases, including type 2 diabetes mellitus. In relation to nutrition, it has also been proposed that an excessive caloric intake leads to obesity and disease because our genomes have been previously selected among others for famine survival. This is known as the **thrifty gene or 'store energy easily' hypothesis** (Neel, 1962, 1999). The argument goes that environmental pressures over time led to the selection of thrifty genes which improve the likelihood of survival after unsuccessful hunts and during famines. This meant that a non-thrifty genotype unable to conserve fuel efficiently would be gradually eliminated from the gene pool. In other words, obesity became an advantage during famines because the more fat an individual possessed, the longer it could survive without food intake. For example, 9 out of the 10 roughly normal-weight Irish hunger strikers died following between 57 and 73 days of starvation. In contrast, a grossly obese individual was reported to have remain fasted for 382 days (Elia, 2000). This demonstrates that the amount of body fat or energy within a human is indeed the key factor that determines survival time during starvation.

Proponents of the 'thrifty genes' hypothesis state that thrifty genes or DNA sequence variations predispose for obesity, type 2 diabetes mellitus and other chronic diseases in the post-industrialization environment. Indeed, the recent history of the island of Nauru provides evidence for this as the dramatic rise in the incidence of type 2 diabetes mellitus on Nauru has been linked to the likely natural selection of thrifty genes during famines followed by a rapid change to a sedentary lifestyle and nutrition-replete environment (see Box 9.1; Diamond, 2003).

BOX 9.1 NAURU: THRIFTY GENES, BAD NUTRITION, SEDENTARY LIFESTYLE AND THE RISE OF CHRONIC DISEASE

Located just south of the Marshall Islands in Micronesia in the South Pacific is Nauru, a 21 km² island that is also the world's smallest republic. As one of the biggest phosphate rock islands in the Pacific Ocean, phosphate mining and exports form one of its most important revenue. Home to just over 9000 islanders, it is not just the rich phosphate deposits that has garnered it much external interest. Nauru also holds the unenviable title of having one of the highest diabetes rates in the world at one time – increasing from 0% to 41% within a brief span of 50 years (1952–2002). This dramatic increase has intrigued doctors and scientists alike. One hypothesis was the double selection theory, which posits that the genes of the ancestral Nauruans underwent an extra bout of selection during the extreme starvation under Japanese occupation in World War 2. This was followed by the relatively sudden switch from a physically active to a sedentary lifestyle after the War as a result of the income and reduced need for manual labour due to the phosphate exports (i.e. non-Nauruan's did the work). It was postulated that these two bouts of natural selection within a short space of time led to the selection of thrifty genes, which ultimately became maladaptive when famines were eventually replaced by abundant food availability after the war. Interestingly, this rise was soon followed by an equally dramatic decrease in the incidence rate of type 2 diabetes mellitus from 17.1 to 7.4 cases per 1000 person-years within just 10 years.

But while the 'thrifty gene' hypothesis can explain at least partially the rise of type 2 diabetes mellitus in Nauru, it is far from clear whether thrifty gene selection is a key contributor to the general rise of type 2 diabetes worldwide, as the frequency of famines and their effect on selection is possibly overestimated by the proponents of the 'thrifty gene' hypothesis. Instead, genetic drift because of reduced selection seems an equally or even more important factor (Speakman, 2008; Prentice et al., 2008). Hopefully, the current extensive use of next-generation sequencing to sequence the genomes of many humans will provide some evidence for or against this hypothesis.

The thrifty gene hypothesis has been extended by stating that both feast–famine and physical activity rest cycles were part of older human lifestyles. A high food intake and sedentary lifestyle breaks these cycles and the consequence is an increased risk of suffering a chronic disease (Chakravarthy and Booth, 2004).

EXERCISE AND TYPE 2 DIABETES MELLITUS

If we have indeed evolved as an high physical activity or even exercise-dependent species, then this implies that a sedentary lifestyle is detrimental and that exercise is the natural treatment for such diseases which are caused by a lack of exercise. Indeed, exercise is an effective intervention for the prevention and treatment of type 2 diabetes mellitus and other chronic diseases. In this section we will now discuss the mechanisms by which exercise improves type 2 diabetes mellitus.

Evidence that exercise is an effective treatment to prevent and treat type 2 diabetes mellitus

The hypothesis that humans have evolved as endurance runners (Bramble and Lieberman, 2004) is consistent with data showing that a sedentary lifestyle is a risk factor for type 2 diabetes mellitus (Hu et al., 2001, 2003) and it can also explain why exercise is an effective treatment to prevent and treat the disease. A key study is the Diabetes Prevention Trial involving over 3000 individuals who were at a high risk of developing type 2 diabetes mellitus (Knowler et al., 2002). The results of this trial show that a lifestyle intervention consisting of intensive exercise and dietary modification is able to reduce the risk of developing type 2 diabetes mellitus in high-risk individuals by 58%, whereas metformin, a common anti-diabetic drug, reduces it by 31% when compared to placebo (Figure 9.2).

Figure 9.2 Effect of exercise/lifestyle intervention on the development of type 2 diabetes mellitus. Cumulative incidence of diabetes in placebo, metformin and lifestyle intervention (lifestyle intervention includes exercise)-treated subjects in a landmark study on the prevention of diabetes mellitus. The data show that a lifestyle intervention with exercise reduces the risk of developing diabetes by 58%. Redrawn after Knowler et al. (2002).

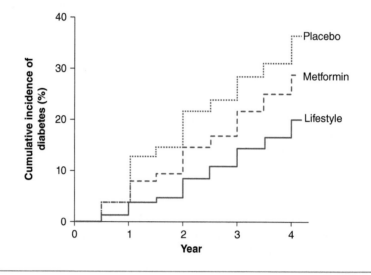

These results are further supported by similar trials performed in other population groups, such as the Da Qing Study and the Finnish Diabetes Prevention Study. Taken together, these studies demonstrate that type 2 diabetes mellitus is a partially preventable disease and that lifestyle modification through physical exercise and exercise can be an effective measure to prevent the disease (Sigal et al., 2006). So what are the molecular mechanisms by which exercise improves insulin resistance, prevents type 2 diabetes mellitus and improves hyperglycaemia?

Exercise stimulates glucose uptake independent of insulin

The effect of insulin on glucose uptake is impaired in type 2 diabetes mellitus as a result of insulin resistance. What about the exercise-stimulated glucose uptake in type 2 diabetes mellitus? Results of *in vivo* studies on type 2 diabetic subjects show that exercise still induces glucose uptake through GLUT4 translocation into the skeletal muscle membrane (Martin et al., 1995; Kennedy et al., 1999). Thus, exercise in these patients will increase glucose uptake even if insulin-stimulated glucose uptake is significantly reduced. For this reason, exercise can be considered as an **insulin mimetic**, a term used to describe a substance or intervention with effects mimicking that of insulin. In general, the higher the exercise intensity and the longer the duration of exercise, the greater the skeletal muscle glucose uptake will be (Rose and Richter, 2005). Moreover, the effects of insulin and exercise are synergistic: in combination, insulin and exercise stimulate glucose uptake more than either one alone (Figure 9.3).

Figure 9.3 Effects of exercise and insulin on glucose uptake. Effects of a high insulin clamp, 30 min of exercise at 40% of $\dot{V}O_{2max}$ only and combined high insulin and exercise on rate of glucose uptake. The figure shows that exercise and insulin together increase glucose uptake more than exercise or insulin alone. Redrawn after DeFronzo et al. (1981).

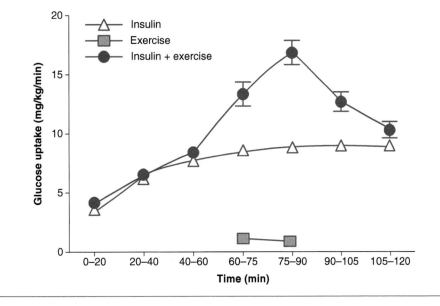

All this suggests that exercise is an effective intervention to increase glucose uptake into skeletal muscle. Thus if a type 2 diabetes mellitus patient were to over-eat, then exercise would be a suitable strategy to lower blood glucose

What are the mechanisms by which exercise and insulin stimulate glucose uptake? In Chapter 8 we have seen how insulin stimulates glucose uptake via a signalling cascade that links the insulin receptor via Akt2 (also known as PKBβ) and AS160 to GLUT4-containing vesicles inside the cell. When insulin triggers this signalling cascade, GLUT4-containing vesicles migrate to the cell membrane and fuse with it. The increased number of GLUT4 transporters in the membrane then increases glucose uptake (Huang and Czech, 2007). Also, glucose delivery and oxidation increase during exercise and contribute to the increased glucose disposal (Romijn et al., 1993).

At a molecular level, exercise is a form of energy consumption which increases the ratios of AMP and ADP to ATP (Xiao et al., 2011). This increase in AMP activates AMPK, which functions as an energy-deficit sensor. AMPK activity is also increased as muscle glycogen, which suppresses AMPK, is lowered with continued exercise (Hudson et al., 2003). One piece of evidence for AMPK involvement comes from transgenic mice studies where both the β1 and β2 AMPK subunits have been knocked out in the skeletal muscle (O'Neill et al., 2011). In these mice, skeletal muscle contraction-induced glucose uptake is greatly reduced when compared to wildtype mice (O'Neill et al., 2011). This suggests that AMPK is a major regulator of exercise-induced glucose uptake.

Muscle contraction is caused by a release of Ca^{2+}, which additionally activates CaMKs (see also Chapter 4). Both CaMKs and AMPK regulate exercise-stimulated glucose uptake independent of insulin (Witczak et al., 2010) by causing the migration of GLUT4-containing vesicles to the outer cell membrane and their subsequent fusion with it. This exercise pathway of glucose uptake is illustrated in more detail in Figure 9.4.

Figure 9.4 Mechanisms by which exercise increases glucose uptake. Exercise increases the glucose uptake of skeletal muscle by moving more GLUT4 transporters from vesicles inside the cell to the cell membrane (Rose and Richter, 2005; Huang and Czech, 2007). (1) Muscle contraction increases the concentrations of AMP and ADP and decreases glycogen, which together result in the activation of AMPK due to the phosphorylation of Thr172 by the upstream kinase LKB1. (2) Ca^{2+} not only triggers muscle contractions but also binds to calmodulin to activate the Ca^{2+}/calmodulin-dependent kinases CaMKK and CaMKII. CaMKK can also phosphorylate AMPK, while CaMKII autophosphorylates itself. (3) The above AMPK and CaMK signalling then inhibits AS160 via phosphorylation. AS160 inhibits Rab protein activation. (4) Thus when AMPK and CaMK are active, intracellular vesicles that contain the glucose transporter GLUT4 migrate to the cell membrane and fuse with it. The result is a higher concentration of GLUT4 transporters in the membrane and an increased glucose uptake into the exercising muscle.

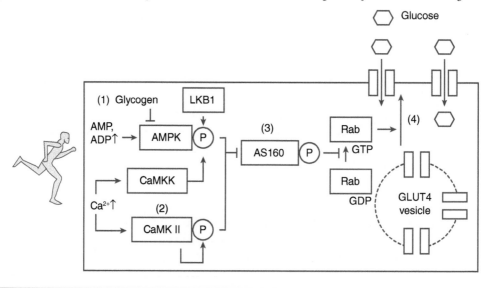

Taken together, acute exercise is an effective tool to reduce blood glucose by increasing glucose delivery, uptake and disposal by oxidation or glycolysis. More importantly, these findings highlight the ability of exercise to stimulate GLUT4 translocation and glucose uptake via AMPK and CaMKs, in contrast to insulin, which stimulates GLUT4 translocation and glucose uptake via Akt2 (PKBβ) and related molecules. The fact that exercise and insulin stimulate glucose disposal via different mechanisms has an important practical implication as exercise can be used to lower blood glucose levels even in insulin-deficient or resistant type 2 diabetic patients.

MECHANISMS BY WHICH EXERCISE IMPROVES INSULIN SENSITIVITY AND TYPE 2 DIABETES MELLITUS

The effect of exercise on glucose uptake and glycaemic control in type 2 diabetes mellitus is not only restricted to acute exercise but insulin sensitivity is also improved for up to

2–3 days (Boule et al., 2005) after endurance exercise (Wang et al., 2009) or resistance exercise (Gordon et al., 2009). How does exercise improve insulin sensitivity? There are at least two types of effects: First, exercise reduces fat and glycogen from muscle and liver through lipolysis and glycogenolysis, respectively. In an oversimplified way, this is equivalent to making space for additional nutrient storage and in the first part of the chapter we discussed how increasing the fat content triggers insulin resistance. Second, exercise activates a signalling cascade centred on AMPK, CaMK, SIRT1 and PGC-1α. The consequences of both processes are adaptations that enhance insulin sensitivity.

Effect of the reduction in fat and glycogen content on insulin sensitivity

Endurance exercise increases the oxidization of intramuscular fat and glycogen (Romijn et al., 1993) and at least temporarily reduces the concentrations of intramuscular fats or triacylglycerides (IMGT) (Watt et al., 2002) and glycogen (Gollnick et al., 1974). Correspondingly, the concentrations of fat and glycogen in the liver and in the adipose tissue will also decrease. As we described earlier in this chapter, a lower fat concentration will improve insulin sensitivity (Kahn et al., 2006; Muoio and Newgard, 2008).

Exercise reduces the concentration of glycogen before recovering to pre-exercise levels and possibly supercompensating to even higher levels (Bergstrom and Hultman, 1966). Reduced glycogen activates glycogen synthase via an unknown mechanism (Danforth, 1965), which increases glycogen synthesis from glucose and may serve as another mechanism of glucose disposal (Manchester et al., 1996). However, the relationship between the glycogen concentration and insulin sensitivity is still unclear, with many unanswered questions remaining. For example, why does insulin sensitivity remain high after exercise even after glycogen levels have recovered and normalized (Maarbjerg et al., 2011)?

Regulation of insulin resistance by the activation of exercise and energy deficit sensors via exercise

AMP, ADP and Ca^{2+} increase and glycogen decreases during exercise. This leads to the activation of the kinases AMPK and CaMK, the deacetylase SIRT1 and the transcriptional co-factor PGC-1α (see Chapter 4). AMPK and PGC-1α in particular trigger several adaptations that contribute to the increased insulin sensitivity. In fact, AMPK has been identified as a treatment target for type 2 diabetes mellitus and the metabolic syndrome for over a decade (Winder and Hardie, 1999; Zhang et al., 2009). Also the anti-diabetic drug metformin is an AMPK activator (Zhou et al., 2001) and increases PGC-1α expression (Suwa et al., 2006). The key anti-diabetic adaptations, which are at least partially dependent on elevated AMPK and/or PGC-1α activity, are as follows:

- AMPK–PGC-1α signalling contributes to the increased expression of the glucose transporter **GLUT4** in skeletal muscle post exercise (Huang and Czech, 2007; Holloszy, 2008). The higher GLUT4 concentration that results likely increases not just the basal but also insulin and exercise-stimulated glucose uptake, as demonstrated in transgenic mice that overexpress skeletal muscle GLUT4 (Hansen et al., 1995).

221

- AMPK promotes fatty acid uptake and oxidation but inhibits enzymes involved in fatty acid synthesis (Thomson and Winder, 2009). This contribute to the increased fat oxidation seen during endurance exercise (Romijn et al., 1993). Consequently, there is a reduction in skeletal muscle and liver fat, which will improve insulin sensitivity in these organs via three mechanisms: the Randle cycle; adipokine actions, and a decrease in the fat metabolites **DAG** (diacylglycerol), **fatty acyl-CoA** and **ceramides** (Kahn et al., 2006; Petersen and Shulman, 2006).

In addition to its effects on skeletal muscle, exercise also causes additional adaptations in the liver, adipose tissue, blood and possibly β-cells. Some of these adaptations include altered concentrations of adipokines and inflammation mediators, modulation of ER stress and O-GlcNac modification.

In summary, the main beneficial effect of exercise on glycaemic control is increased glucose uptake acutely via AMPK/CaMK–AS160 signalling in insulin-resistant and type 2 diabetic individuals. There is also improved insulin sensitivity for up to 2–3 days (Boule et al., 2005). This is partially due to the glycogen- and fat-lowering effects of exercise and increased AMPK–PGC-1α signalling, which upregulates GLUT4 transporters.

EXERCISE MIMETICS AS A THERAPEUTIC AGENT IN TYPE 2 DIABETES MELLITUS

Exercise activates AMPK, CaMK, SIRT1 and PGC-1α signalling in target organs, including skeletal muscle, and this in turn increases glucose uptake and improves insulin sensitivity. One possible treatment option for sedentary subjects is to try to target these anti-diabetic exercise pathways using exercise-mimetic drugs, which activate the signalling molecules listed above (Narkar et al., 2008). Such exercise mimetics or pills seem an attractive idea, but while they might activate signalling molecules in ways similar to that of exercise, they fail to increase energy expenditure and lower the fat and carbohydrate content of the body. Metformin and thiazolidinediones are two classes of anti-diabetic agents which stimulate AMPK (Hardie et al., 2003) and are thus exercise mimetics. Another AMPK activator is α-lipoic acid, a naturally occurring short-chain fatty acid and a powerful antioxidant, which has been shown to improve glucose disposal as well as decrease oxidative stress. Currently, its main indication for use in type 2 diabetes mellitus is diabetic neuropathy (Singh and Jialal, 2008). Finally, although animal studies with resveratrol (a phytoalexin enriched in red grape skin and a constituent of red wine) showed a potential glucose-lowering effect (Baur et al., 2006), the equivalent dosing for humans (400 mg/kg per day or 30 g/day for a 75 kg person) may limit its therapeutic use in humans. Nevertheless, the ongoing development of small-molecule SIRT1 activators, with a biopotency 1000 times higher than that of resveratrol, mean that a new class of anti-diabetic agents may soon emerge (Milne et al., 2007; Houtkooper et al., 2010).

SUMMARY

Type 2 diabetes mellitus is a disease involving many organs, signal transduction pathways, DNA sequence variations and epigenetics (Goh and Sum, 2010). It is characterized by chronic hyperglycaemia resulting from **insulin resistance** in primarily the skeletal muscle and liver and is often associated with relative insulin deficiency as a result of β-**cell dysfunction**. This is due to the high concentrations of circulating fatty acids and fat-related metabolites, perhaps lower numbers of mitochondria, the signalling of adipokines, pro-inflammatory mediators, ER stress and O-GlcNAc modification. The possible causes of β-cell dysfunction include ER stress and a loss of β-cells. Type 2 diabetes mellitus is partially inherited because a family history of type 2 diabetes increases an individual's risk of developing type 2 diabetes mellitus. Specific mutations have been associated with maturity-onset diabetes of the young (MODY) and GWAS studies have identified some polymorphisms that explain a small fraction of the heritability of type 2 diabetes. Finally, low birth weight increases the risk of developing type 2 diabetes mellitus, which is believed to be due to epigenetic regulation. The fact that high fat content in certain visceral organs and a sedentary lifestyle cause type 2 diabetes can be traced back to human evolution as there is good evidence that humans have evolved as a physically active species for a lifestyle characterized by physical activity–rest and feast–famine cycles. Unfortunately, modern lifestyles are characterized by low physical activity and overeating, which lead to genome–lifestyle divergence that may explain the rise of type 2 diabetes mellitus. In addition, it is likely that disease-predisposing alleles have increased in the human genome because modern medicine has allowed carriers of such alleles to survive and reproduce. If a sedentary lifestyle causes type 2 diabetes mellitus and other diseases then physical activity should be a suitable treatment, which is indeed the case. Exercise improves glucose metabolism in two ways. First, exercise increases GLUT4-mediated glucose uptake directly even if insulin-stimulated GLUT4-mediated glucose uptake is defective. Second, exercise increases insulin sensitivity in the days after exercise via its actions on AMPK, CaMK, SIRT1 and PGC-1α signalling.

REVIEW QUESTIONS

1   Define insulin resistance and β-cell dysfunction, summarize their causes and explain how these two factors contribute to the development of type 2 diabetes mellitus.
2   Discuss the relationships and putative mechanisms that link human evolution, fetal nutrition (birth weight) and type 2 diabetes mellitus.
3   What is meant by lifestyle–genome divergence in the context of type 2 diabetes mellitus?
4   What is the signal transduction mechanism by which exercise stimulates glucose uptake into skeletal muscle? Compare and contrast it to insulin-stimulated glucose uptake.

FURTHER READING

Abdul-Ghani MA and DeFronzo RA (2010). Pathogenesis of insulin resistance in skeletal muscle. *J Biomed Biotechnol* 2010, 476279.
Bramble DM and Lieberman DE (2004). Endurance running and the evolution of homo. *Nature* 432, 345–352.

Kahn SE, Hull RL and Utzschneider KM (2006). Mechanisms linking obesity to insulin resistance and type 2 diabetes. *Nature* 444, 840–846.

Knowler WC, Barrett-Connor E, Fowler SE, Hamman RF, Lachin JM, Walker EA, et al. (2002). Reduction in the incidence of type 2 diabetes with lifestyle intervention or metformin. *N Engl J Med* 346, 393–403.

Muoio DM and Newgard CB (2008). Mechanisms of disease: molecular and metabolic mechanisms of insulin resistance and beta-cell failure in type 2 diabetes. *Nat Rev Mol Cell Biol* 9, 193–205.

Rose AJ and Richter EA (2005). Skeletal muscle glucose uptake during exercise: how is it regulated? *Physiology (Bethesda)* 20, 260–270.

REFERENCES

Abdul-Ghani MA and DeFronzo RA (2010). Pathogenesis of insulin resistance in skeletal muscle. *J Biomed Biotechnol* 2010, 476279.

Amati F, Dubé JJ, Alvarez-Carnero E, Edreira MM, Chomentowski P, Coen PM, et al. (2011). Skeletal muscle triglycerides, diacylglycerols, and ceramides in insulin resistance: another paradox in endurance-trained athletes? *Diabetes* 60, 2588–2597.

ADA (American Diabetes Association) (2013). Standards of medical care in diabetes – 2013. *Diabetes Care* 36 Suppl 1, S11–S66.

Archer E and Blair SN (2011). Physical activity and the prevention of cardiovascular disease: from evolution to epidemiology. *Prog Cardiovasc Dis* 53, 387–396.

Barker DJ (1997). Maternal nutrition, fetal nutrition, and disease in later life. *Nutrition* 13, 807–813.

Baur JA, Pearson KJ, Price NL, Jamieson HA, Lerin C, Kalra A, et al. (2006). Resveratrol improves health and survival of mice on a high-calorie diet. *Nature* 444, 337–342.

Bergstrom J and Hultman E (1966). Muscle glycogen synthesis after exercise: an enhancing factor localized to the muscle cells in man. *Nature* 210, 309–310.

Bersaglieri T, Sabeti PC, Patterson N, Vanderploeg T, Schaffner SF, Drake JA, et al. (2004). Genetic signatures of strong recent positive selection at the lactase gene. *Am J Hum Genet* 74, 1111–1120.

Billings LK and Florez JC (2010). The genetics of type 2 diabetes: what have we learned from GWAS? *Ann N Y Acad Sci* 1212, 59–77.

Blair SN, Kampert JB, Kohl HW, III, Barlow CE, Macera CA, Paffenbarger RS, Jr., et al. (1996). Influences of cardiorespiratory fitness and other precursors on cardiovascular disease and all-cause mortality in men and women. *JAMA* 276, 205–210.

Boule NG, Weisnagel SJ, Lakka TA, Tremblay A, Bergman RN, Rankinen T, et al. (2005). Effects of exercise training on glucose homeostasis: the HERITAGE Family Study. *Diabetes Care* 28, 108–114.

Bramble DM and Lieberman DE (2004). Endurance running and the evolution of homo. *Nature* 432, 345–352.

Cerling TE, Wynn JG, Andanje SA, Bird MI, Korir DK, Levin NE, et al. (2011). Woody cover and hominin environments in the past 6 million years. *Nature* 476, 51–56.

Chakravarthy MV and Booth FW (2004). Eating, exercise, and "thrifty" genotypes: connecting the dots toward an evolutionary understanding of modern chronic diseases. *J Appl Physiol* 96, 3–10.

Cnop M, Foufelle F and Velloso LA (2011). Endoplasmic reticulum stress, obesity and diabetes. *Trends Mol Med* 18, 59–68.

Danaei G, Finucane MM, Lu Y, Singh GM, Cowan MJ, Paciorek CJ, et al. (2011). National, regional, and global trends in fasting plasma glucose and diabetes prevalence since 1980: systematic analysis of health examination surveys and epidemiological studies with 370 country-years and 2.7 million participants. *Lancet* 378, 31–40.

Danforth WH (1965). Glycogen synthetase activity in skeletal muscle. Interconversion of two forms and control of glycogen synthesis. *J Biol Chem* 240, 588–593.

DeFronzo RA, Ferrannini E, Sato Y, Felig P and Wahren J (1981). Synergistic interaction between exercise and insulin on peripheral glucose uptake. *J Clin Invest* 68, 1468–1474.

Diamond J (2003). The double puzzle of diabetes. *Nature* 423, 599–602.

Dolinoy DC (2008). The agouti mouse model: an epigenetic biosensor for nutritional and environmental alterations on the fetal epigenome. *Nutr Rev* 66 Suppl 1, S7–11.

Elia M (2000). Hunger disease. *Clin Nutr* 19, 379–386.

Ferrannini E, Barrett EJ, Bevilacqua S and DeFronzo RA (1983). Effect of fatty acids on glucose production and utilization in man. *J Clin Invest* 72, 1737–1747.

Finucane MM, Stevens GA, Cowan MJ, Danaei G, Lin JK, Paciorek CJ, et al. (2011). National, regional, and global trends in body-mass index since 1980: systematic analysis of health examination surveys and epidemiological studies with 960 country-years and 9.1 million participants. *Lancet* 377, 557–567.

George S, Rochford JJ, Wolfrum C, Gray SL, Schinner S, Wilson JC, et al. (2004). A family with severe insulin resistance and diabetes due to a mutation in AKT2. *Science* 304, 1325–1328.

Gluckman PD, Hanson MA, Buklijas T, Low FM and Beedle AS (2009). Epigenetic mechanisms that underpin metabolic and cardiovascular diseases. *Nat Rev Endocrinol* 5, 401–408.

Goh KP and Sum CF (2010). Connecting the dots: molecular and epigenetic mechanisms in type 2 diabetes. *Curr Diabetes Rev* 6, 255–265.

Gollnick PD, Piehl K and Saltin B (1974). Selective glycogen depletion pattern in human muscle fibres after exercise of varying intensity and at varying pedalling rates. *J Physiol* 241, 45–57.

Gordon BA, Benson AC, Bird SR and Fraser SF (2009). Resistance training improves metabolic health in type 2 diabetes: a systematic review. *Diabetes Res Clin Pract* 83, 157–175.

Gresham D, Desai MM, Tucker CM, Jenq HT, Pai DA, Ward A, et al. (2008). The repertoire and dynamics of evolutionary adaptations to controlled nutrient-limited environments in yeast. *PLoS Genet* 4, e1000303.

Hansen PA, Gulve EA, Marshall BA, Gao J, Pessin JE, Holloszy JO and Mueckler M (1995). Skeletal muscle glucose transport and metabolism are enhanced in transgenic mice overexpressing the Glut4 glucose transporter. *J Biol Chem* 270, 1679–1684.

Hardie DG, Scott JW, Pan DA and Hudson ER (2003). Management of cellular energy by the AMP-activated protein kinase system. *FEBS Lett* 546, 113–120.

Hart GW, Slawson C, Ramirez-Correa G and Lagerlof O (2011). Cross talk between O-GlcNAcylation and phosphorylation: roles in signaling, transcription, and chronic disease. *Annu Rev Biochem* 80, 825–858.

Hill AM, Fleming JA and Kris-Etherton PM (2009). The role of diet and nutritional supplements in preventing and treating cardiovascular disease. *Curr Opin Cardiol* 24, 433–441.

Holloszy JO (2008). Regulation by exercise of skeletal muscle content of mitochondria and GLUT4. *J Physiol Pharmacol* 59 Suppl 7, 5–18.

Holloszy JO (2009). Skeletal muscle 'mitochondrial deficiency' does not mediate insulin resistance. *Am J Clin Nutr* 89, 463S–466S.

Houtkooper RH, Canto C, Wanders RJ and Auwerx J (2010). The secret life of NAD+: an old metabolite controlling new metabolic signaling pathways. *Endocr Rev* 31, 194–223.

Hu FB, Stampfer MJ, Solomon C, Liu S, Colditz GA, Speizer FE, et al. (2001). Physical activity and risk for cardiovascular events in diabetic women. *Ann Intern Med* 134, 96–105.

Hu FB, Li TY, Colditz GA, Willett WC and Manson JE (2003). Television watching and other sedentary behaviors in relation to risk of obesity and type 2 diabetes mellitus in women. *JAMA* 289, 1785–1791.

Huang S and Czech MP (2007). The GLUT4 glucose transporter. *Cell Metab* 5, 237–252.

Hudson ER, Pan DA, James J, Lucocq JM, Hawley SA, Green KA, et al. (2003). A novel domain in AMP-activated protein kinase causes glycogen storage bodies similar to those seen in hereditary cardiac arrhythmias. *Curr Biol* 13, 861–866.

Isbell LA and Young TP (1996). The evolution of bipedalism in hominids and reduced group size in chimpanzees: alternative responses to decreasing resource availability. *J Hum Evol* 30, 289–297.

Kahn SE, Hull RL and Utzschneider KM (2006). Mechanisms linking obesity to insulin resistance and type 2 diabetes. *Nature* 444, 840–846.

Kahn SE, Zraika S, Utzschneider KM and Hull RL (2009). The beta cell lesion in type 2 diabetes: there has to be a primary functional abnormality. *Diabetologia* 52, 1003–1012.

Kennedy JW, Hirshman MF, Gervino EV, Ocel JV, Forse RA, Hoenig SJ, et al. (1999). Acute exercise induces GLUT4 translocation in skeletal muscle of normal human subjects and subjects with type 2 diabetes. *Diabetes* 48, 1192–1197.

Kim JK, Gavrilova O, Chen Y, Reitman ML and Shulman GI (2000). Mechanism of insulin resistance in A-ZIP/F-1 fatless mice. *J Biol Chem* 275, 8456–8460.

Knowler WC, Barrett-Connor E, Fowler SE, Hamman RF, Lachin JM, Walker EA, et al. (2002). Reduction in the incidence of type 2 diabetes with lifestyle intervention or metformin. *N Engl J Med* 346, 393–403.

Levy-Marchal C, Arslanian S, Cutfield W, Sinaiko A, Druet C, Marcovecchio ML, et al. (2010). Insulin resistance in children: consensus, perspective, and future directions. *J Clin Endocrinol Metab* 95, 5189–5198.

Lopez AD, Mathers CD, Ezzati M, Jamison DT and Murray CJ (2006). Global and regional burden of disease and risk factors, 2001: systematic analysis of population health data. *Lancet* 367, 1747–1757.

Lowell BB and Shulman GI (2005). Mitochondrial dysfunction and type 2 diabetes. *Science* 307, 384–387.

Maarbjerg SJ, Sylow L and Richter EA (2011). Current understanding of increased insulin sensitivity after exercise – emerging candidates. *Acta Physiol (Oxford)* 202, 323–335.

Manchester J, Skurat AV, Roach P, Hauschka SD and Lawrence JC, Jr. (1996). Increased glycogen accumulation in transgenic mice overexpressing glycogen synthase in skeletal muscle. *Proc Natl Acad Sci U S A* 93, 10707–10711.

Martin IK, Katz A and Wahren J (1995). Splanchnic and muscle metabolism during exercise in NIDDM patients. *Am J Physiol* 269, E583–E590.

Milne JC, Lambert PD, Schenk S, Carney DP, Smith JJ, Gagne DJ, et al. (2007). Small molecule activators of SIRT1 as therapeutics for the treatment of type 2 diabetes. *Nature* 450, 712–716.

Mootha VK, Lindgren CM, Eriksson KF, Subramanian A, Sihag S, Lehar J, et al. (2003). PGC-1alpha-responsive genes involved in oxidative phosphorylation are coordinately downregulated in human diabetes. *Nat Genet* 34, 267–273.

Muoio DM and Newgard CB (2008). Mechanisms of disease: molecular and metabolic mechanisms of insulin resistance and beta-cell failure in type 2 diabetes. *Nat Rev Mol Cell Biol* 9, 193–205.

Myatt JP, Schilling N and Thorpe SK (2011). Distribution patterns of fibre types in the triceps surae muscle group of chimpanzees and orangutans. *J Anat* 218, 402–412.

Narkar VA, Downes M, Yu RT, Embler E, Wang YX, Banayo E, et al. (2008). AMPK and PPARdelta agonists are exercise mimetics. *Cell* 134, 405–415.

Neel JV (1962). Diabetes mellitus: a "thrifty" genotype rendered detrimental by "progress"? *Am J Hum Genet* 14, 353–362.

Neel JV (1999). The "thrifty genotype" in 1998. *Nutr Rev* 57, S2–S9.

O'Neill HM, Maarbjerg SJ, Crane JD, Jeppesen J, Jorgensen SB, Schertzer JD, et al. (2011). AMP-activated protein kinase (AMPK) beta1beta2 muscle null mice reveal an essential role for AMPK in maintaining mitochondrial content and glucose uptake during exercise. *Proc Natl Acad Sci U S A* 108, 16092–16097.

Patti ME, Butte AJ, Crunkhorn S, Cusi K, Berria R, Kashyap S, et al. (2003). Coordinated reduction of genes of oxidative metabolism in humans with insulin resistance and diabetes: Potential role of PGC1 and NRF1. *Proc Natl Acad Sci U S A* 100, 8466–8471.

Petersen KF and Shulman GI (2006). Etiology of insulin resistance. *Am J Med* 119, S10–S16.

Petrie JR, Pearson ER and Sutherland C (2011). Implications of genome wide association studies for the understanding of type 2 diabetes pathophysiology. *Biochem Pharmacol* 81, 471–477.

Prentice AM, Hennig BJ and Fulford AJ (2008). Evolutionary origins of the obesity epidemic: natural selection of thrifty genes or genetic drift following predation release? *Int J Obes (Lond)* 32, 1607–1610.

Randle PJ, Garland PB, Hales CN and Newsholme EA (1963). The glucose fatty-acid cycle. Its role in insulin sensitivity and the metabolic disturbances of diabetes mellitus. *Lancet* 1, 785–789.

Reasner CA and DeFronzo RA (2001). Treatment of type 2 diabetes mellitus: a rational approach based on its pathophysiology. *Am Fam Physician* 63, 1687–2, 1694.

Romijn JA, Coyle EF, Sidossis LS, Gastaldelli A, Horowitz JF, Endert E, et al. (1993). Regulation of endogenous fat and carbohydrate metabolism in relation to exercise intensity and duration. *Am J Physiol* 265, E380–E391.

Rose AJ and Richter EA (2005). Skeletal muscle glucose uptake during exercise: how is it regulated? *Physiology (Bethesda)* 20, 260–270.

Scholz MN, D'Aout K, Bobbert MF and Aerts P (2006). Vertical jumping performance of bonobo (Pan paniscus) suggests superior muscle properties. *Proc Biol Sci* 273, 2177–2184.

Sigal RJ, Kenny GP, Wasserman DH, Castaneda-Sceppa C and White RD (2006). Physical activity/exercise and type 2 diabetes: a consensus statement from the American Diabetes Association. *Diabetes Care* 29, 1433–1438.

Singh U and Jialal I (2008). Alpha-lipoic acid supplementation and diabetes. *Nutr Rev* 66, 646–657.

Sinha R, Fisch G, Teague B, Tamborlane WV, Banyas B, Allen K, et al. (2002). Prevalence of impaired glucose tolerance among children and adolescents with marked obesity. *N Engl J Med* 346, 802–810.

Soares P, Alshamali F, Pereira JB, Fernandes V, Silva NM, Afonso C, et al. (2012). The expansion of mtDNA haplogroup L3 within and out of Africa. *Mol Biol Evol* 29, 915–927.

Speakman JR (2008). Thrifty genes for obesity, an attractive but flawed idea, and an alternative perspective: the 'drifty gene' hypothesis. *Int J Obes (Lond)* 32, 1611–1617.

Stumvoll M, Goldstein BJ and van Haeften TW (2005). Type 2 diabetes: principles of pathogenesis and therapy. *Lancet* 365, 1333–1346.

Suwa M, Egashira T, Nakano H, Sasaki H and Kumagai S (2006). Metformin increases the PGC-1alpha protein and oxidative enzyme activities possibly via AMPK phosphorylation in skeletal muscle in vivo. *J Appl Physiol* 101, 1685–1692.

The Wellcome Trust Case Control Consortium (2007). Genome-wide association study of 14,000 cases of seven common diseases and 3,000 shared controls. *Nature* 447, 661–678.

Thomson DM and Winder WW (2009). AMP-activated protein kinase control of fat metabolism in skeletal muscle. *Acta Physiol (Oxford)* 196, 147–154.

Ungar PS and Sponheimer M (2011). The diets of early hominins. *Science* 334, 190–193.

Wang Y, Simar D and Fiatarone Singh MA (2009). Adaptations to exercise training within skeletal muscle in adults with type 2 diabetes or impaired glucose tolerance: a systematic review. *Diabetes Metab Res Rev* 25, 13–40.

Ward CV (2002). Interpreting the posture and locomotion of Australopithecus afarensis: where do we stand? *Am J Phys Anthropol* Suppl 35, 185–215.

Watt MJ, Heigenhauser GJ and Spriet LL (2002). Intramuscular triacylglycerol utilization in human skeletal muscle during exercise: is there a controversy? *J Appl Physiol* 93, 1185–1195.

Winder WW and Hardie DG (1999). AMP-activated protein kinase, a metabolic master switch: possible roles in type 2 diabetes. *Am J Physiol* 277, E1–10.

Witczak CA, Jessen N, Warro DM, Toyoda T, Fujii N, Anderson ME, et al. (2010). CaMKII regulates contraction- but not insulin-induced glucose uptake in mouse skeletal muscle. *Am J Physiol Endocrinol Metab* 298, E1150–E1160.

Xiao B, Sanders MJ, Underwood E, Heath R, Mayer FV, Carmena D, et al. (2011). Structure of mammalian AMPK and its regulation by ADP. *Nature* 472, 230–233.

Zhang BB, Zhou G and Li C (2009). AMPK: an emerging drug target for diabetes and the metabolic syndrome. *Cell Metab* 9, 407–416.

Zhou G, Myers R, Li Y, Chen Y, Shen X, Fenyk-Melody J, et al. (2001). Role of AMP-activated protein kinase in mechanism of metformin action. *J Clin Invest* 108, 1167–1174.

Zimmet P, Alberti KG and Shaw J (2001). Global and societal implications of the diabetes epidemic. *Nature* 414, 782–787.

10 Molecules, ageing and exercise

Henning Wackerhage

INTRODUCTION

The question 'why do we age and die?' is a question not only of interest to philosophers, gerontologists and physicians but also exercise physiologists. This is because with age our maximal heart rate, aerobic capacity ($\dot{V}O_{2max}$), muscle strength, speed and power all decline. We will start this chapter by reviewing the major ageing hypotheses. These suggest that we age principally because molecules, cells and tissues become damaged during ageing, and ageing programmes such as cell division counters trigger a decline in function. All these molecular and cellular changes increase the risk of becoming ill (morbidity) or of dying (mortality). Specifically we will discuss three major hypotheses of ageing:

- the oxidative stress hypothesis as a damage hypothesis,
- the Hayflick limit or telomere hypothesis and
- the mTOR hypothesis as programmed ageing hypotheses.

Ideally, you should read the section on the mTOR pathway in Chapter 6 before starting this chapter as we assume this knowledge here.

After an introduction to the general ageing hypotheses we will ask the question 'Why does our exercise capacity decrease with advancing age?' First, we will discuss the normal decline of our muscle mass and strength during ageing, which is called sarcopenia (Rosenberg, 1997). Second, we will examine why our endurance capacity or $\dot{V}O_{2max}$ drops and try to identify molecular causes for these ageing events.

GENERAL AGEING MECHANISMS

Currently we are unable to give a definite answer to the question 'Why do we age?' This is not because there is a lack of ageing hypotheses: Back in 1990 Medvedev was already able to list 300 theories of ageing (Medvedev, 1990). Since then, the emerging picture is that ageing is not just because of one mechanism. What is also evident is that some mechanisms appear more important to the ageing process than others. In general, we can divide ageing mechanisms or hypotheses into one of two types:

- damage (or error) hypotheses;
- programmed ageing hypotheses.

In this section we will discuss three general ageing mechanisms or hypotheses. First, the oxidative stress hypothesis as a major damage hypothesis and then the Hayflick limit/ telomere and mTOR hypotheses as two major programmed ageing hypotheses.

THE OXIDATIVE STRESS HYPOTHESIS AS AN EXAMPLE OF A DAMAGE HYPOTHESIS

One of the first ageing hypotheses, the **rate of living hypothesis**, states that the higher the metabolic rate the shorter the lifespan (Pearl, 1928). To put it simply 'if you live and metabolize fast, you will die young'. Data on houseflies support this hypothesis (Figure 10.1). Sohal and Buchan (1981) placed houseflies in differently sized environments and compared flies with and without wings. The flies without wings and those housed within a small bottle moved less and had a lower metabolic rate than those with wings or those living in a larger bottle. The authors found that the higher the estimated physical activity and metabolic rate of the flies the shorter their lifespan (Sohal and Buchan, 1981).

These data support Pearl's rate of living hypothesis and make bad reading for Tour de France cyclists and other endurance athletes who, if this applied to humans, should all age more quickly and die younger than their sedentary counterparts. Luckily, this does not seem to be the case, probably because human ageing is different than fly ageing.

Figure 10.1 Relationship between different treatments that affect physical activity and lifespan in houseflies. The authors state 'The median and the maximum life spans of the de-winged flies were about 25 per cent longer (Fig. 1) as compared to the normal winged flies ($p < 0.05$) which suggested that flight activity has a deleterious effect on the life span'. Redrawn after Sohal and Buchan (1981).

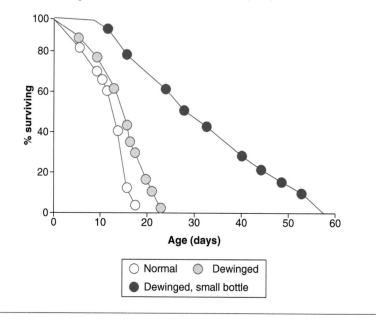

The rate of living hypothesis was modified and made more mechanistic when Harman proposed the **free radical theory** (Harman, 1956), now usually referred to as the **oxidative stress hypothesis**. In its simplest form, the oxidative stress hypothesis states that reactive oxygen species (ROS) and reactive nitrogen species (RNS; RONS for reactive oxygen and nitrogen species together) react with biomolecules such as lipids, proteins and nucleic acids, which results in damage and ultimately leads to ageing. This theory supposes that there is a direct relationship between RONS and metabolic rate, whereby a higher metabolic rate leads to higher RONS production.

However, several exceptions to this rule have been reported. The level of RONS produced is, for example, dependent on the mitochondrial state and it has been demonstrated that elevated, so-called state 3 mitochondrial respiration actually produces less RONS per mole oxygen compared to a lower, state 4 mitochondrial respiration (Sanz et al., 2006). For example, fewer RONS result if a high metabolic rate is due to increased mitochondrial uncoupling which refers to heat rather than ATP production by mitochondria. Indeed, mice with higher than average metabolic rates had more uncoupled skeletal muscle mitochondria and an associated significant increase in lifespan (Speakman et al., 2004). The important information for exercise physiologists is that the rate of living hypothesis and the oxidative stress hypothesis are two different hypotheses because a high rate of living does not automatically mean a high amount of RONS.

What is the difference between free radicals, RONS, the antioxidative defence and oxidative stress?

In relation to the Harman free radical hypothesis the terms 'free radicals', 'RONS', 'antioxidants' and 'oxidative stress' are all used. Here we wish to explain and compare these terms. **Free radicals** are atoms or molecules that have an unpaired electron (an electron is a particle with a negative electric charge and is abbreviated as 'e$^-$'). A single unpaired electron usually pairs up with an unpaired electron from another molecule, resulting in a more stable molecule. Free radicals are highly reactive and can lead to uncontrolled reactions that damage molecules such as lipids, proteins or even DNA. Free radicals, however, are not the only highly reactive molecules within our bodies, as other highly reactive molecules not containing unpaired electrons also exist. Because of this the term 'reactive species' is now preferred over the term 'free radical', even if 'free radical' is still widely used.

Reactive species usually contain oxygen or nitrogen and they are referred to as **reactive oxygen species (ROS), reactive nitrogen species (RNS)** or **reactive oxygen and nitrogen species (RONS)**. To give some examples of molecules, free radical ROS are the **superoxide anion ($O_2 \cdot^-$**; the '$\cdot$' behind the O_2 denotes the unpaired electron) and the **hydroxyl radical (HO$\cdot$)**.

Hydrogen peroxide (H_2O_2) is also highly reactive but it is not a free radical because it does not carry an unpaired electron. Major reactive nitrogen species (RNS) are the **nitric oxide (NO$\cdot$)** free radical and the **peroxynitride anion (NONOO$^-$)** which is not a free radical but again a highly reactive molecule ROS are produced mainly by the electron transport chain of the mitochondria and NO is synthesized by the neural, inducible and endothelial isoforms of the enzyme **nitric oxide synthase (nNOS, iNOS or eNOS)**. NO is a major signalling molecule as it regulates vasodilation, showing that not all RONS are always bad for the organism.

Our bodies are not defenceless against RONS as we have an **antioxidant defence**. The antioxidant defence removes RONS either (a) via antioxidant enzymes that catalyse controlled reactions where RONS are substrates or (b) by antioxidants which are molecules such as vitamins or proteins that prevent other molecules from reacting with RONS. Three examples of important antioxidant enzymes are:

- **superoxide dismutase** (SOD), which removes the superoxide anion $O_2 \cdot^-$ by catalysing $2O_2 \cdot^- + 2H^+ \rightarrow O_2 + H_2O_2$;
- **catalase** (Cat), which gets rid of H_2O_2 by catalysing the following reaction: $2H_2O_2 \rightarrow 2H_2O + O_2$;
- **glutathione peroxidase** (GPx), which removes H_2O_2 via the following reaction: $2GSH + H_2O_2 \rightarrow GS\text{-}SG + 2H_2O$, where GSH is glutathione and GS-SG is glutathione disulfide.

Non-enzymatic antioxidants are either taken up with the diet or directly produced by the body. Major examples are (Powers and Jackson, 2008):

- **vitamins A, C and E** as well as α-**lipoic acid**, which are taken up with the diet; and
- **glutathione**, which is a peptide comprising three amino acids, **biliburin**, a breakdown product of haemoglobin, and **uric acid**, a by-product of purine metabolism, which are all produced by the body.

The RONS that are not removed by the antioxidant defence cause **oxidative stress**. The higher the oxidative stress, the higher the molecular and cellular damage and the faster the rate of ageing according to the oxidative stress hypothesis. Thus the rate of living is not necessarily proportional to RONS and oxidative stress as the extent of mitochondrial uncoupling and the antioxidant defence also play a role.

For the researcher, the problem is that it is nigh impossible to measure RONS or oxidative stress directly using a straightforward chemical assay. This is because of the high reactivity and short half-life time of RONS. Thus indirect methods are usually used to estimate oxidative stress. In practice, researchers can use four types of assays to measure oxidative stress and the antioxidant defence (Powers and Jackson, 2008):

- **Oxidants** are the ROS that researchers would like to measure directly in the best case scenario. They are, however, very difficult to measure.
- **Oxidation products** are molecules 'damaged' by RONS. Examples are protein carbonyls (marker for protein oxidation/damage), markers for lipid peroxidation, or 8-hydroxy-2' deoxyguanosine (8-OH-dG) as a marker for DNA oxidation/damage.
- **Antioxidants**. Examples are glutathione or the total antioxidant capacity.
- **Antioxidant/pro-oxidant balance**. An example is the ratio of oxidized to total glutathione (GSH/TGSH).

Because none of the above assays is free of limitations, it is best to measure several of them in a research project to convincingly demonstrate the level of oxidative stress. Also, when judging published papers it is important to look at the oxidative stress markers that were actually measured in order to decide whether the measurements were robust.

So what about exercise and oxidative stress? Single bouts of aerobic and anaerobic exercise generally increase oxidative stress, which has been measured using several of the aforementioned oxidative stress assays (Powers and Jackson, 2008; Fisher-Wellman and Bloomer, 2009). According to the aforementioned 'rate of living' and 'free radical' and 'oxidative stress' hypotheses of ageing (Harman, 1956) exercise should shorten the lifespan, especially of elite endurance athletes. Is there any evidence for this? No. In a recent review of studies on this topic the opposite was concluded, namely that 'elite endurance athletes tend to survive longer than people in the general population' (Teramoto and Bungum, 2010). What is the explanation for this conundrum? Are the 'rate of living', 'free radical' and 'oxidative stress' hypotheses all wrong or are there mechanisms by which oxidative stress is reduced in elite endurance athletes? Here, we discuss four points to try to answer this question.

1 Many epidemiological studies show that exercise reduces disease risk or morbidity for cardiovascular disease, stroke, some forms of cancer, type 2 diabetes, obesity and osteoporosis (Blair and Morris, 2009). Because most of these diseases increase mortality, endurance and other athletes will reduce mortality simply by reducing morbidity even if RONS production is high.

2 While elite endurance athletes probably generate high levels of RONS, their oxidative defence will adapt. The literature is not entirely clear but many studies suggest that endurance exercise increases the activity of the antioxidant enzymes superoxide dismutase (SOD), glutathione peroxidases (GPx) and also the concentration of

non-enzyme antioxidants such as glutathione (GSH) (Powers and Jackson, 2008). Thus, while the RONS production of an elite endurance athlete might be high it is probably better dealt with because the skeletal muscle antioxidant defence increases. which reduces the amount of oxidative stress.

3   While RONS cause damage, they are also beneficial as they trigger adaptations that are beneficial for health (Powers and Jackson, 2008). For this reason antioxidant supplementation can be detrimental if it reduces RONS to an extent that these adaptations are prevented. This has, for example, been demonstrated in a study where the researchers compared two groups of healthy young males. One group did exercise only and the other group did the same exercise but also received the anti-oxidants vitamin C and E. The researchers found that the doses of antioxidant supplementation used prevented adaptations to exercise such as an increased ability to take up glucose by muscle (Ristow et al., 2009). Together with other studies where antioxidant supplementation with or without exercise led to negative results or lack of effect (Powers and Jackson, 2008), this suggests that at least moderately increased RONS are actually beneficial for health and thus probably also for longevity. This has been termed the hormesis or **mitohormesis hypothesis**. Hormesis describes that low exposures to stressors and toxins can have beneficial effects. Thus during some training sessions and in some races elite endurance athletes may produce detrimentally high concentrations of RONS but during normal training the level and resulting oxidative stress might be beneficial or at least not too detrimental. Thus the relationship between oxidative stress generated by exercise and longevity might be best described by an inverted 'U' curve: low levels are bad, medium levels optimal and high levels are again detrimental. Finally, very high amounts of nutritional antioxidants can reduce adaptations to exercise and have other detrimental effects so athletes, coaches and exercise physiologists should read the literature in order not to overdose with these supplements.

4   The RONS production of at least some elite endurance athletes might have a negative effect on longevity but other positive effects might outweigh the negative effects. For example, endurance exercise activates the energy status sensor AMPK in skeletal muscle (see Chapter 4), which in turn inhibits the longevity-reducing mTOR pathway (Inoki et al., 2003) which we describe below.

To conclude, the relationship between oxidative stress generated by exercise and longevity is complex and probably resembles an inverted 'U' curve as some oxidative stress triggers beneficial adaptations. In addition, many other mechanisms link exercise to ageing. These include a reduced morbidity, telomere shortening and AMPK–mTOR signalling. Thus the 'oxidative stress' hypothesis alone is too simplistic to explain the link between exercise and ageing.

PROGRAMMED AGEING HYPOTHESES: THE TELOMERASE–P53 PATHWAY AS AN EXAMPLE

In this section we will discuss **the replicative or cellular senescence hypothesis of ageing** as the first example of a programmed ageing hypothesis. The replicative senescence hypothesis of ageing is considered to be a programmed ageing hypothesis because it proposes a programme or mechanism that drives the ageing process.

What is replicative senescence? Cells in culture can only divide a limited number of times. For this, Leonard Hayflick is usually quoted. He showed that human fibroblasts derived from embryonic tissues could divide ≈50 times in a culture dish (Hayflick, 1965). Because of this limit the number of divisions that a cell line can undergo is often referred to as the **Hayflick limit**. This is perhaps unfair because Weismann had originally first published the idea that the number of cell divisions is limited some 70 years earlier (Weismann, 1892) so it should probably be the Weismann–Hayflick limit!

Replicative senescence can be linked to overall ageing by hypothesizing that ageing is a result of more and more cells reaching their Hayflick limit and becoming senescent. How is replicative senescence regulated? A key regulator is the **telomere**, which lies at the end of chromosomes and which shortens during each cell division. Telomeres are made up of single strands of repeated TTAGGG segments. Together with some specific proteins it forms a cap that protects the end of chromosomes. Telomeres are elongated in germ cells by an enzyme which has been termed **telomerase**. Telomerase is a ribonucleoprotein enzyme that is expressed only in germ cells and some stem cells but not in other body cells. Because telomerase elongates telomeres, the cells in an embryo start out with long telomeres. Telomeres then shorten during each cell division, which is termed 'telomere erosion'. This was first shown in cultured cells *in vitro* (Harley et al., 1990). In addition, oxidative stress also contributes to telomere shortening (von Zglinicki, 2002), which connects these two fundamental ageing mechanisms.

At the Hayflick limit, telomeres reach a critically short length, which triggers signalling events that result in cellular senescence or, in other words, the cells stop dividing. But is this just a circumstantial event or is telomere erosion required for replicative senescence? To test this, the telomere-lengthening enzyme telomerase was overexpressed in cultured human epithelial cells and fibroblasts. With the added telomerase these cells could divide at least another 20 rounds on top of what normal cells could do (Bodnar et al., 1998). In contrast, many immortal tumour cells have an increased telomerase activity and thus longer telomeres. If telomerase is inhibited in these cells then progressive telomere shortening and cell senescence or death occurs (Herbert et al., 1999). To summarize, germ cells and embryos start with long telomeres. In somatic cells telomeres then shorten with each division and after a set number of cell divisions the Hayflick limit is reached and signalling events trigger cellular senescence. Immortal cells such as tumour cells often have increased telomerase activity which ensures long telomeres and delayed senescence.

Relatively little is known about the relationship between the length of telomeres and exercise. However, in one study (Figure 10.2) researchers found that telomeres were longer in white blood cells (leukocytes) of people with high physical activity levels than in people with less physical activity (Cherkas et al., 2008).

These data suggest that physical activity may prevent leukocyte telomere shortening, demonstrating a relationship between physical activity and telomere length in at least one cell type. However, it is important not to over-interpret this finding, as this is only a cross-sectional study and the effect might be limited to leukocytes.

One question that we have not addressed so far is the mechanism by which telomere erosion triggers replicative senescence. A key player is the **tumour suppressor** protein

Figure 10.2 Relationship between physical activity levels and telomere length in leukocytes of human volunteers. White bars are adjusted for age, sex and extraction year, black bars additionally for BMI, smoking and socioeconomic status. Redrawn after Cherkas et al. (2008). Note that this is only an association and does not constitute proof that exercise actually limits the shortening of telomeres in leukocytes or other cells.

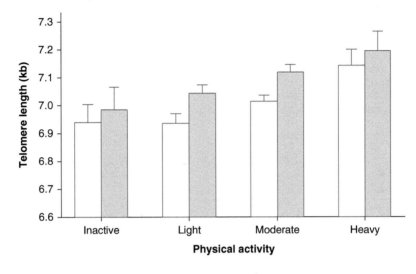

p53 ('p' stands for protein and '53' for the weight in kilodaltons), which is one of the most frequently mutated genes in tumours and has been dubbed the 'guardian of the genome'. p53 is not only one of the most researched proteins in cancer research, as it frequently defective in cancer cells, but is also a factor in ageing. Unique insight into the function of p53 comes from an experiment where researchers had generated transgenic mice where either p53 activity was reduced (indicated by a '–'), where p53 activity was normal (indicated by a '+'; wildtype) or where p53 was more hyperactive (abbreviated as 'm') (Tyner et al., 2002). The researchers then studied tumours or cancer, ageing and survival of these mice (Figure 10.3). Given that p53 is a tumour suppressor it is not surprising that mice with low p53 activity died early, mainly because >80% of them developed tumours. The wildtype p53 mice survived longer but >45% still had tumours as mice 'naturally' have a higher rate of tumours than humans. In the mice with highly active p53 only 6% developed tumours, but surprisingly these mice still died younger than their wildtype counterparts as they were ageing prematurely. How can these findings be explained? The telomere–p53 system regulates the number of cell divisions that cells can undergo before senescence. If the system responds early then cells can proliferate less. As potentially tumour-causing mutations can occur during each cell division, the result is a lower tumour risk, but because cells become senescent early, the whole body ages prematurely. In contrast, if the telomere–p53 system allows many cell divisions, then ageing is delayed but the frequency of tumours in high-proliferating cells increases. This experiment uniquely demonstrates the relationship between telomere–p53 signalling, cell division, cancer and ageing.

To summarize, replicative senescence is a mechanism that protects against cancer as it limits the number of cell divisions before cells become senescent. While this is

Figure 10.3 p53, cancer and ageing in mice. (a) Survival and tumour incidence of mice with low p53, normal p53 and high p53 activity. Mice with high p53 activity develop few tumours but age prematurely. In contrast, >80% of mice with low p53 activity develop tumours. (b) Phenotype of skinned mice with normal and high p53 activity. The mice with high p53 activity display signs of premature ageing. Redrawn after Tyner et al. (2002).

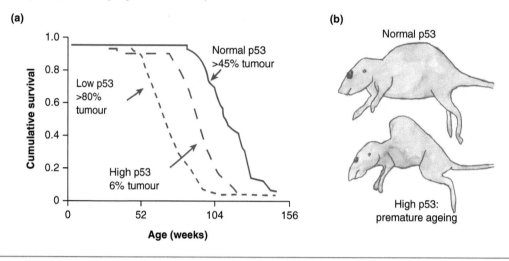

beneficial in mid-life it is detrimental at older age because if the telomere–p53 system is too active then cells become senescent prematurely and the organism ages (Shawi and Autexier, 2008). But can the telomere–p53 system potentially be targeted as an anti-ageing intervention? Two high-profile studies suggest so: In the first study, investigators asked whether telomerase reactivation can not only halt but also reverse ageing processes. They generated knock-in mice in which telomerase could be expressed by giving tamoxifen to the animal. They found that telomerase expression in old animals reversed the ageing of several organs (Jaskelioff et al., 2011). Related to this, in another study researchers developed a transgenic mouse that allowed them to eliminate senescent cells which are marked by the expression of p16[Ink4a]. They found that the removal of such senescent cells delayed ageing in several organs, showing that the presence of senescent cells alone contributes to the ageing phenotype (Baker et al., 2011). Thus one future strategy to slow down ageing could be to develop treatments that allow the removal of senescent cells.

PROGRAMMED AGEING HYPOTHESES: THE mTOR HYPOTHESIS AS A SECOND EXAMPLE

In Chapter 6 we described how the mTOR pathway mediates skeletal muscle hypertrophy in response to resistance exercise. Here, we reveal a darker side of this pathway, namely its inverse relationship with longevity. There is now compelling evidence that an activation mTOR pathway shortens lifespan. Does this mean that bodybuilders should die young as their skeletal muscle hypertrophy is driven by the frequent activation of the mTOR pathway?

The negative effect of components of the mTOR pathway on lifespan was initially shown in yeast, worms and flies (Kapahi et al., 2010) and more recently in mice (Selman et al., 2009; Johnson et al., 2013). In these studies scientists deleted individual genes in this pathway and showed that these deletions led to increased lifespan and an improved healthspan in mice. In addition, in two mouse studies (Figure 10.4), the mTOR inhibitor rapamycin was given to female and male mice later in life and this extended lifespan (Harrison et al., 2009; Chen et al., 2009), confirming that an inhibition of the mTOR pathway increases lifespan.

Figure 10.4 mTOR inhibition and lifespan. (a) Survival of untreated female mice (continuous line) and mice treated with rapamycin (dotted line, the arrow indicates the start of rapamycin treatment). (b) Survival of wildtype mice (continuous line) and mice where S6k1 (also known as p70 S6k) was knocked out. Redrawn after Harrison et al. (2009) and Selman et al. (2009). These results suggest that an inhibition of members of the mTOR pathway can increase lifespan in mammals.

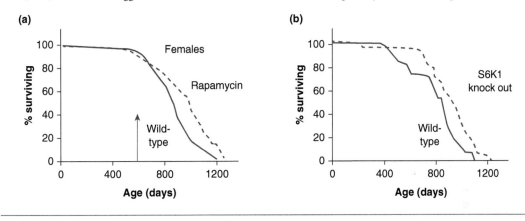

In another study, p70 S6k (in that paper abbreviated as S6K1), which signals downstream of mTOR, was knocked out. This led to increased survival in female mice (Selman et al., 2009). The S6K1 knockout mice weighed less, had a reduced fat mass and several age-related disease variables, such as insulin resistance, were delayed. All this demonstrates that whole-body high mTOR pathway activity shortens lifespan in many species, including mammals.

In Chapter 6, we discussed how the activity of the mTOR pathway is influenced by different modes of exercise: resistance exercise increases mTOR signalling whereas endurance exercise inhibits mTOR activity, especially during acute exercise because AMPK inhibits mTOR, for example, via AMPK–TSC2 signalling (Inoki et al., 2003) and AMPK–Raptor (Gwinn et al., 2008) links. Thus is there any evidence that AMPK activation by exercise or other means increases longevity? As we have discussed already, exercise is associated with lower morbidity and mortality rates in humans, whether people do moderate of exercise (Blair and Morris, 2009) or are elite athletes (Teramoto and Bungum, 2010). No-one so far has tested the lifespan of mice that express constitutively active AMPK, either in the whole body or in individual tissues such as skeletal muscle. However, one study showed that treatment with the AMPK activator metformin increased the lifespan of female mice (Anisimov et al., 2008), but not in Fischer-344 rats (Smith et al., 2010). Another chemical that activates AMPK, albeit indirectly, is resveratrol, which is a polyphenol found, for example, in the skins of red grapes. Initially, it was identified as an activator of sirtuins, but it is now believed that resveratrol activates AMPK (Fullerton and Steinberg, 2010). It was found that resveratrol treatment increased health and survival of mice on a high-calorie but not normal diet (Baur et al., 2006). Not only that, in another study it was demonstrated that resveratrol treatment increased the aerobic capacity of mice on a high-fat diet, increased mitochondrial biogenesis and also the running time to exhaustion (Lagouge et al., 2006). Thus resveratrol not only stimulates some exercise-like adaptations but also increases lifespan, at least in some animal models.

Taken together, there is compelling evidence that an inhibition of the mTOR pathway increases lifespan in many species, including mammals. This can be achieved by inhibiting parts of the mTOR pathway directly, or by activating AMPK, which inhibits mTOR. The AMPK inhibition of mTOR is potentially a key mechanism to explain why endurance exercise increases longevity. However, what about bodybuilders who increase mTOR activation with resistance exercise and nutrients? Generally, elite strength and power athletes usually have a normal life expectancy but not the increased life expectancy seen in elite endurance athletes (Teramoto and Bungum, 2010). In contrast to the mTOR inhibition studies, bodybuilders increase mTOR activity only in their skeletal muscles but not the whole body. Currently it is unknown whether the effect of mTOR activity on lifespan is dependent on certain organs and whether skeletal muscle is one of these organs. Thus we cannot answer the question whether increased mTOR activity in skeletal muscle reduces lifespan in humans or, put simply, whether bodybuilders should die young because of that.

Before ending the section on global ageing hypotheses we wish to discuss one intervention that can be used for increasing lifespan in many species, including primates (Colman et al., 2009). It is **caloric restriction**, which is a reduction of nutrient intake without causing malnutrition. The result is weight loss, and because of the link between the body mass index (BMI) and morbidity this will reduce the incidence of diseases where an increased BMI is a risk factor and thus morbidity. The second effect is that the mTOR pathway will be less activated. After each mixed meal, carbohydrates, which are digested to glucose, trigger an increased release of insulin. Proteins, which are digested to amino acids, activate the mTOR pathway. Thus if the nutrient intake is reduced then the insulin release and the amino acids will be lower and the mTOR pathway will be less activated. This can explain part of the longevity effect of caloric restriction.

Summary

Many mechanisms contribute to ageing. The oxidative stress hypothesis is an example of a key damage hypothesis and the replicative senescence/telomere and mTOR pathway hypotheses are examples of major programmed ageing hypotheses. Exercise is linked to all three aforementioned hypotheses, and endurance exercise in particular increases lifespan via many potential mechanisms, including an increase in the antioxidant defence and an inhibition of the mTOR pathway due to AMPK activation. Resistance exercise theoretically could have a detrimental effect on lifespan as it activates the mTOR pathway specifically in skeletal muscle, but the evidence for this is limited. Figure 10.5 is a schematic diagram that gives an overview of major ageing hypotheses and the link to endurance and resistance exercise.

PHYSICAL FITNESS AND ORGAN-SPECIFIC AGEING

In the second half of this chapter we move from global ageing mechanisms to the ageing of two organs that are associated with the loss of exercise capacity during normal ageing. The first one is the normal decline of skeletal muscle mass and strength during ageing, defined as sarcopenia (Rosenberg, 1997). The second is the drop in maximal oxygen uptake, which is to a large extent due to a drop in the maximal heart rate, roughly described by Fitt's '$HR_{max} = 220 - age$' formula. Of these two phenomena, sarcopenia

Figure 10.5 Summary of how exercise might affect longevity. (1) Endurance exercise increases oxidative phosphorylation (and the 'rate of living') which leads to an increase in the production of RONS. However, the expression of uncoupling protein 3 (UCP3) also increases after endurance exercise, resulting in increased uncoupling which limits the RONS and oxidative stress increase after exercise in line with the 'uncoupling to survive' hypothesis (Speakman et al., 2004). Also, exercise increases the expression of some components of the enzymatic and non-enzymatic antioxidant defence (Powers and Jackson, 2008) which also reduces oxidative stress. (2) Oxidative stress reduces longevity but also induces adaptation, which may reduce morbidity and possibly mortality (Ristow et al., 2009). (3) Physical activity is associated with longer telomeres in white blood cells but the effect on telomeres in other tissues and stem cells is unknown. The telomere–p53 system affects longevity via the incidence of cancers and the rate of ageing. (4) Resistance exercise leads to a skeletal muscle-specific activation of the mTOR pathway. It is unknown whether this reduces longevity as mTOR activation reduces longevity in many species. Activation of AMPK by resveratrol and metformin will reduce mTOR activity and this is the likely mechanism by which resveratrol and metformin increase longevity at least under some conditions. Endurance exercise also activates AMPK in skeletal muscle but it is unknown whether this contributes the increased longevity experienced by endurance athletes and individuals engaging in endurance exercise.

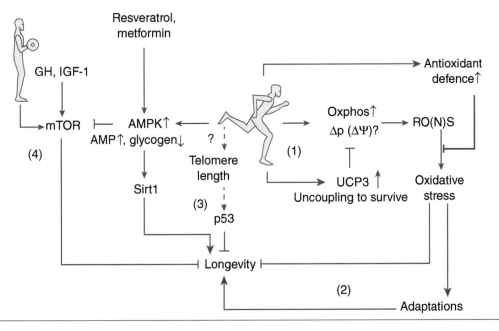

has been more intensely studied, whereas the drop of the maximal heart rate during ageing has been almost ignored, despite having such a major effect on the exercise capacity. Other relevant changes that affect the exercise capacity are ageing-induced neurodegeneration and a loss of bone mass which we will not cover here.

Organ-specific ageing and exercise: sarcopenia

Arguably the most important ageing-related change that affects our exercise capacity is sarcopenia, defined as the loss of muscle mass, strength and power that occurs both in males and females during normal ageing. The effect of ageing on muscle size and strength is illustrated in Figure 10.6.

Figure 10.6 shows three things: First, there is a large variation in human muscle mass, size and strength at all ages in men and women, as indicated by the scatter of the points at each given age. In Chapter 7 we have already discussed that DNA sequence variations can explain much of the observed variation of muscle mass and strength within a population at a given age. Second, women on average have a lower muscle mass, size and strength than men. Third, muscle size declines at a greater rate after around 60 years of age.

Why is sarcopenia so important? It is important because a decreased muscle mass and or strength has been associated with a reduced ability to perform normal life tasks such as rising from a chair, walking at a reasonable speed and ascending and descending stairs (Ploutz-Snyder et al., 2002), frailty and old age disability (Rantanen et al., 1999), falls

Figure 10.6 Sarcopenia. (a) MRI scans of the legs at mid thigh of a young (top) and sarcopenic, elderly (bottom) subject. Skeletal muscle can be seen in grey. The cross-sectional muscle area is smaller in the sarcopenic subject, and fat, seen in white, infiltrates the muscle. (b) Changes in calf muscle cross-sectional area and knee extension torque (strength) during normal ageing in women (filled circles) and men (open circles) (Lauretani et al., 2003). Note the large variation at any given age and the average decline with age in males and females.

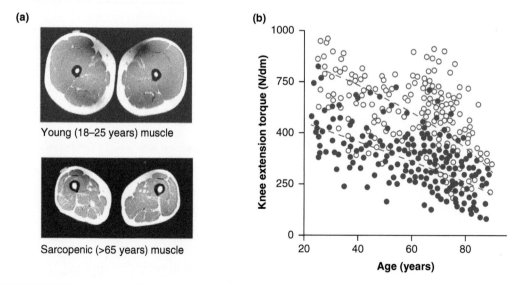

(a)

Young (18–25 years) muscle

Sarcopenic (>65 years) muscle

(b)

(Moreland et al., 2004), poor recovery after hip fracture or major disease (Wolfe, 2006) and even increased mortality (Ruiz et al., 2008). Sarcopenia thus profoundly affects lifestyle, morbidity and even mortality and thus affects the quality of life and health at old age more than the drop in the $\dot{V}O_{2max}$ which we will discuss later in this chapter.

During sarcopenia many things change, but three changes seem particularly important:

- **α-Motor neuron and muscle fibre loss and type II atrophy**: During ageing and especially after around 60 years of age we lose muscle α-motor neurons and muscle fibres (possibly predominantly type II fibres (Trappe et al., 1995)) and especially fast type II fibres atrophy (Lexell et al., 1988).
- **Anabolic resistance**: Old muscle responds with decreased protein synthesis to anabolic stimuli such as proteins/amino acids (Cuthbertson et al., 2005), insulin (Rasmussen et al., 2006) and resistance exercise (Kumar et al., 2009) compared with young muscle. This hypothesis is not, however, universally accepted.
- **Muscle regeneration defects**: Muscle regeneration after injury or atrophy is impaired in old age (Carlson et al., 2009). This is associated with reduced satellite cell numbers and probably caused by impaired satellite responses to injury (Carlson et al., 2009). This dysfunction is well described for mice but it is unclear whether satellite cell function is impaired in human muscle (Alsharidah et al., 2013).

The above changes are well described in the literature (Narici and Maffulli, 2010) but the underlying mechanisms are incompletely researched. We first summarize the changes that contribute to sarcopenia in Figure 10.7 and after that we will discuss the major changes and mechanisms in detail.

Muscle fibre loss and type II fibre atrophy

Why do we lose muscle fibres when we age? As it turns out, the loss of muscle fibres is at least partially due to the loss of α-motor neurons, which innervate muscle fibres (Lexell, 1997; Aagaard et al., 2010). The loss of α-motor neurons has been observed in human cadavers (Tomlinson and Irving, 1977; Mittal and Logmani, 1987) but the causes are incompletely understood. The motor neurons that survive ageing appear to sprout to some denervated muscle fibres, resulting in **large motor units** (a motor unit is an α-motor neuron and the muscle fibres innervated by it). This may also explain the **fibre grouping** within a muscle that is another characteristic of old muscle (Lexell and Downham, 1991). There is some evidence that the faster FF α-motor neurons in particular are lost and decline in size, while the slower S α-motor neurons are preserved and increase in size during ageing (Kanning et al., 2010). Furthermore, ageing also affects the function of axons and motor endplates (Aagaard et al., 2010).

What is causing this **neurodegeneration** or death of α-motor neurons and the related loss of muscle fibres? The exact cause is unknown but there are some intriguing similarities between the neurodegeneration during ageing and the neurodegeneration that occurs during amyotrophic lateral sclerosis (ALS, also known as Lou Gehrig's disease). In 1993 it was reported that some patients with ALS have a mutation in their Cu/Zn superoxide dismutase (SOD) gene (Rosen et al., 1993), which we reviewed earlier in

Figure 10.7 Three major factors in sarcopenia. (1) Muscle fibre loss. Possibly RONS cause a loss of α-motor neurons and muscle fibres. (2) Anabolic resistance: Many but not all researchers find that in response to insulin, resistance exercise and amino acids (protein), muscle protein synthesis is reduced in old muscle when compared to young. This might be related to changed concentrations of signal transduction proteins that mediate muscle hypertrophy and atrophy. (3) Satellite cells decrease by ≈50% during ageing. The remaining satellite cells regenerate muscle poorly, which is partially due to circulatory factors. This has been proposed to be related to increased TGFβ–Smad signalling and reduced Notch-activation of satellite cells, at least in mice. As a consequence, muscle regenerates poorly after atrophy and injury, and as a further consequence the frequency of fat or connective tissue infiltrations increases (Carlson et al., 2009). It is unclear, however, whether satellite cell function is impaired in human skeletal muscle (Alsharidah et al., 2013).

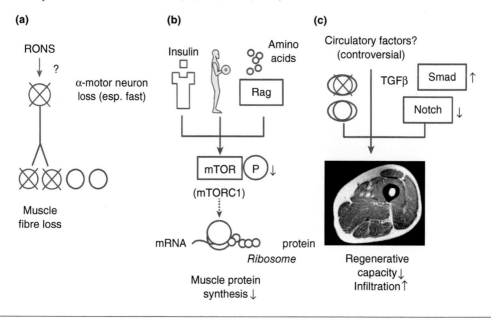

this chapter as one of the antioxidant enzymes. Researchers have created SOD1 knock-out mice and mice with a mutation that mimics the mutation found in the affected families. The SOD1 knockout suffer from increased muscle loss, especially in faster muscles during ageing which resembles sarcopenia (Jang et al., 2010). In mice that carry the SOD1 mutation seen in families with ALS, the number of motor units declines (Shefner et al., 1999), while the mice who have a SOD1 knockout have more peripheral defects in the neuromuscular junctions (Jang et al., 2010). All this suggests indirectly that RONS may be involved in the neurodegeneration and loss of muscle fibres and fibre grouping that is a hallmark of sarcopenia. The mechanism by which RONS or other triggers kill motor neurons and/or muscle fibres is unknown.

Anabolic resistance

A second contributor to sarcopenia is **anabolic resistance** (i.e. the decreased response of muscle protein synthesis to anabolic stimulus). This is found by many but not all researcher groups that are active in this field. Anabolic resistance of old muscle has been reported for the muscle protein synthesis response to amino acids (Cuthbertson et al.,

Figure 10.8 Effect of essential amino acid feeding on muscle protein synthesis in young (28 ± 6 years) and elderly (70 ± 6 years) men. The data show that elderly subjects increase their muscle protein synthesis less in response to a given stimulation by amino acids than young men. Of practical relevance, the data suggest that 10 g of essential amino acids, which equate to 20 g of protein, are sufficient to stimulate protein synthesis near maximally in young and elderly men. Redrawn after Cuthbertson et al. (2005).

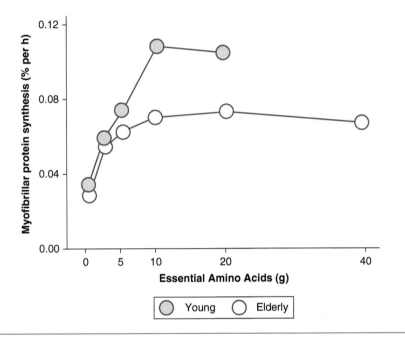

2005), insulin (Rasmussen et al., 2006) and resistance exercise (Kumar et al., 2009). Anabolic resistance can explain especially the atrophy of predominantly type II fibres during sarcopenia (Lexell et al., 1988). Figure 10.8 shows anabolic resistance for amino acid-stimulated myofibrillar (i.e. myosin, actin and other proteins of the sarcomere) protein synthesis. In this study, fasted protein synthesis is similar in young and old men. However, when stimulated with essential amino acids then old muscle synthesizes less myofibrillar protein than young muscle.

What is the mechanism that causes anabolic resistance? Again this is incompletely understood. Given that the stimulation is similar (same amount of amino acids, insulin and resistance exercise) the problem is that the input appears to be amplified less in older, sarcopenic subjects than in young. This could potentially be due to a lower expression or function of signalling proteins in the mTOR pathway that links anabolic stimuli to protein synthesis. Indeed, the concentrations of the anabolic proteins mTOR and p70 S6k are roughly half of the levels in young muscle, while catabolic NF-κB, an atrophy mediator, is ≈4-fold higher in the elderly than in the young (Cuthbertson et al., 2005). However, the mTOR signal transduction network is extensive (Caron et al., 2010) and a mathematical modelling or systems biology approach is required to determine how the observed changes in signal transduction protein concentrations affect the muscle protein synthesis output in response to a given input signal.

Muscle regeneration defects

Old skeletal muscle regenerates less well than young muscle. This has been perhaps best demonstrated by showing that after immobility-induced muscle atrophy the muscle of young subjects (≈20 years) recovered normally while old (≈70 years) muscle recovered incompletely with scar tissue formation (Carlson et al., 2009). Also old satellite cells proliferate less during recovery from atrophy in old than young muscle, which again reduces repair (Carlson et al., 2009) although some studies suggest that there is no difference between young and old muscle satellite cells and other myoblasts (Alsharidah et al., 2013).

What are the mechanisms responsible for the decreased regenerative capacity of old skeletal muscle? Several causes have been identified in mouse studies. One possible cause is that the number of satellite cells is decreased by ≈50% in old when compared to young human muscle by quantifying satellite cells on the basis of the three satellite cell markers Pax7, NCAM and M-cadherin (Carlson et al., 2009). In addition, old satellite cells seem less capable of regeneration, at least in mice. There are several potential causes:

- In old mouse skeletal muscle satellite cells proliferate more and they are closer to senescence which reduces ability to contribute to regeneration and muscle maintenance (Chakkalakal et al., 2012).
- Old serum has higher concentrations of factors that impair the function of satellite cells (Conboy et al., 2005; Carlson et al., 2009). However, this is controversial because the sera of young and elderly subjects had no effect on the behaviour of myoblasts, which resemble activated satellite cells, in culture (George et al., 2010).
- The Notch and TGFβ–Smad pathways have been suggested to be specific causes of the impaired regenerative capacity of old satellite cells (Carlson et al., 2008, 2009).

While differences in the proliferative and regenerative capacity are well documented for mice, it is unclear whether human satellite cells function less in the elderly. In one study the authors reported that young and old human satellite cells and other precursors proliferate and differentiate similarly in culture, suggesting that there is no defect in the old satellite cells and other precursors (Alsharidah et al., 2013). For this reason the cause of the regeneration effect of old human muscles (Carlson et al., 2009) is still incompletely understood.

Organ-specific ageing and exercise: $\dot{V}O_{2max}$

After sarcopenia, the drop of $\dot{V}O_{2max}$ is the second most important effect of ageing on our exercise capacity. The $\dot{V}O_{2max}$ depends mainly on the maximal cardiac output (Ekblom, 1968), which in turn is a product of heart rate × stroke volume. One of the most striking effects of human ageing is the drop of the maximal heart rate. This was described by Fitt's best guess formula 'HR_{max} (in beats per minute) = 220 − age' or as a best-fit line for actual experimental data (Tanaka et al., 2001) (Figure 10.9).

So what is the mechanism that reduces the maximal heart rate with age and via that the $\dot{V}O_{2max}$? One would imagine that such an important question was intensely researched, but surprisingly this is not the case. There is only a limited number of studies

Figure 10.9 Factors associated with the drop of the maximal heart rate during normal ageing. The maximal heart rate, which declines with age, depends on (a) the intrinsic heart rate and (b) the response of the heart to β-agonists, which are drugs that work in a similar manner than for example adrenaline. The authors also demonstrated that the intrinsic heart rate is lower in elderly subjects and that the drop of the intrinsic heart rate with ageing and the response of the heart to β-agonists explains together much of the drop of the maximal heart rate during ageing. Redrawn after Christou and Seals (2008).

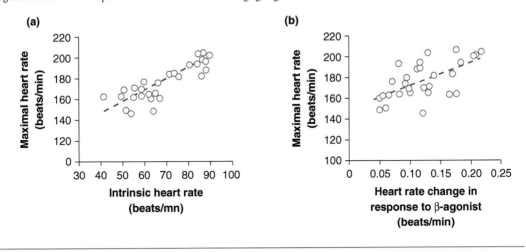

and these studies do not identify the causative mechanisms. Here, we summarize this research and discuss possible mechanisms for the drop of the maximal heart rate with ageing.

The heart rate depends on pacemaker cells within the so-called sinoatrial node (SA node) in the right atrium of the heart (Monfredi et al., 2010). The SA node will even generate an intrinsic heart rate in the absence of hormones and nerves. The catecholamines adrenaline and noradrenaline (termed epinephrine and norepinephrine in the USA) and sympathetic nerves increase the activity while parasympathic nerves such as the nervus vagus reduce it. So is the maximal heart rate decreased because elderly subjects have less sympathetic activation? This does not appear to be the case, as young and elderly appear to have similar adrenaline and noradrenaline concentrations during a graded exercise test (Lehmann and Keul, 1986). What else could it be? In an elegant study, Christou and Seals (2008) found that both the **intrinsic heart rate** and the β-**adrenergic responsiveness**, which is the increase in heart rate in response to adrenaline and noradrenaline, are both reduced and explain much of the decline of the maximal heart rate during ageing (Christou and Seals, 2008).

Both **intrinsic heart rate** and β-**adrenergic responsiveness** are vague terms, and for molecular exercise physiologists the aim is to identify the underlying cellular and molecular changes. As we have stated above, the likely culprit for the drop of maximal heart rate during ageing is a less well-performing SA node. Here, ageing could either:

- affect the number and size especially of pacemaker cells in the SA node or
- change the expression of the ion channels and other proteins that drive the action potentials.

Is there any evidence for this happening? There is some evidence that the collagen content of the SA node increases during ageing (Monfredi et al., 2010) and that the relative SA nodal cell volume decreases (Shiraishi et al., 1992). However, even if these changes occur, it is unclear whether they contribute to the drop of the maximal heart rate with ageing. A second factor could be that the SA node cells change their pace-making characteristics so that a quick firing necessary for a high maximal heart rate cannot be achieved. Is there any evidence for that? In one study the hearts of young (25 ± 1 years), middle-aged (50 ± 2) and elderly (66 ± 2 years) individuals were investigated using various electrophysiological methods. The authors found a slowing of the conduction velocity and sinus node recovery time, which was interpreted as a reduction of sinus node function, with age (Kistler et al., 2004).

Although we do not know what is causing the drop of the intrinsic heart rate and the loss of β-adrenergic responsiveness with ageing there are some candidate genes which might be involved. As an example, inhibition of the kinase CaMKII by the inhibitor AC3-I does reduce the increase of the heart rate in response to β-adrenergic stimulation with isoproterenol (Wu et al., 2009). Thus a reduction of CaMKII expression with age is one of the many possible molecular mechanisms that can explain the drop in maximal heart rate and thus $\dot{V}O_{2max}$ with ageing.

We will now stop the speculation about the cause of the drop in HR_{max} during ageing here and summarize. During ageing, the $\dot{V}O_{2max}$ drops, which is mainly due to a fall in the maximal heart rate with age, which can be estimated with the popular 'HR_{max} = 220 – age' formula or with formulas obtained from trend lines fitted through measured data (Tanaka et al., 2001). The drop in the maximal heart rate is due to a reduction of the **intrinsic heart rate** and a diminished heart rate increase in response to adrenaline and noradrenaline (i.e. **β-adrenergic responsiveness** (Christou and Seals, 2008)). Both the intrinsic heart rate and β-adrenergic responsiveness are linked to the pacemaker cells within the SA node. In the SA node the volumes of cells and the electrophysiology changes with age. The pacemaker cell membrane clock depends on ion channels, transporters and regulatory proteins. Proteins such as CaMKII affect the β-adrenergic responsiveness and changes in the expression of such proteins could potentially explain the drop in the maximal heart rate with age. Finally, the drop in the $\dot{V}O_{2max}$ with ageing also depends on many other factors, such as muscle strength (Frontera et al., 1990), pulmonary function, blood oxygen transport capacity capillarization and metabolic enzymes within the exercising muscle.

SUMMARY

Ageing hypotheses can be subdivided into damage or wear-and-tear hypotheses and programmed ageing hypotheses. An example of a damage hypothesis states that excessive amounts of RONS and the resulting oxidative stress result in ageing. In contrast, the cellular senescence hypothesis, which is a programmed ageing hypothesis, states that critical telomere shortening causes cellular senescence. A key component of this pathway is p53, which has been dubbed the guardian of the genome. Too little p53 activity increases the likelihood of cancer as cells proliferate more and are more likely to pick up cancer-inducing mutations. On the other hand, too much p53 activity drives cells into senescence too early, which results in premature ageing. Finally, ageing is affected by the activity of the mTOR pathway. This pathway promotes muscle growth but increased activity of this pathway over a long time causes reduced lifespan, while a reduction of pathway activity increases lifespan. All the above mechanisms can be related to endurance exercise, which generally increases lifespan.

Sarcopenia is the loss of muscle mass and strength during normal ageing. Key mechanisms are the loss of muscle fibres, which is probably secondary to the loss of α-motor neurons. Elderly people also appear to develop anabolic resistance, which is the reduced response to anabolic signals, such as resistance exercise, amino acids and insulin. Finally, old muscle has a decreased regenerative capacity, which is probably due to a lower number and impaired function of satellite cells, although this is not universally accepted and might differ between humans and mice. During ageing the $\dot{V}O_{2max}$ also decreases, which is due to a decline of the maximal cardiac output which in turn is due to a drop in the HR_{max}. This important phenomenon is under researched but a potentially key factor appears to be a reduced intrinsic heart rate and decreased β-adrenergic responsiveness. The causes of this are unknown but changes in the signalling or electrophysiology of the SA node are a possible cause.

REVIEW QUESTIONS

1   According to the 'rate of living' hypothesis, endurance athletes should die earlier than sedentary subjects. Why is this not the case?
2   Why would an excessive antioxidant intake be detrimental?
3   Is it probable that bodybuilders die young because of increased mTOR activation?
4   What is sarcopenia? Discuss two mechanisms that contribute to sarcopenia?
5   Why does our $\dot{V}O_{2max}$ decline with ageing? Speculate about possible cellular and molecular mechanisms.

FURTHER READING

Johnson SC, Rabinovitch PS and Kaeberlein M (2013). mTOR is a key modulator of ageing and age-related disease. *Nature* 493, 338–345.
Narici MV and Maffulli N (2010). Sarcopenia: characteristics, mechanisms and functional significance. *Br Med Bull.* 95, 139–159
Powers SK and Jackson MJ (2008). Exercise-induced oxidative stress: cellular mechanisms and impact on muscle force production. *Physiol Rev* 88, 1243–1276.
Shawi M and Autexier C (2008). Telomerase, senescence and ageing. *Mech Ageing Dev* 129, 3–10.
Tyner SD, Venkatachalam S, Choi J, Jones S, Ghebranious N, Igelmann H, et al. (2002). p53 mutant mice that display early ageing-associated phenotypes. *Nature* 415, 45–53.

REFERENCES

Aagaard P, Suetta C, Caserotti P, Magnusson SP and Kjaer M (2010). Role of the nervous system in sarcopenia and muscle atrophy with aging: strength training as a countermeasure. *Scand J Med Sci Sports* 20, 49–64.

Alsharidah M, Lazarus NR, George TE, Agley CC, Velloso CP and Harridge SD (2013). Primary human muscle precursor cells obtained from young and old donors produce similar proliferative, differentiation and senescent profiles in culture. *Aging Cell* 12, 333–344.

Anisimov VN, Berstein LM, Egormin PA, Piskunova TS, Popovich IG, Zabezhinski MA, et al. (2008). Metformin slows down aging and extends life span of female SHR mice. *Cell Cycle* 7, 2769–2773.

Baker DJ, Wijshake T, Tchkonia T, LeBrasseur NK, Childs BG, van de Sluis B, et al. (2011). Clearance of p16Ink4a-positive senescent cells delays ageing-associated disorders. *Nature* 479, 232–236.

Baur JA, Pearson KJ, Price NL, Jamieson HA, Lerin C, Kalra A, et al. (2006). Resveratrol improves health and survival of mice on a high-calorie diet. *Nature* 444, 337–342.

Blair SN and Morris JN (2009). Healthy hearts – and the universal benefits of being physically active: physical activity and health. *Ann Epidemiol* 19, 253–256.

Bodnar AG, Ouellette M, Frolkis M, Holt SE, Chiu CP, Morin GB, et al. (1998). Extension of life-span by introduction of telomerase into normal human cells. *Science* 279, 349–352.

Carlson ME, Hsu M and Conboy IM (2008). Imbalance between pSmad3 and Notch induces CDK inhibitors in old muscle stem cells. *Nature* 454, 528–532.

Carlson ME, Suetta C, Conboy MJ, Aagaard P, Mackey A, Kjaer M, et al. (2009). Molecular aging and rejuvenation of human muscle stem cells. *EMBO Mol Med* 1, 381–391.

Caron E, Ghosh S, Matsuoka Y, Ashton-Beaucage D, Therrien M, Lemieux S, et al. (2010). A comprehensive map of the mTOR signaling network. *Mol Syst Biol* 6, 453.

Chakkalakal JV, Jones KM, Basson MA and Brack AS (2012). The aged niche disrupts muscle stem cell quiescence. *Nature* 490, 355–360.

Chen C, Liu Y, Liu Y and Zheng P (2009). mTOR regulation and therapeutic rejuvenation of aging hematopoietic stem cells. *Sci Signal* 2, ra75.

Cherkas LF, Hunkin JL, Kato BS, Richards JB, Gardner JP, Surdulescu GL, et al. (2008). The association between physical activity in leisure time and leukocyte telomere length. *Arch Intern Med* 168, 154–158.

Christou DD and Seals DR (2008). Decreased maximal heart rate with aging is related to reduced {beta}-adrenergic responsiveness but is largely explained by a reduction in intrinsic heart rate. *J Appl Physiol* 105, 24–29.

Colman RJ, Anderson RM, Johnson SC, Kastman EK, Kosmatka KJ, Beasley TM, et al. (2009). Caloric restriction delays disease onset and mortality in rhesus monkeys. *Science* 325, 201–204.

Conboy IM, Conboy MJ, Wagers AJ, Girma ER, Weissman IL and Rando TA (2005). Rejuvenation of aged progenitor cells by exposure to a young systemic environment. *Nature* 433, 760–764.

Cuthbertson D, Smith K, Babraj J, Leese G, Waddell T, Atherton P, et al. (2005). Anabolic signaling deficits underlie amino acid resistance of wasting, aging muscle. *FASEB J* 19, 422–424.

Ekblom B (1968). Effect of physical training on oxygen transport system in man. *Acta Physiol Scand Suppl* 328, 1–45.

Fisher-Wellman K and Bloomer RJ (2009). Acute exercise and oxidative stress: a 30 year history. *Dyn Med* 8, 1.

Frontera WR, Meredith CN, O'Reilly KP and Evans WJ (1990). Strength training and determinants of VO_{2max} in older men. *J Appl Physiol* 68, 329–333.

Fullerton MD and Steinberg GR (2010). SIRT1 takes a backseat to AMPK in the regulation of insulin sensitivity by resveratrol. *Diabetes* 59, 551–553.

George T, Velloso CP, Alsharidah M, Lazarus NR and Harridge SD (2010). Sera from young and older humans equally sustain proliferation and differentiation of human myoblasts. *Exp Gerontol* 45, 875–881.

Gwinn DM, Shackelford DB, Egan DF, Mihaylova MM, Mery A, Vasquez DS, et al. (2008). AMPK phosphorylation of raptor mediates a metabolic checkpoint. *Mol Cell* 30, 214–226.

Harley CB, Futcher AB and Greider CW (1990). Telomeres shorten during ageing of human fibroblasts. *Nature* 345, 458–460.

Harman D (1956). Aging: A theory based on free radical and radiation chemistry. *J Gerontol* 11, 298–300.

Harrison DE, Strong R, Sharp ZD, Nelson JF, Astle CM, Flurkey K, et al. (2009). Rapamycin fed late in life extends lifespan in genetically heterogeneous mice. *Nature* 460, 392–395.

Hayflick L (1965). The limited in vitro lifetime of human diploid cell strains. *Exp Cell Res* 37, 614–636.

Herbert B, Pitts AE, Baker SI, Hamilton SE, Wright WE, Shay JW and Corey DR (1999). Inhibition of human telomerase in immortal human cells leads to progressive telomere shortening and cell death. *Proc Natl Acad Sci U S A* 96, 14276–14281.

Inoki K, Zhu T and Guan KL (2003). TSC2 mediates cellular energy response to control cell growth and survival. *Cell* 115, 577–590.

Jang YC, Lustgarten MS, Liu Y, Muller FL, Bhattacharya A, Liang H, et al. (2010). Increased superoxide in vivo accelerates age-associated muscle atrophy through mitochondrial dysfunction and neuromuscular junction degeneration. *FASEB J* 24, 1376–1390.

Jaskelioff M, Muller FL, Paik JH, Thomas E, Jiang S, Adams AC, et al. (2011). Telomerase reactivation reverses tissue degeneration in aged telomerase-deficient mice. *Nature* 469, 102–106.

Johnson SC, Rabinovitch PS and Kaeberlein M (2013). mTOR is a key modulator of ageing and age-related disease. *Nature* 493, 338–345.

Kanning KC, Kaplan A and Henderson CE (2010). Motor neuron diversity in development and disease. *Annu Rev Neurosci* 33, 409–440.

Kapahi P, Chen D, Rogers AN, Katewa SD, Li PW, Thomas EL and Kockel L (2010). With TOR, less is more: a key role for the conserved nutrient-sensing TOR pathway in aging. *Cell Metab* 11, 453–465.

Kistler PM, Sanders P, Fynn SP, Stevenson IH, Spence SJ, Vohra JK, et al. (2004). Electrophysiologic and electroanatomic changes in the human atrium associated with age. *J Am Coll Cardiol* 44, 109–116.

Kumar V, Selby A, Rankin D, Patel R, Atherton P, Hildebrandt W, et al. (2009). Age-related differences in the dose-response relationship of muscle protein synthesis to resistance exercise in young and old men. *J Physiol* 587, 211–217.

Lagouge M, Argmann C, Gerhart-Hines Z, Meziane H, Lerin C, Daussin F, et al. (2006). Resveratrol improves mitochondrial function and protects against metabolic disease by activating SIRT1 and PGC-1alpha. *Cell* 127, 1109–1122.

Lauretani F, Russo CR, Bandinelli S, Bartali B, Cavazzini C, Di IA, et al. (2003). Age-associated changes in skeletal muscles and their effect on mobility: an operational diagnosis of sarcopenia. *J Appl Physiol* 95, 1851–1860.

Lehmann M and Keul J (1986). Age-associated changes of exercise-induced plasma catecholamine responses. *Eur J Appl Physiol Occup Physiol* 55, 302–306.

Lexell J (1997). Evidence for nervous system degeneration with advancing age. *J Nutr* 127, 1011S–1013S.

Lexell J and Downham DY (1991). The occurrence of fibre-type grouping in healthy human muscle: a quantitative study of cross-sections of whole vastus lateralis from men between 15 and 83 years. *Acta Neuropathol (Berl)* 81, 377–381.

Lexell J, Taylor CC and Sjostrom M (1988). What is the cause of the ageing atrophy? Total number, size and proportion of different fiber types studied in whole vastus lateralis muscle from 15- to 83-year-old men. *J Neurol Sci* 84, 275–294.

Medvedev ZA (1990). An attempt at a rational classification of theories of ageing. *Biol Rev Camb Philos Soc* 65, 375–398.

Mittal KR and Logmani FH (1987). Age-related reduction in 8th cervical ventral nerve root myelinated fiber diameters and numbers in man. *J Gerontol* 42, 8–10.

Monfredi O, Dobrzynski H, Mondal T, Boyett MR and Morris GM (2010). The anatomy and physiology of the sinoatrial node – a contemporary review. *Pacing Clin Electrophysiol* 33, 1392–1406.

Moreland JD, Richardson JA, Goldsmith CH and Clase CM (2004). Muscle weakness and falls in older adults: a systematic review and meta-analysis. *J Am Geriatr Soc* 52, 1121–1129.

Narici MV and Maffulli N (2010). Sarcopenia: characteristics, mechanisms and functional significance. *Br Med Bull* 95, 139–159.

Pearl R (1928). *The rate of living.* University of London Press, London.

Ploutz-Snyder LL, Manini T, Ploutz-Snyder RJ and Wolf DA (2002). Functionally relevant thresholds of quadriceps femoris strength. *J Gerontol A Biol Sci Med Sci* 57, B144–B152.

Powers SK and Jackson MJ (2008). Exercise-induced oxidative stress: cellular mechanisms and impact on muscle force production. *Physiol Rev* 88, 1243–1276.

Rantanen T, Guralnik JM, Foley D, Masaki K, Leveille S, Curb JD, et al. (1999). Midlife hand grip strength as a predictor of old age disability. *JAMA* 281, 558–560.

Rasmussen BB, Fujita S, Wolfe RR, Mittendorfer B, Roy M, Rowe VL and Volpi E (2006). Insulin resistance of muscle protein metabolism in aging. *FASEB J* 20, 768–769.

Ristow M, Zarse K, Oberbach A, Kloting N, Birringer M, Kiehntopf M, et al. (2009). Antioxidants prevent health-promoting effects of physical exercise in humans. *Proc Natl Acad Sci U S A* 106, 8665–8670.

Rosen DR, Siddique T, Patterson D, Figlewicz DA, Sapp P, Hentati A, et al. (1993). Mutations in Cu/Zn superoxide dismutase gene are associated with familial amyotrophic lateral sclerosis. *Nature* 362, 59–62.

Rosenberg IH (1997). Sarcopenia: origins and clinical relevance. *J Nutr* 127, 990S–991S.

Ruiz JR, Sui X, Lobelo F, Morrow JR, Jr., Jackson AW, Sjostrom M, et al. (2008). Association between muscular strength and mortality in men: prospective cohort study. *BMJ* 337, a439.

Sanz A, Pamplona R and Barja G (2006). Is the mitochondrial free radical theory of aging intact? *Antioxid Redox Signal* 8, 582–599.

Selman C, Tullet JM, Wieser D, Irvine E, Lingard SJ, Choudhury AI, et al. (2009). Ribosomal protein S6 kinase 1 signaling regulates mammalian life span. *Science* 326, 140–144.

Shawi M and Autexier C (2008). Telomerase, senescence and ageing. *Mech Ageing Dev* 129, 3–10.

Shefner JM, Reaume AG, Flood DG, Scott RW, Kowall NW, Ferrante RJ, et al. (1999). Mice lacking cytosolic copper/zinc superoxide dismutase display a distinctive motor axonopathy. *Neurology* 53, 1239–1246.

Shiraishi I, Takamatsu T, Minamikawa T, Onouchi Z and Fujita S (1992). Quantitative histological analysis of the human sinoatrial node during growth and aging. *Circulation* 85, 2176–2184.

Smith DL, Jr., Elam CF, Jr., Mattison JA, Lane MA, Roth GS, Ingram DK, et al. (2010). Metformin supplementation and life span in Fischer-344 rats. *J Gerontol A Biol Sci Med Sci* 65, 468–474.

Sohal RS and Buchan PB (1981). Relationship between physical activity and life span in the adult housefly, Musca domestica. *Exp Gerontol* 16, 157–162.

Speakman JR, Talbot DA, Selman C, Snart S, McLaren JS, Redman P, et al. (2004). Uncoupled and surviving: individual mice with high metabolism have greater mitochondrial uncoupling and live longer. *Aging Cell* 3, 87–95.

Tanaka H, Monahan KD and Seals DR (2001). Age-predicted maximal heart rate revisited. *J Am Coll Cardiol* 37, 153–156.

Teramoto M and Bungum TJ (2010). Mortality and longevity of elite athletes. *J Sci Med Sport* 13, 410–416.

Tomlinson BE and Irving D (1977). The numbers of limb motor neurons in the human lumbosacral cord throughout life. *J Neurol Sci* 34, 213–219.

Trappe SW, Costill DL, Fink WJ and Pearson DR (1995). Skeletal muscle characteristics among distance runners: a 20-yr follow-up study. *J Appl Physiol* 78, 823–829.

Tyner SD, Venkatachalam S, Choi J, Jones S, Ghebranious N, Igelmann H, et al. (2002). p53 mutant mice that display early ageing-associated phenotypes. *Nature* 415, 45–53.

von Zglinicki T (2002). Oxidative stress shortens telomeres. *Trends Biochem Sci* 27, 339–344.

Weismann A (1892). *Über Leben und Tod.* Gustav Fischer, Jena.

Wolfe RR (2006). The underappreciated role of muscle in health and disease. *Am J Clin Nutr* 84, 475–482.

Wu Y, Gao Z, Chen B, Koval OM, Singh MV, Guan X, et al. (2009). Calmodulin kinase II is required for fight or flight sinoatrial node physiology. *Proc Natl Acad Sci U S A* 106, 5972–5977.

11 Molecular neuroscience and exercise

Peer Wulff and Henning Wackerhage

At the end of the chapter you should be able to:

- List the major cells and anatomical structures within the spinal cord and brain and explain how they function.

- Explain how movement is regulated by neurons within the spinal cord and in the brain. Explain how the firing of α-motor neurons is regulated during contractions of different intensity (Henneman size principle), during the stretch reflex, cyclic movements such as walking or swimming and by higher motor centres in the brain.

- Explain long-term potentiation, changes of dendritic spines and neurogenesis as forms of neural plasticity. Give examples of their importance during motor skill learning and explain some of the underlying molecular mechanisms.

INTRODUCTION

When unicellular organisms evolved into multicellular organisms, systems co-evolved that allowed all the cells within the organism to function as a team rather than as individuals. These systems are the endocrine system and the nervous system. In the context of sport, the key function of the nervous system is to learn and execute movements and to respond with movement to specific situations, such as a high ball in football or a sound during dancing. The complex movement sequences of whole limbs in gymnasts or very fine movements of small parts of the body, as in concert pianists, demonstrate the precision and complexity that can be achieved if talent (i.e. the genetic variants that determine trainability for motor skills) and intense training are combined. However, some neural circuits interfere with the motor systems of the brain and spinal cord: during penalties in a World Cup Final, emotion and crowd noise turn a simple penalty motor task into a difficult task.

Because neuroscience is not always taught in sport and exercise courses, we will first review neuroanatomy and cellular neuroscience in relation to sport and exercise to introduce this topic. First we introduce the cells and major systems within the nervous system. After that we review the neural circuits that regulate motor skill execution by starting with the regulation of muscle fibres by α-motor neurons, and we end with the regulation of complex movements by neural circuits in the brain. Finally, we will discuss neural networks that control motor learning and the molecular mechanisms that facilitate memory formation and learning.

MOLECULAR AND CELLULAR NEUROSCIENCE IN A NUTSHELL

The nervous system can be divided into two systems (Figure 11.1):

- the **central nervous system** (CNS), comprising the brain and spinal cord protected by skull and spine, and the
- **peripheral nervous system** (PNS), comprising the neuronal tissue outside the central nervous system.

The peripheral nervous system can be further divided into the

- **somatic nervous system** and the
- **autonomic** (sympathetic, parasympathetic) **nervous system**.

In the somatic, peripheral nervous system **sensory afferent neurons** sense signals (e.g. pain, muscle length, visual signals or sounds) and transmit this information from the periphery to the central nervous system. In contrast, **efferent α-motor neurons** receive inputs from the central nervous system and execute these by innervating muscle fibres within muscles.

The **autonomic** nervous system controls and orchestrates the organs and glands within the body. Its sympathetic division is activated during excitement or stress, causing a '**fight or flight**' response with increased blood levels of **adrenaline** (also called epinephrine), increased heart rate and blood pressure, and widened bronchial tubes for better air flow to the lungs. In '**rest and digest**' situations the parasympathetic division takes over, reversing the actions of the sympathetic system. Although the autonomic nervous system is relevant for physical and mental performance, we will focus in this chapter on the somatic part of the peripheral nervous system and the motor systems of the central nervous system.

THE CELLS OF THE NERVOUS SYSTEM

There are two major cell types within the nervous system:

- **Neurons** are cells that can receive, convey, integrate and pass on information. They signal via changes in their membrane potential (e.g. action potentials) and release neurotransmitters.
- **Glia** (Greek for 'glue') are cells that nurture, protect and insulate neurons.

Figure 11.1 Central and peripheral nervous system. (a) MRI image of the brain or central nervous system with the (1) cortex and (2) cerebellum visible. (b) MRI image of the (2) cerebellum for comparison with (a) and (3) the spinal cord, which is a key part of the peripheral central nervous system. Images were kindly provided by Ourania Varsou and Prof. Christian Schwarzbauer.

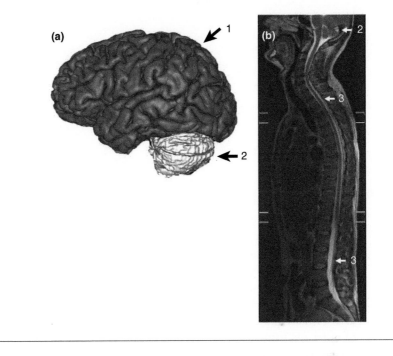

In the human brain there are probably roughly ≈100 billion neurons (Williams and Herrup, 1988). All have similar components (Figure 11.2):

- **Dendrites** sense **neurotransmitters** released from other neurons via **neurotransmitter receptors**. These open or close ion channels or transporters, changing the **membrane potential** (increased or decreased membrane potentials are known as excitatory or inhibitory postsynaptic potentials or **EPSP** or **IPSP**, respectively).
- The **soma** or **cell body** contains the nucleus and the machinery to maintain the neuron. It is the 'central decision maker' as it receives dendritic inputs (i.e. EPSPs and IPSPs) which combine to change the membrane potential. If, as a consequence, the membrane potential rises from normally ≈−70 mV to the threshold of ≈−30 mV, an **action potential** (a rapid spike of the membrane potential to positive values) is triggered.
- Action potentials travel down the **axon** of the neuron to reach the **synapse**, which is a point of contact with other neurons. Calcium entry into the synapse leads to the release of **neurotransmitters**, which can activate or inhibit downstream neurons.

Essentially, neurons do what mathematical equations do (London and Hausser, 2005) as they concentrate several inputs (EPSPs, IPSPs) into one (action potential or not).

Figure 11.2 Schematic drawing of a neuron highlighting the dendrites, nucleus, axon, synapses and neurotransmitters. Neurons come in many shapes and sizes and α-motor neurons, which span from the spinal cord to distant muscles, are the longest cells of the human body.

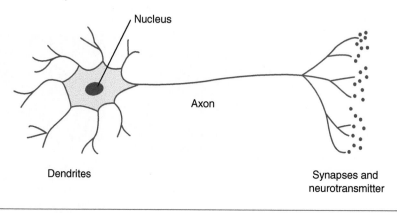

We will now discuss neurotransmitters, as they are key regulators in the nervous system. Neurotransmitters are molecules released by neurons that act on target cells which can be either other neurons or cells such as muscle fibres. Neurotransmitters are released by synapses, diffuse through the synaptic cleft and then bind to specific receptors on target neurons, muscle fibres and other cells. In the brain, **glutamate** is the main excitatory neurotransmitter, whereas **gamma amino butyric acid (GABA)** and **glycine** are the main inhibitory neurotransmitters. The effects of other neurotransmitters, such as dopamine, noradrenaline or serotonin, can be excitatory or inhibitory, depending on the respective receptor subtype they act on. The most important neurotransmitters, their receptors and some of their functions are summarized in Table 11.1.

Table 11.1 Some important neurotransmitters

NEUROTRANSMITTER	RECEPTORS AND SOME FUNCTIONS
Glutamate, aspartate	NMDA and non-NMDA receptors; important for LTP and LTD
GABA	$GABA_A$ and $GABA_B$; inhibitory neurotransmitter
Glycine	Gly receptor (GlyR); inhibitory neurotransmitter
Serotonin (5-HT)	Serotoninergic (5-HT) receptors are classified as 5-HT_1, 5-HT_2 and 5-HT_3 receptors
Acetylcholine	Nicotinic N_1 or N_2 or muscarinic M_1–M_5. Acetylcholine is released, for example, from α-motor neuron synapses and binds to N_2 receptors at the neuromuscular junction on muscle fibres
Dopamine	Dopaminergic receptors D_1–D_5. Dopamine has been linked to central fatigue (Foley and Fleshner, 2008)
Noradrenaline (norepinephrine)	Adrenergic α- and β-receptors (important for the fight-or-flight response)
Endorphins, enkephalins	Endorphin–enkephalin (opioid) receptors (many forms labelled μ_1, μ_2, δ_1, δ_2, κ_1, κ_2, κ_3, σ). Has been linked to the runner's high hypothesis (Boecker et al., 2008)

From Porter and Kaplan (2010).

One of the most fascinating hypotheses involving neurotransmitters in sport is the so-called 'runner's high' hypothesis, which states that endorphins, which are morphines produced in the body, induce euphoria when they rise during exercise. To test this hypothesis, researchers have infused the opiodergic receptor analogue [^{18}F]FDPN into their subjects and searched for its binding in the brain with positron emission tomography (PET) after endurance running or at rest. The researchers additionally performed tests for euphoria (Boecker et al., 2008) to test whether running leads to euphoria in their subjects and whether this is linked to endorphin receptor binding. The researchers reported a reduced binding of [^{18}F]FDPN in some brain regions after endurance running compared to rest, suggesting that opiodergic peptides such as endorphins were occupying the receptors after running. This also correlated with euphoria ratings, suggesting a causal relationship (Boecker et al., 2008) and that the 'runner's high' might be a real phenomenon.

FROM BOTTOM TO TOP: THE MAMMALIAN MOTOR SYSTEM AND MOTOR LEARNING

After the quick introduction to the nervous system and its cells we will now focus on how the motor system controls movement. We will first discuss how the motor system works and then analyse how the motor system can be modulated by training e.g. while learning movements.

MOTOR NEURON TYPES AND DEVELOPMENT

All instructions that result in movement are sent as action potentials to skeletal muscles via motor neurons. There are various types of motor neuron (Kanning et al., 2010):

- α-**Motor neurons** innervate skeletal muscle fibres. They vary in size, dendrite branching, axon size and conduction velocity, depending on whether they innervate slow type I, intermediate type IIa or fast type IIb/x muscle fibres (Kanning et al., 2010).
- γ-**Motor neurons** innervate so-called intrafusal fibres within length-sensing muscle spindles (they modulate the sensitivity of muscle spindles).
- β-**Motor neurons** innervate both muscle fibres and intrafusal fibres.

The cell bodies (soma) of motor neurons are located in the spinal cord. From there their axons go to all skeletal muscles where their synapses release **acetylcholine** to innervate the muscle fibres to which they connect. The acetylcholine then binds to N_2 nicotinic receptors at the **neuromuscular junction** of muscle fibres. It has been estimated that there are 120 000 motor neurons in the spinal cord, which innervate 100 million muscle fibres (Kanning et al., 2010).

Motor neurons and the fibres to which they connect form a so-called **motor unit**. Different classes of motor units have been identified in experiments where researchers have electrically stimulated motor neurons and measured the resultant force output, which depends both on the properties of the motor unit and on the properties of the muscle fibres that are innervated by the motor neuron. On the basis of these experiments three types of motor units have been proposed (Burke et al., 1973):

- **FF (fast-fatiguing; a related class is termed Fint; intermediate fatiguability)** motor units comprise large, hard-to-excite α-motor neurons with large, fast-conducting axons that innervate fast but easily fatiguing type IIb and IIx fibres*;
- **FR (fatigue-resistant)** motor units comprise α-motor neurons that innervate type intermediate type IIa fibres. The α-motor neuron properties are likely between those of FF and S α-motor neurons.
- **S (slow)** motor units comprise small α-motor neurons with a low excitation threshold and small axons that innervate slow type I fibres.

*The above relationships between motor unit class, α-neuron class and fibre type have not conclusively been demonstrated as reliable molecular markers, especially for the different α-motor neurons, are lacking (Kanning et al., 2010).

The different α-motor neurons are schematically depicted in Figure 11.3.

Figure 11.3 (a) Schematic drawing of different α-motor neurons/motor units found in the spinal cord. From top to bottom: FF (innervating IIb/x fibres; note that intermediate Fint motor neurons are similar), FR (innervates IIa fibres) and S (innervates type I fibres) motor unit. Redrawn after Kanning et al. (2010). (b) Microscopic image of a mouse α-motor neuron ending kindly provided by Dr Abraham Acevedo-Arozena, MRC Harwell.

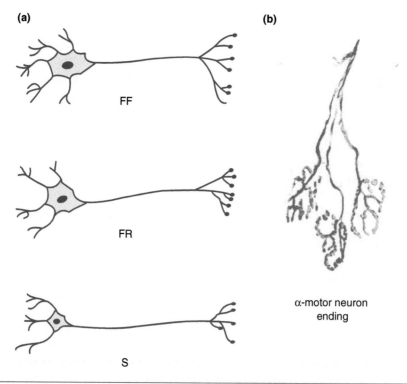

Thus to summarize, there are different muscle fibre types (I, IIa, IIx and IIb, note that IIb do not occur regularly in humans) and the different fibre types are innervated by specific α-motor neurons (S, FR and FF) to form S, FR and FF motor units. It is an intriguing question whether the type of muscle fibre is determined by the type of α-motor neuron or vice versa or whether the types of muscle fibre and α-motor neuron are regulated independently.

Finally, we will discuss how α-motor neurons and motor units develop. In this process so-called early spinal progenitor cells first stop dividing and then differentiate into mature α-motor neurons that find their way to and then connect to muscle fibres (Dalla Torre di Sanguinetto et al., 2008). At the end of development the α-motor neurons in the spinal cord are organized broadly into **motor columns** and within these motor columns **motor neuron pools** develop. The α-motor neurons within a motor neuron pool innervate the fibres of one skeletal muscle. For example, in the so-called lateral motor column ≈50 distinct motor neuron pools have been identified (Dalla Torre di Sanguinetto et al., 2008). The location and type of α-motor neurons depend especially on the transcription factors which are expressed within the α-motor neuron (Dalla Torre di Sanguinetto et al., 2008). The factors that regulate the expression of different transcription factors within α-motor neurons, and thereby their type and location, are currently incompletely understood.

HENNEMAN SIZE PRINCIPLE

In the last section we have described how motor neuron pools comprise FF, FR and S motor units, or α-motor neurons that innervate the type IIx/b, type IIa and type I fibres of the target muscle, respectively. We have also discussed that α-motor neurons form motor neuron pools, which connect to one muscle. Within a motor neuron pool there is a mix of α-motor neuron types, as skeletal muscles generally comprise a mix of muscle fibres. Functionally, a key question is: What determines the innervation or recruitment of S, FR, FF α-motor neurons or motor units within a motor neuron pool and thereby the contraction of type I, IIa, IIx and IIb muscle fibres? Imagine we decide to contract our vastus lateralis with increasing intensity from a very mild contraction to maximal. During this motor task the neurons that connect the higher motor centres in the brain to the vastus lateralis motor neuron pool in the spinal cord will increase the rate of action potential firing from low to maximal. As a consequence, more and more α-motor neurons in the vastus lateralis motor neuron pool will reach their threshold and will start to fire action potentials, which travel to muscle fibres in the vastus lateralis and innervate these. Crucially, S, FR and FF α-motor neurons have different innervations thresholds, with S α-motor neurons having the lowest, FR having an intermediate and FF α-motor neurons having the highest threshold. Therefore, if an upstream neuron fires action potentials at a low frequency into a motor neuron pool then only small S α-motor neurons will reach their threshold and innervate the type I muscle fibres to which they connect. If the upstream neurons increase the action potential firing rate, then additionally FR and, at very high rates also FF α-motor neurons will be recruited. This is known as the **Henneman size principle** or as **ramp-like recruitment** (Mendell, 2005). This system ensures that the faster but fatiguing α-motor neurons or motor units

Figure 11.4 Henneman size principle or ramp-like recruitment. Schematic drawing depicting how the firing rate of a descending neuron determines the firing of S, FR and FF α-motor neurons or motor units. (a) If the descending neuron fires at a low rate (indicated by a only three spikes in the graph showing voltage (V) against time) then only S α-motor neurons are recruited as they have the lowest threshold for recruitment. This ensures that during low-intensity contractions, such as maintaining one's posture or walking, only type I muscle fibres will be recruited. (b) If the descending neuron fires at near maximal rates, as is the case during maximal contractions, then all α-motor neurons will be recruited and type I, IIa, IIb/x muscle fibres contract. This is known as the Henneman size principle or ramp-like recruitment.

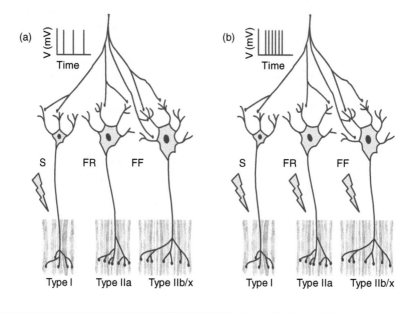

are only additionally recruited when high forces need to be generated. The function of this system has been demonstrated in freely moving cats (Walmsley et al., 1978) and indirectly by glycogen-depletion patterns during different exercise intensities in fast and slow muscle fibres in humans (Gollnick et al., 1974).

The recruitment of different motor units is additionally affected by fatigue and this explains why FR and FF muscle fibres glycogen deplete, for example, at the end of a marathon race. A marathon runner will initially innervate mainly S α-motor neurons and slow type I fibres, as the contraction forces are relatively low. However, over time the slow type I fibres will lose glycogen (Gollnick et al., 1974) and as a consequence fatigue and produce less force. If that happens then the only way to maintain the running pace is for the upstream neurons to fire action potentials at a higher rate so that more α-motor neurons are recruited. The more the type I and other fibres fatigue, the more the upstream neurons need to fire action potentials, and at one point type FR and eventually also FF α-motor neurons will be recruited and type IIa and IIx and IIb muscle fibres will contract and use up their glycogen (Gollnick et al., 1974).

The effect of fatigue also applies to resistance or strength training, where athletes sometimes try to recruit as many α-motor neurons as possible in order to increase their neuromuscular activation. One might think that the only way to do so is to lift a maximal weight. However, an alternative way to innervate all fibres that can be voluntarily innervated is to lift a lower weight to failure. In the latter case the innervated muscle fibres will fatigue more and more with each repetition and again the only way to maintain force output is to recruit additional α-motor neurons until the weight can no longer be lifted, despite maximal α-motor neuron recruitment (Jungblut, 2009). This means that if a strength athlete wishes to train with maximal neuromuscular activation, then she or he can either perform maximal contractions or alternatively perform sets with submaximal contractions to failure.

To summarize, the Henneman size principle is key to understanding how S, FR and FF α-motor neurons or type I, IIa and IIx muscle fibres are innervated by upstream neurons. In general, S α-motor neurons have the lowest innervations threshold, followed by FR and FF α-motor neurons. As a consequence, S α-motor neurons are the only neurons that are innervated during low-intensity exercise. If muscle fibres fatigue either during a marathon but also during a submaximal set of resistance exercise, then the upstream neurons need to increase their firing frequency so that more and more α-motor neurons are recruited, which is necessary to maintain exercise intensity. This means that even FR and FF α-motor neurons can be recruited during low-intensity exercise if the exercise duration is long enough. Also, maximal neuromuscular activation can be achieved during an exercise which is performed to failure.

REFLEXES AND CENTRAL PATTERN GENERATORS

In the previous section we have outlined how motor neuron pools are organized in the spinal cord, how their identity is regulated by transcription factors and how the firing of S, FR and FF α-motor neurons within a motor neuron pool is regulated. We now wish to discuss neural circuits that regulate α-motor neuron firing within the spinal cord. The first type of neural circuits that we are going to discuss are simple reflexes, and the second type of neural circuits are known as central pattern generators. We will mainly explain how these neural circuits work, but we will additionally review some of the molecular biology that controls their development and function.

Reflexes are important for immediate response to sensory input, in particular for sensing potentially dangerous signals such as muscle stretch, heat or injury. Depending on the input and output pathways and the number of neurons involved, mono- and polysynaptic reflexes can be distinguished. The best-known spinal reflex is the stretch reflex (Figure 11.5) which functions as follows:

- Stretch, especially rapid stretch of a muscle spindle leads to the activation of afferent **Ia sensory neurons**. These then start to fire action potentials which activates α-motor neurons via one excitatory synapse.
- The activated α-**motor neurons** in turn innervate the muscle fibres that belong to their motor unit within the stretched muscle.

How important are the cells within the stretch reflex neural circuit for the normal functioning of the motor system? Molecular studies have helped to answer this question. Knocking out the genes for Runx3, ErbB2, Nrg1 and Egr3, either from birth or in the adult, reduces or eliminates muscle spindles or inhibits their connection to α-motor

Figure 11.5 Schematic drawing depicting the monosynaptic stretch reflex. (1) Rapid stretch causes depolarization of muscle spindles and firing of IIa neurons. (2) IIa Neurons are connected via a single synapse to α-motor neurons which (3) innervates the muscle fibres to which it is connected. This short circuit with only one synapse allows rapid responses to a potentially damaging stimulus. However, note that many other neurons connect to α-motor neurons. Such neurons can override the stretch reflex, for example in ballet dancers, martial artists or hurdlers.

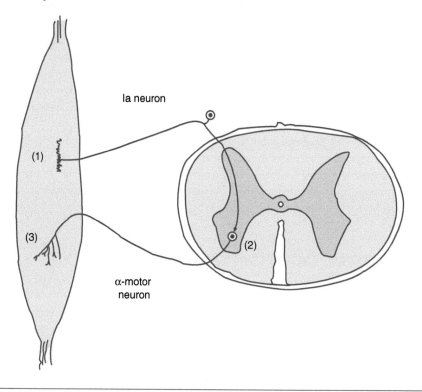

neurons. As a consequence muscle length perception and the stretch reflex work poorly or not at all. If such a gene knockout is induced then the mice suffer from ataxia (i.e. poor muscle coordination) and/or severe movement disorders (Chen et al., 2003). In an extreme example, the knockdown of ErbB2 and muscle spindles results in a mouse where the limbs are all extended (Figure 11.6), demonstrating the importance of the muscle spindle–Ia sensory neuron system and muscle length perception for normal movement (Andrechek et al., 2002).

In sport, the stretch reflex can be utilized to try to increase the neuromuscular activation during a contraction. Such training is known as **plyometric training**. A typical exercise is a drop jump, where the athlete jumps from a box onto the ground followed by another jump. When the athlete lands on the ground, the leg extensors will be stretched, and as a consequence the stretch reflex is triggered and Ia neurons will fire. Thus, during the subsequent jump the α-motor neurons will be innervated by both Ia neurons and upstream neurons, and this might help the athlete to activate his or her muscles. To conclude, spinal reflexes connect proprioceptive organs, such as the muscle spindle and Golgi tendon organ, directly or via very few interneurons to pools of α-motor neurons. When triggered in an emergency, such as during a rapid stretch, they cause the

Figure 11.6 Schematic drawing demonstrating the importance for muscle spindles for the control of muscle contractions. (a) Posture of a wildtype mouse compared to (b) which depicts a transgenic mouse where ErbB2 and, via that, muscle spindles have been knocked out. The transgenic mice presumably have no or only little sensation about their muscle length and as a consequence the limbs are, in severely affected mice, extended. Redrawn after Andrechek et al. (2002).

(a)　　　　　　　　　　　**(b)**

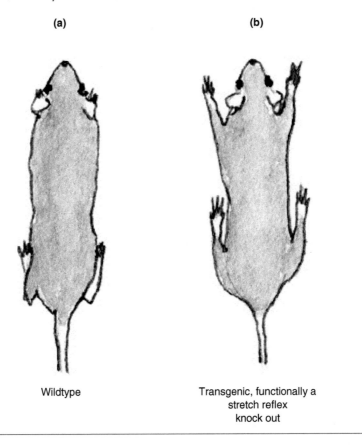

Wildtype

Transgenic, functionally a
stretch reflex
knock out

immediate firing of α-motor neurons and thereby prevent damage. In addition, these reflex circuits are important for muscle tone and length.

Central pattern generators are more complex neural circuits which are, like reflexes, located in the **spinal cord**. They regulate rhythmic movements, such as walking, running or swimming. Central pattern generators function even if the spinal cord is severed, without interference from the brain. They can regulate the following functions (Kiehn, 2006):

- **left–right** coordination;
- **flexor–extensor alternation**;
- **rhythm** generation.

The neural circuits that form central pattern generators work by connecting the α-motor neuron pools that innervate the muscles involved in rhythmic movements. The connecting neurons are termed **interneurons**. Such interneurons express specific transcription factors and can be subdivided into V0, V1, V2, V3 and Hb9 interneurons (Kiehn,

Figure 11.7 Simplified schematic model of a vertebrate central pattern generator that may regulate, for example, rhythmic movements such as walking. In this model filled circles indicate excitatory synapses whereas open circles indicate inhibitory synapses. (1) Rhythm-generating cells (rh) fire rhythmically and (2) innervate α-motor neurons (mn), which cause rhythmic muscle contractions. (3) Rhythm-generating cells additionally innervate inhibitory interneurons (in) which ensure that flexors or antagonists do not contract when extensors or agonists do. (4) Additionally, commissural, inhibitory interneurons, which span from the left to the right side of the spinal cord or vice versa, ensure that the right muscles do not contract when the left muscles do. Higher brain centres switch the central pattern generators on and off and modulate their intensity (slow, fast walking, running) and reflex arcs override the central pattern generators, for example when a muscle is stretched unexpectedly or when we step onto a sharp object.

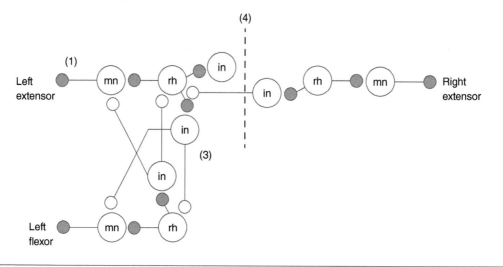

2010). In a nutshell, rhythm-generating neurons initiate the movement, and inhibitory interneurons ensure that the left leg does not contract when the right left does and that antagonists do not contract when antagonists do. Figure 11.7 gives a simplified but still complex overview over a central pattern generator and shows how rhythm-generating ability, left–right and flexor–extensor coordination may be linked to each other.

We will now review transgenic mouse models that support the existence of central pattern generators. In one transgenic mouse strain the transcription factor Dbx1, which is expressed in V0 commissural interneurons was knocked out. The researchers found that these mice had a tendency towards hopping, where the right and left muscles contract at the same time rather than walking, where left and right muscle contractions alternate. This suggests that the knockout has interneurons, which regulate left–right alternation (Lanuza et al., 2004). Similarly, ephrin A4 and ephrinB3 knockout mice also struggle to perform left–right alternation (Figure 11.8).

Transgenic mice were also used to study **rhythm** generation by glutamate-releasing neurons. In these mice, the rhythm-generating neurons can be switched on as these neurons express a light-sensitive ion channel (Figure 11.9). The consequence of such activation is rhythmic α-motor neuron firing, which is consistent with a central pattern generator as there is an alternating ipsilateral and left–right firing pattern. In other words, antagonists relax when agonists contract and also the same muscles on the right and left contract in an alternating fashion (Hagglund et al., 2010).

Figure 11.8 The walking patterns of (a) a wildtype mouse and (b) transgenic ephrin A4 and ephrin B3 knockout mice. The traces show the firing of the right and left muscles. Wildtype mice are capable of a normal alternating gait, whereas the transgenic mice perform a synchronous, hopping gait. This suggests that the ablated ephrins and neurons are important for alternating left–right contractions. Redrawn after Kullander et al. (2003).

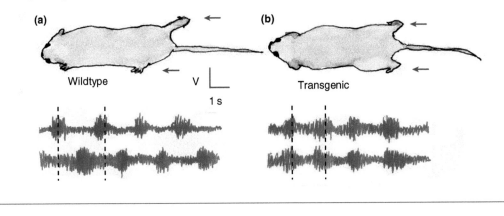

Figure 11.9 Experimental demonstration of a central pattern generator. Researchers have created a transgenic mouse in which glutamate-releasing neurons can be activated by light. These neurons in the spinal cord (depicted in 1) were exposed to light and the output on the left (left L2, left L5) and right side (right L5) of lumbar segments was recorded. (2) The researchers found alternating firing between left left L2 and left L5, presumably depicting agonist and antagonist firing, and between left L5 and right L5, presumably demonstrating left–right firing. This suggests that glutamine-releasing neurons (a) can drive the rhythm, (b) coordinate alternative ipsilateral pools of α-motor neurons (e.g. flexor, extensor) and (c) coordinate the alternating left–right contraction. Thus, such glutamine-releasing neurons appear to kickstart the central pattern generator for alternating movements such as walking. Redrawn after Hagglund et al. (2010).

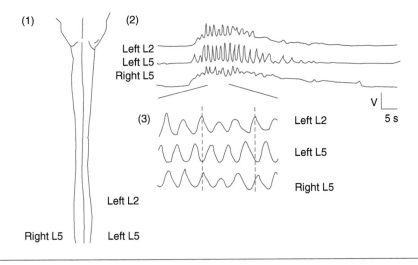

To summarize, reflexes and central pattern generators are neural circuits within the spinal cord that regulate emergency reactions and relatively complex movements such as walking or running. Central pattern generators comprise largely interneurons that connect the motor neuron pools responsible for rhythm generation, left–right as well as extensor–flexor coordination. Many of these interneurons are inhibitory and ensure that the right or antagonist α-motor neurons do not fire when the left or agonist α-motor neurons do.

HIGHER MOTOR CENTRES

In the previous section we summarized research which shows that the brain is not essential for the generation of relatively complex movements such as walking or running because of reflex arcs and central pattern generators. But what is the function of the brain or of higher motor areas in the control of movement? Some of the functions include the following:

- Starting or stopping central pattern generators via neurons in the spinal cord or brainstem (Hagglund et al., 2010), which again have connections to higher brain centres (Goulding, 2009).
- Overriding reflex arcs: For example, ballet dancers and martial arts athletes, such as Bruce Lee, override the stretch reflex e.g. during fast, high kicks.
- Overriding central pattern generators: Breathing depends on central pattern generators in the medulla, which is part of the brainstem (Garcia-Campmany et al., 2010). This central pattern generator operates even when we are unconscious or sleeping but it is no problem to consciously override it, for example, when we decide to do a breath hold dive.
- **Fine finger and hand movements** depend on connections from the motor cortex in the brain to motor neurons (Courtine et al., 2007; Lemon, 2008). These neural tracts are only present in primates and humans but not other species and are believed to be important for the high finger and hand dexterity of humans.
- Regulating movement in relation to (a) sensory inputs, especially visual inputs (e.g. tennis player reacting to a ball), (b) emotion (e.g. penalty in a World Cup Final), (c) cognition (e.g. planned movements) and (d) memory (i.e. learned movement).

Here we will only very briefly discuss the overall locomotor system in vertebrates (Figure 11.10).

We will now discuss the structure of the locomotor system shown in Figure 11.10 in more detail. (1) α-Motor neuron pools in relation with central pattern generators within the spinal cord (2) innervate muscle fibres within skeletal muscles according to the Henneman size principle (Mendell, 2005). Ia (and II) sensory neurons from stretch-sensing muscle spindles and Ib sensory neurons from tension-sensing Golgi tendon organs run back to the spinal cord and influence the activity of α-motor neurons. This is termed **proprioception** (meaning 'one's own sensation'). (3) A key function of the cerebellum is to compare a copy of a movement programme going down through the spinal cord via descending pathways with proprioception. This allows the locomotor system to check whether a movement feels as it was planned. (4) Descending pathways, especially the reticulospinal pathway, which is activated by the mesencephalic locomotor region (MLR), initiate locomotion and help to control it. (5) The motor cortex can directly innervate α-motor neurons that regulate the fingers of the hand in primates and

Figure 11.10 Schematic drawing of the structure of the locomotor system in vertebrates. (1) Pools of α-motor neurons are connected via reflexes such as the stretch reflex and part of spinal cord networks such as central pattern generators (CPG). They innervate muscle fibres and at the same time receive feedback about muscle length or tension. (2) Higher motor centres in the brain additionally regulate α-motor neuron pools in the spinal cord via descending pathways (see text below). (4) This is then sent via descending pathways to motor neuron pools. A key function of the cerebellum is not to initiate movements but to check whether the executed movements match the planned movements. Thus if a cerebellum is injured, movements become slow and erratic. Redrawn after Goulding (2009).

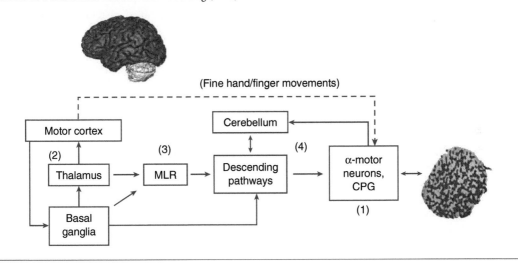

humans (Lemon, 2008). Alternatively, in combination with higher brain regions called the thalamus and basal ganglia, the motor cortex can also control locomotion.

MOTOR LEARNING AND NEURAL PLASTICITY

After introducing the nervous system and the parts that regulate muscle contractions and movement, we will now focus on the molecular mechanisms that enable us to learn new movements or motor skills. We learn such motor skills throughout life: After birth, children learn to sit, crawl, stand and walk. Even speaking is a motor skill as it requires us to contract muscles to produce sounds. Later in life we learn to swim, cycle or to perform a wide variety of other motor skills. In order to be able to perform complicated motor skills, both many years of training and a high level of talent are required. The importance of talent or of being the carrier of advantageous DNA sequence variations for motor learning have been demonstrated in twin studies. Such studies have shown that both neuromuscular coordination (Fox et al., 1996; Missitzi et al., 2004) and motor learning (Fox et al., 1996) are highly inherited. This matches what we see when a group of individuals attempt to learn a difficult motor task, such as a somersault. Some group members will learn the task quickly while others will struggle to learn it at all. The exact DNA sequence variations that are responsible for the heritability of motor learning skills are largely unknown.

In general, short- and long-term learning and memory formation can be distinguished. For short-term learning and memory formation no protein synthesis is required and the memory is not very stable. This is different from long-term learning and memory formation, which requires the synthesis of new proteins (Costa-Mattioli et al., 2009). On a cellular level, learning means that the neurons involved need to increase or decrease their effect on each other or new connections between neurons need to be made or existing connections need to be removed. This is termed **neural plasticity**. There is evidence for neural plasticity both in brain areas such as the motor cortex, which can be measured with functional MRI in humans (Karni et al., 1995), and in the spinal cord, which can be demonstrated by measuring the so-called H-reflex (Thompson et al., 2009). A remarkable feature of motor skill learning is that learned skills, unlike the endurance or strength built by training, are well retained throughout life. This explains why we can swim, cycle or ski even if we have not done so for over 10 years, implying that the new skill is 'engraved' somewhere within the brain or spinal cord. For molecular exercise physiologists, the challenge is to identify the molecular mechanisms that cause long-term motor memories. We will now attempt to answer this question.

In 1894 the Spanish neuroanatomist Ramon y Cajal was one of the first scientists to suggest that memories might be stored by strengthening the connections between neurons (Cajal, 1894). This idea was further developed in the 1940s by the Canadian psychologist Donald Hebb, who suggested that strengthening of connections between active neurons might underlie learning (Hebb, 1949). It took a long time before researchers managed to demonstrate this experimentally. The evidence was produced in 1973 when Bliss and Lømo used so-called 'conditioning trains', which are repeated electrical shocks of certain neurons, and observed that these increased the amplitude of excitatory postsynaptic potentials (EPSPs) in postsynaptic or downstream neurons by nearly 3-fold and that these changes persisted for several hours afterwards (Bliss and Lømo, 1973; Lomo, 2003). They termed this effect **long-term potentiation** (**LTP**; Figure 11.11).

BOX 11.1 TERJE LØMO ON THE DISCOVERY OF LONG-TERM POTENTIATION

Long-term potentiation (LTP) was first observed in 1966 (Lømo, 1966) when I worked for my doctorate in medicine with Per Andersen in Oslo. The discovery initiated a follow-up study with Tim Bliss in Andersen's laboratory in 1968–1969, which resulted in the first description of some of the basic features of LTP (Bliss and Lømo, 1973). Interest in LTP, initially feeble, exploded in the mid-1980s after three important discoveries: its presence in slices of the hippocampus maintained *in vitro* (Schwartzkroin and Wester, 1975), its dependence on activation of NMDA receptors (Collingridge et al., 1983) and its dependence on an increase in intracellular levels of Ca^{2+} (Lynch et al., 1983). Since then LTP has been demonstrated in most, if not all, parts of the central nervous system. Many of the underlying molecular mechanisms have been revealed in amazing detail and, today, LTP appears necessary for both hippocampus-dependent (declarative) and -independent (procedural) forms of learning and memory. A more recent discovery is PKMζ, a protein kinase C isoform, which becomes constitutively active in postsynaptic dendritic spines after LTP-inducing stimulation or natural learning (Sacktor, 2011). Remarkably, such activity is a prerequisite for the long-term maintenance of LTP and natural memories. Thus, local injection of a specific inhibitor of

PKMζ into relevant cortical regions erases both LTP and natural memories, even when the injection occurs several months after the learning event. LTP occurs in the sensorimotor cortex in association with learning new motor skills and also the injection of the PKMζ inhibitor erases the skill (von Kraus et al., 2010). Importantly, however, such erasure only brings the relevant synapses back to their basal condition and does not interfere with the learning of new skills after the disappearance of the inhibitor. Evidently, LTP is a necessary form of synaptic plasticity that allows the formation of new circuits in the brain where new facts and skills can be stored and later recalled.

Terje Lømo, 2012

Figure 11.11 Long-term potentiation as a form of neural plasticity. Redrawn after Bliss and Lømo (1973). (a) Recording of neurons before and after 'conditioning trains', which are electrical stimulations of 10 s with 15 Hz, or 15 electrical shocks per second. One can see that the amplitude of the recording of the conditioned neurons is larger. (b) Plot of the excitatory postsynaptic potential (EPSP) before and after four conditioning trains. The EPSP amplitude increases in the conditioned neurons (filled circle) but not in the control neurons (open circle).

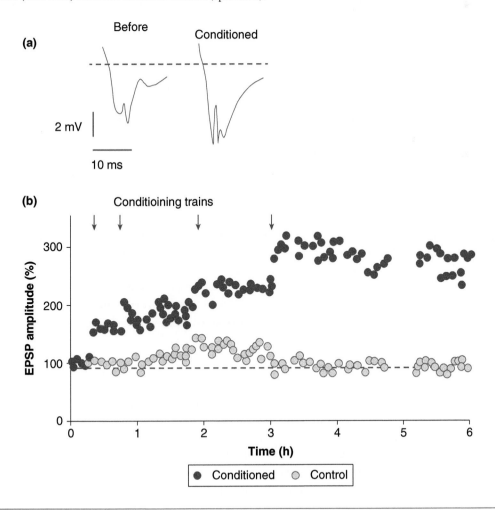

Neuroscientists became excited about the discovery of LTP because they realized that this was the first time that a neuronal 'memory' had been demonstrated at the cellular level and because it occurred in a part of the hippocampus that was important for the storage of memories for facts and events (Kim and Linden, 2007).

Is LTP the universal cellular mechanism that makes neurons change their behaviour in response to activity and thus responsible for all learning and memory? Probably not. Since the discovery of LTP, researchers have identified many mechanisms by which the neuron activity changes the strength of existing connections, leads to the formation of new connections, alters the overall responsiveness of neurons and even leads to the addition of neurons. Thus neural plasticity or changes in the nervous system can occur because of various mechanisms, with LTPs being one of them. Below are three key forms of neuronal plasticity:

- **Long-term potentiation** (LTP) (Bliss and Lømo, 1973) or **long-term depression** (Lynch et al., 1977) of glutamatergic (i.e. glutamate-releasing), GABAergic (GABA-releasing) and other synapses can develop in response to suitable stimuli. LTPs increase the EPSP and LTDs decrease the EPSP of postsynaptic neurons, respectively (Kim and Linden, 2007).
- **Structural plasticity**, which includes the formation or elimination of **dendritic spines** (Alvarez and Sabatini, 2007; Yu and Zuo, 2011), can strengthen or weaken the connections between neurons. New dendritic spine connections with existing axons strengthen the connection between two neurons, while dendritic spine connections with new axons change the processing of information within a neural circuit.
- Emergence of new neurons, which is termed **neurogenesis**, can occur as a consequence of the division and/or differentiation of neural stem cells (van Praag, 2009; Shors et al., 2011).

Thus, to summarize, neurons can increase or decrease the strength of postsynaptic activation (EPSP) or inhibition (IPSP), rewire as a result of the formation of new dendritic spine–axon connections or entirely new neurons can emerge as a result of neural stem cell action. These three types of mechanisms are believed to be responsible for motor skill and other learning and the formation of memory. In the following section we will give examples where the three mechanisms have been related to motor learning and then discuss the underlying molecular mechanisms.

EVIDENCE FOR LTP AND LTD DURING MOTOR SKILL LEARNING

LTP and LTD have long been hypothesized to be one mechanism by which motor skill and other leaning occurs. However, good evidence for this hypothesis was only published in 2000. The experiment involved training rats in a motor skill task followed by measuring LTP and LTD in the motor cortex of the hemisphere of the brain that was regulating the motor skill, using the other hemisphere as a control (Rioult-Pedotti et al., 2000). The idea was that if motor learning induced LTP in motor cortex neurons then experimenters would struggle to increase the LTP further with electrical stimulation, as these neurons should be closer to their LTP ceiling than control neurons. This is what the team observed and it was interpreted as indirect evidence that LTPs occur during motor skill learning in the motor cortex.

EVIDENCE FOR DENDRITIC SPINE PLASTICITY DURING MOTOR SKILL LEARNING

The second mechanism of neural plasticity is the formation or elimination of dendritic spines. Dendritic spines are protrusions from dendrites that receive input via a synapse from an axon. They come in various shapes and sizes but the majority (>95% at least in some neurons) have a connection to an axon (Arellano et al., 2007). Dendritic spines are not stable but constantly form and disappear. Do dendritic spines change during motor learning? To answer this question researchers trained mice in a task where the animal had to grasp a single seed with the forelimb as a model for motor skill learning. The researchers then imaged dendritic spines in neurons of the motor cortex that were made detectable by overexpressing a yellow fluorescent protein (Xu et al., 2009). The authors observed two phenomena: first motor skill learning rapidly increased the formation of new dendritic spines and second motor skill learning made new dendritic spines survive for longer when compared to controls (Figure 11.12). This suggests that the formation of dendritic spines and their survival is changed in response to motor skill learning, so that overall new dendritic spines occur when a motor skill is learned.

Figure 11.12 Dendritic spines and motor learning. (a) Drawing of dendritic spines protruding from a dendrite. (b) Dendritic spine formation is increased during motor learning. The authors also found (not shown) that newly formed dendritic spines are more stabilized in the period after motor learning when compared to controls. This suggests that dendritic spines may be linked to the strengthening of neuron connections during motor learning. Redrawn after Xu et al. (2009).

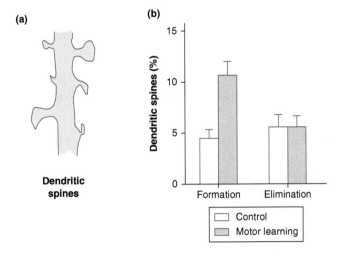

EVIDENCE FOR NEUROGENESIS DURING EXERCISE

Adult **neurogenesis** describes the emergence of new neurons due to division and/or differentiation of neural stem and progenitor cells in the adult brain. For many decades it had been assumed that the adult brain cannot generate new neurons. However, using cell division markers such as BrdU, researchers discovered that there are areas in the brain where dividing cells exist. These are the subventricular zone of the lateral ventricle wall and the subgranular zone of the hippocampal dentate gyrus (Ma et al., 2009). The hippocampus, where the dentate gyrus is located, is important for learning and memory formation and one can see how new neurons might help in encoding new memories. Related to this, researchers tested whether endurance exercise increased neurogenesis in the dentate gyrus of mice by using BrdU, which is incorporated into dividing cells. The researchers found that it does and therefore demonstrated that exercise can lead to the formation of new neurons (Pereira et al., 2007).

The next question is: are these new neural stem cell-derived neurons important? To test this questions researchers created a knockout mouse where the transcription factor TLX, which plays an important role in neural stem cells, was knocked out. This reduced stem cell proliferation and spatial learning but did not affect, for example, locomotion (Zhang et al., 2008). Taken together, this suggests that exercise increases the production of new neurons, presumably from neural stem cells, and this might be important for some forms of learning.

NEURAL PLASTICITY: SIGNALLING MECHANISMS

The above has shown that during motor skill learning LTP, LTD and denditric spines form and that neurogenesis or neuron formation occurs depending on the task learned and on the area of the brain involved. The challenge for molecular exercise physiologists is to identify the underlying molecular mechanisms. The research in this area has led to the identification of several molecular mechanisms. Here, we focus on two mechanisms illustrated in Figure 11.13 and explained in more detail in the text below:

- Ca^{2+}–CaMKII–AMPA receptor signalling and the regulation of short-term LTP;
- protein synthesis and the regulation of long-term LTPs.

CA^{2+}–CAMKII–AMPA RECEPTOR SIGNALLING AND SHORT-TERM LTP

Glutamate is the neurotransmitter in many but not all neurons that can produce LTPs (Kim and Linden, 2007). When glutamate is released from the synapse it binds to so-called **NMDA** receptors on dendrites of the postsynaptic neuron. They are termed NMDA receptors because they can be selectively activated by the drug NMDA (*N*-methyl D-aspartate). When stimulated, NMDA receptors open and Ca^{2+} flows into the postsynaptic neuron. This Ca^{2+} then binds to the calcium sensor calmodulin, which in turn activates the Ca^{2+}/calmodulin-dependent protein kinase

Figure 11.13 Examples of mechanisms responsible for short-term (a) and long-term (b) LTP and memory. (a) (1) Suitable stimuli in glutamatergic synapses lead to an activation of the NMDA receptor (NMDA-R). This causes a Ca^{2+} influx into the postsynaptic neuron. Ca^{2+} then binds to calmodulin, which activates CaMKII, which in turn phosphorylates itself (Lisman et al., 2002). CaMKII also phosphorylates AMPA receptors which move to the postsynaptic membrane and respond to glutamate with an influx of ions such as Ca^{2+}. As there are now more glutamate-activated channels at the postsynaptic membrane, a given amount of glutamate will lead to an LTP of the neuron. (b) (3) eIF2α dephosphorylation by a phosphatase (PP) regulates the translation of ATF4 which in turn represses CREB, a known repressor of LTP and memory. Thus dephosphorylation of eIF2α favours the formation of long-term LTPs and memory (Costa-Mattioli et al., 2007).

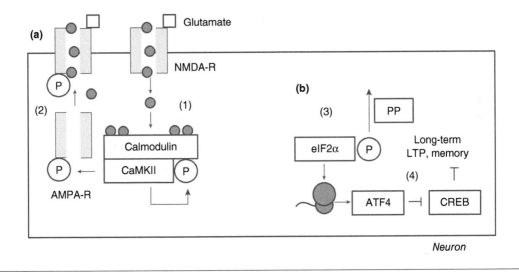

CaMKII by autophosphorylation at threonine residues (Lee et al., 2009). CaMKII in turn phosphorylates a subunit of the AMPA receptors at Ser831. AMPA receptors then become inserted into the cell membrane (Lisman et al., 2002). Since AMPA receptors are channels that are activated by glutamate and allow the influx of sodium into the cell, the increased presence of AMPA receptors in the postsynaptic membrane will lead to increased responses upon subsequent exposure to glutamate, and an LTP results. Thus, not surprisingly, the knockout of CaMKII and other parts of this signalling pathway in mice greatly impairs the ability to produce LTP (Silva et al., 1992) and also a variety of behavioural and learning defects result (Lisman et al., 2002). Now, while the glutamate → NMDA receptor → Ca^{2+}/calmodulin → CaMKII → AMPA receptor signalling pathway is a key pathway for the formation of LTP and LTD, it is not the only one responsible for how the activity of neurons affects neural plasticity in the short and long term (Kim and Linden, 2007; Flavell and Greenberg, 2008).

PROTEIN SYNTHESIS AND CONTROL OF LONG-TERM LTP

Earlier in this chapter we had indicated that both LTP and memories can be subdivided into **short-term** and **long-term**. Long-term memories can last a lifetime. For example,

movement skills such as swimming, cycling, skiing or ice skating are relatively difficult to learn, but when learned and stored as memory they are never fully lost (Costa-
Mattioli et al., 2009). For long-term LTP and memories, the mTOR pathway and
protein synthesis are required, as treatment with protein synthesis inhibitors prevents
the formation of late LTPs and long-term learning (Tang et al., 2002; Lynch, 2004;
Costa-Mattioli et al., 2009).

In this section we review the molecular mechanisms that underly long-term LTPs
and memory. At the core of the molecular mechanism lies the protein synthesis regulator **eIF2α** and its phosphorylation at a crucial Ser51 site by the protein kinase **GCN2**
and other upstream kinases. Phosphorylated eIF2α then selectively stimulates the
translation or synthesis of the transcription factor ATF4 which, when present, inhibits CREB-dependent gene expression which is crucial for memory formation (Costa-
Mattioli et al., 2009). So how important is eIF2α for long-term LTP and memory
formation? To test this, researchers generated transgenic mice in which the crucial serine
51 (Ser51) of eIF2α was mutated to an alanine, which can no longer be phosphorylated (the mutated allele was termed eIF2α S51A). Because mice homozygous for this
mutation did not survive, the researchers only used heterozygous eIF2α S51A mice, in
which one allele was wildtype eIF2α and the other was eIF2α S51A. In these heterozygous mice the researchers found that it was much easier to induce LTP (Figure 11.14)
and that learning was facilitated in several different tests (Costa-Mattioli et al., 2007).

Figure 11.14 Mechanisms of LTP formation. A single 100 Hz train of stimulation in wildtype mice (open circles) produces
an early-phase LTP, but after ≈2 h it returns to baseline. In contrast, in heterozygous mice with one wildtype and one eIF2α
S51A allele (filled circles; the mutated allele cannot be phosphorylated) the LTP remains beyond 3 h. These data suggest that
LTP is regulated via the Ser51 phosphorylation site of eIF2α. Redrawn after Costa-Mattioli et al. (2007).

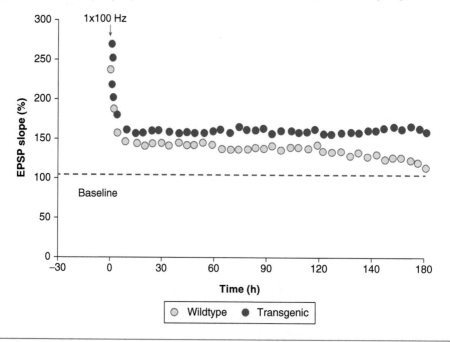

Thus this demonstrates that this signalling mechanism is involved in LTP and memory formation and that can it manipulated in a way that memory formation is improved.

Another pathway that regulates protein synthesis in response to LTP-inducing stimuli involves the kinase mTOR, which can via 4E-BP1 enhance translation of sets of mRNA that increase the capacity for protein synthesis. To test the relevance of this pathway for synaptic plasticity and learning, researchers inhibited mTOR. These experiments revealed that mTOR activity was necessary for both LTP and memory formation in different behavioural tests (Tang et al., 2002; Tischmeyer et al., 2003; Costa-Mattioli et al., 2009). Taken together, these data show that long-term LTP and memory formation depend on protein synthesis, which is regulated via several pathways. It is likely that similar mechanisms operate during actual motor learning.

SUMMARY

The two parts of the nervous system are the central nervous system (CNS; brain and spinal cord) and the peripheral nervous system, which lies outside the CNS. Key cells within the nervous system are neurons, which receive, convey and integrate information. They signal via action potentials, which lead to the release of neurotransmitters at the synapse at the end of the axon. For example, endorphins are released during running and bind to their receptors, which suggests that this might contribute to the 'runner's high'. In the motor system α-motor neurons originate in the spinal cord and their axons innervate and form a motor unit, usually with several muscle fibres. Motor units can be subdivided into fast-fatiguing (FF), fatigue-resistant (FR) and slow (S) motor units, or α-motor neurons that innervate type IIx/b, type IIa and type I fibres, respectively. The innervation threshold of different α-motor neuron types differs and is highest in FF, intermediate in FR and lowest in S α-motor neurons. This is the basis for the Henneman size principle or ramp-like recruitment. As a consequence S motor units are active during mild activity such as standing or walking, while FR and FF motor units are only recruited during medium- and high-intensity exercise or when S motor units are fatigued. Pools of α-motor neurons in the spinal cord are connected to each other via excitatory or inhibitory interneurons and form reflex arcs, such as the stretch reflex or central pattern generators. Such central pattern generators control (a) the rhythm, (b) left–right coordination and (c) agonist–antagonist firing during cyclic movements such as walking or running. Such neural networks in the spine can be started and stopped, modified, overridden or bypassed by higher motor centres in the brain.

Motor learning can be subdivided into short-term and long-term motor learning and some motor skills such as swimming are retained throughout life once learned. Long-term potentiation (LTP) or long-term depression describes how inputs increase or decrease the signalling to postsynaptic neurons. One mechanism is the formation or elimination of dendritic spines. Such dendritic spines form during motor learning and are maintained better than when not learning. Finally, some types of learning can lead to the emergence of new neurons, which is termed neurogenesis. The molecular mechanisms involved include the regulation of translation or protein synthesis.

REVIEW QUESTIONS

1   What is a motor unit and how is the recruitment of slow (S), fatigue-resistant (FR) and fast-fatiguing (FF) motor units or α-motor neurons regulated during light and intense muscle contraction?

2   Describe the three key components of central pattern generators. Give an example of a transgenic mouse model that has increased our understanding of one of the components of central pattern generator function.

3   What is long-term potentiation and long-term depression? What is the link to motor learning?

4   What are dendritic spines and how do they function during motor learning?

5   Discuss one signal transduction mechanism that has been linked to memory formation.

FURTHER READING

Costa-Mattioli M, Sossin WS, Klann E and Sonenberg N (2009). Translational control of long-lasting synaptic plasticity and memory. *Neuron* 61, 10–211.

Kandel ER, Schwartz JH, Jessell TM and Hudspeth AJ (2012). *Principles of neural science*, 5th edn. McGraw-Hill, New York.

Kanning KC, Kaplan A and Henderson CE (2010). Motor neuron diversity in development and disease. *Annu Rev Neurosci* 33, 409–440.

Kim SJ and Linden DJ (2007). Ubiquitous plasticity and memory storage. *Neuron* 56, 582–592.

Lomo T (2003). The discovery of long-term potentiation. *Philos Trans R Soc Lond B Biol Sci* 358, 617–620.

Mendell LM (2005). The size principle: a rule describing the recruitment of motoneurons. *J Neurophysiol* 93, 3024–3026.

REFERENCES

Alvarez VA and Sabatini BL (2007). Anatomical and physiological plasticity of dendritic spines. *Annu Rev Neurosci* 30, 79–97.

Andrechek ER, Hardy WR, Girgis-Gabardo AA, Perry RL, Butler R, Graham FL, et al. (2002). ErbB2 is required for muscle spindle and myoblast cell survival. *Mol Cell Biol* 22, 4714–4722.

Arellano JI, Benavides-Piccione R, Defelipe J and Yuste R (2007). Ultrastructure of dendritic spines: correlation between synaptic and spine morphologies. *Front Neurosci* 1, 131–143.

Bliss TV and Lømo T (1973). Long-lasting potentiation of synaptic transmission in the dentate area of the anaesthetized rabbit following stimulation of the perforant path. *J Physiol* 232, 331–356.

Boecker H, Sprenger T, Spilker ME, Henriksen G, Koppenhoefer M, Wagner KJ, et al. (2008). The runner's high: opioidergic mechanisms in the human brain. *Cereb Cortex* 18, 2523–2531.

Burke RE, Levine DN, Tsairis P and Zajac FE, III (1973). Physiological types and histochemical profiles in motor units of the cat gastrocnemius. *J Physiol* 234, 723–748.

Cajal R (1894). The Croonian lecture: la fine structure des centres nerveux. *Proc R Soc Lond B Biol Sci* 55, 444–468.

Chen HH, Hippenmeyer S, Arber S and Frank E (2003). Development of the monosynaptic stretch reflex circuit. *Curr Opin Neurobiol* 13, 96–102.

Collingridge GL, Kehl SJ and McLennan H (1983). The antagonism of amino acid-induced excitations of rat hippocampal CA1 neurons in vitro. *J Physiol* 334, 19–31.

Costa-Mattioli M, Gobert D, Stern E, Gamache K, Colina R, Cuello C, et al. (2007). eIF2alpha phosphorylation bidirectionally regulates the switch from short- to long-term synaptic plasticity and memory. *Cell* 129, 195–206.

Costa-Mattioli M, Sossin WS, Klann E and Sonenberg N (2009). Translational control of long-lasting synaptic plasticity and memory. *Neuron* 61, 10–26.

Courtine G, Bunge MB, Fawcett JW, Grossman RG, Kaas JH, Lemon R, et al. (2007). Can experiments in nonhuman primates expedite the translation of treatments for spinal cord injury in humans? *Nat Med* 13, 561–566.

Dalla Torre di Sanguinetto S, Dasen JS and Arber S (2008). Transcriptional mechanisms controlling motor neuron diversity and connectivity. *Curr Opin Neurobiol* 18, 36–43.

Flavell SW and Greenberg ME (2008). Signaling mechanisms linking neuronal activity to gene expression and plasticity of the nervous system. *Annu Rev Neurosci* 31, 563–590.

Foley TE and Fleshner M (2008). Neuroplasticity of dopamine circuits after exercise: implications for central fatigue. *Neuromolecular Med* 10, 67–80.

Fox PW, Hershberger SL and Bouchard TJ, Jr. (1996). Genetic and environmental contributions to the acquisition of a motor skill. *Nature* 384, 356–358.

Garcia-Campmany L, Stam FJ and Goulding M (2010). From circuits to behaviour: motor networks in vertebrates. *Curr Opin Neurobiol* 20, 116–125.

Gollnick PD, Piehl K and Saltin B (1974). Selective glycogen depletion pattern in human muscle fibres after exercise of varying intensity and at varying pedalling rates. *J Physiol* 241, 45–57.

Goulding M (2009). Circuits controlling vertebrate locomotion: moving in a new direction. *Nat Rev Neurosci* 10, 507–518.

Hagglund M, Borgius L, Dougherty KJ and Kiehn O (2010). Activation of groups of excitatory neurons in the mammalian spinal cord or hindbrain evokes locomotion. *Nat Neurosci* 13, 246–252.

Hebb DO (1949). *Organization of behaviour: a neuropsychological theory.* John Wiley, New York.

Jungblut S (2009). The correct interpretation of the size principle and its practical application to resistance training. *Medicina Sportiva* 13, 203–209.

Kanning KC, Kaplan A and Henderson CE (2010). Motor neuron diversity in development and disease. *Annu Rev Neurosci* 33, 409–440.

Karni A, Meyer G, Jezzard P, Adams MM, Turner R and Ungerleider LG (1995). Functional MRI evidence for adult motor cortex plasticity during motor skill learning. *Nature* 377, 155–158.

Kiehn O (2006). Locomotor circuits in the mammalian spinal cord. *Annu Rev Neurosci* 29, 279–306.

Kiehn O (2010). Development and functional organization of spinal locomotor circuits. *Curr Opin Neurobiol* 21, 100–109.

Kim SJ and Linden DJ (2007). Ubiquitous plasticity and memory storage. *Neuron* 56, 582–592.

Kullander K, Butt SJ, Lebret JM, Lundfald L, Restrepo CE, Rydstrom A, et al. (2003). Role of EphA4 and EphrinB3 in local neuronal circuits that control walking. *Science* 299, 1889–1892.

Lanuza GM, Gosgnach S, Pierani A, Jessell TM and Goulding M (2004). Genetic identification of spinal interneurons that coordinate left-right locomotor activity necessary for walking movements. *Neuron* 42, 375–386.

Lee SJ, Escobedo-Lozoya Y, Szatmari EM and Yasuda R (2009). Activation of CaMKII in single dendritic spines during long-term potentiation. *Nature* 458, 299–304.

Lemon RN (2008). Descending pathways in motor control. *Annu Rev Neurosci* 31, 195–218.

Lisman J, Schulman H and Cline H (2002). The molecular basis of CaMKII function in synaptic and behavioural memory. *Nat Rev Neurosci* 3, 175–190.

Lømo T (1966). Frequency potentiation of excitatory synaptic activity in the dentate area of the hippocampal formation. *Acta Physiol Scand* 68, 128.

Lømo T (2003). The discovery of long-term potentiation. *Philos Trans R Soc Lond B Biol Sci* 358, 617–620.

London M and Hausser M (2005). Dendritic computation. *Annu Rev Neurosci* 28, 503–532.

Lynch GS, Dunwiddie T and Gribkoff V (1977). Heterosynaptic depression: a postsynaptic correlate of long-term potentiation. *Nature* 266, 737–739.

Lynch G, Larson J, Kelso S, Barrionuevo G, and Schottler F (1983). Intracellular injections of EGTA block induction of hippocampal long-term potentiation. *Nature* 305, 719–721.

Lynch MA (2004). Long-term potentiation and memory. *Physiol Rev* 84, 87–136.

Ma DK, Bonaguidi MA, Ming GL and Song H (2009). Adult neural stem cells in the mammalian central nervous system. *Cell Res* 19, 672–682.

Mendell LM (2005). The size principle: a rule describing the recruitment of motoneurons. *J Neurophysiol* 93, 3024–3026.

Missitzi J, Geladas N and Klissouras V (2004). Heritability in neuromuscular coordination: implications for motor control strategies. *Med Sci Sports Exerc* 36, 233–240.

Pereira AC, Huddleston DE, Brickman AM, Sosunov AA, Hen R, McKhann GM, et al. (2007). An in vivo correlate of exercise-induced neurogenesis in the adult dentate gyrus. *Proc Natl Acad Sci U S A* 104, 5638–5643.

Porter RS and Kaplan LS (2010). *The Merck Manual*, 19 edn. Merck Sharp & Dohme Corp., White house Station, New Jersey.

Rioult-Pedotti MS, Friedman D and Donoghue JP (2000). Learning-induced LTP in neocortex. *Science* 290, 533–536.

Sacktor TC (2011). How does PKMzeta maintain long-term memory? *Nat Rev Neurosci* 12, 9–15.

Schwartzkroin PA and Wester K (1975). Long-lasting facilitation of a synaptic potential following tetanization in the in vitro hippocampal slice. *Brain Res* 89, 107–119.

Shors TJ, Anderson ML, Curlik DM and Nokia MS (2011). Use it or lose it: How neurogenesis keeps the brain fit for learning. *Behav Brain Res* 227, 450–458.

Silva AJ, Stevens CF, Tonegawa S and Wang Y (1992). Deficient hippocampal long-term potentiation in alpha-calcium-calmodulin kinase II mutant mice. *Science* 257, 201–206.

Tang SJ, Reis G, Kang H, Gingras AC, Sonenberg N and Schuman EM (2002). A rapamycin-sensitive signaling pathway contributes to long-term synaptic plasticity in the hippocampus. *Proc Natl Acad Sci U S A* 99, 467–472.

Thompson AK, Chen XY and Wolpaw JR (2009). Acquisition of a simple motor skill: task-dependent adaptation plus long-term change in the human soleus H-reflex. *J Neurosci* 29, 5784–5792.

Tischmeyer W, Schicknick H, Kraus M, Seidenbecher CI, Staak S, Scheich H, et al. (2003). Rapamycin-sensitive signalling in long-term consolidation of auditory cortex-dependent memory. *Eur J Neurosci* 18, 942–950.

van Praag H (2009). Exercise and the brain: something to chew on. *Trends Neurosci* 32, 283–290.

von Kraus LM, Sacktor TC and Francis JT (2010). Erasing sensorimotor memories via PKMzeta inhibition. *PLoS ONE* 5, e11125.

Walmsley B, Hodgson JA and Burke RE (1978). Forces produced by medial gastrocnemius and soleus muscles during locomotion in freely moving cats. *J Neurophysiol* 41, 1203–1216.

Williams RW and Herrup K (1988). The control of neuron number. *Annu Rev Neurosci* 11, 423–453.

Xu T, Yu X, Perlik AJ, Tobin WF, Zweig JA, Tennant K, et al. (2009). Rapid formation and selective stabilization of synapses for enduring motor memories. *Nature* 462, 915–919.

Yu X and Zuo Y (2011). Spine plasticity in the motor cortex. *Curr Opin Neurobiol* 21, 169–174.

Zhang CL, Zou Y, He W, Gage FH and Evans RM (2008). A role for adult TLX-positive neural stem cells in learning and behaviour. *Nature* 451, 1004–1007.

At the end of this chapter you should be able to:

- Understand the basic physiology of the immune system and how it protects us from bacteria and pathogens.

- Explain the J-shaped curve relating the amount and/or intensity of exercise performed and the risk of subsequent upper respiratory tract infection.

- Discuss the changes in the immune system that have been observed after an acute bout of exercise and whether they link to infection risk.

- Explain nutritional strategies that can be employed to improve immune function after exercise.

- Explain that skeletal muscle acts as an endocrine organ that secretes so-called myokines (muscle hormones).

- Discuss how factors within the immune system (e.g. NF-κB) can affect skeletal muscle mass and exercise performance.

- Detail how T cells can modulate the function of satellite cells.

12 Molecular exercise immunology

Stuart Gray and Henning Wackerhage

INTRODUCTION

Athletes sometimes ask 'I have a cold, should I still train?' or 'Why do I often come down with a cough or a cold after a period of heavy training?' The answer is related to the function of the immune system and how it is affected by exercise. We will begin this chapter by reviewing the immune system in relation to exercise. The immune system is by nature a molecular and cellular subject. This is because the main players are cells that can be identified by molecules, rather than organs such as the heart in the cardiovascular system. In fact it is often said that molecular biology is a major driving force behind immunology research, with a prime example being the ability to identify and clone the numerous cytokines involved in immune regulation. A recent example of molecular methods in immunology is seen in the awarding of one half of the 2011 Nobel Prize in Physiology or Medicine to Jules Hoffmann and Bruce Beutler. In 1996 Hoffmann found that when infected with bacteria or fungi, fruitflies with Toll gene mutations died because they were unable to mount an effective immune defence. Two years later, Bruce Beutler found that a gene in mice very similar to Toll, thus named the Toll-like receptor, was the receptor for lipopolysaccharide, an endotoxin secreted by bacteria. These two discoveries were the starting point in identifying the way in which pathogens activate the innate immune system and are great examples of the importance of molecular techniques in immunology research.

From our review of the immune system we will see that there appears to be an increase in infection risk after periods of heavy exercise and potentially a reduced risk in those who regularly perform moderate intensity exercise. On top of that we will show that certain aspects of the immune system are suppressed after a single bout of exercise.

Although we will describe the immune system, due to its complexity we are not able to cover more detailed aspects. After this we will then look at how nutritional intake can potentially restore immune function after exercise and reduce the risk of subsequent infection. We will then review molecular exercise immunology research. In this part we will discuss some research showing how molecules of the immune system can control aspects of physiology relevant to exercise and health. These include controlling glucose and fat metabolism and also the size of skeletal muscle mass.

THE IMMUNE SYSTEM IN A NUTSHELL

Our world is inhabited by many **microbes** (i.e. microscopic organisms) and some infect the human body. An **infection** is when a parasite such as a microbe enters and colonizes a host organism. Some infections, such as the infection of our gut with beneficial microbes, are symbiotic (i.e. good for parasite and host), while others cause harm and can even be lethal (e.g. the *Mycobacterium tuberculosis* bacterium or the HIV virus); these are termed **pathogens**.

This is where the **immune system** comes in: It is an organ system that keeps pathogens from entering the body and identifies and kills any pathogens that do enter. As can be seen in the general overview in Figure 12.1 the immune system has two, interacting branches:

- the **innate** immune system and
- the **acquired** (adaptive) immune system.

The innate immune system is the first defence system. It is supported by the slower acquired, or adaptive, immune system, which fights against microbes that may have evaded the innate system.

Figure 12.1 General overview of the cells of the innate and acquired immune system. The functions of several of the cells shown here are explained in the text.

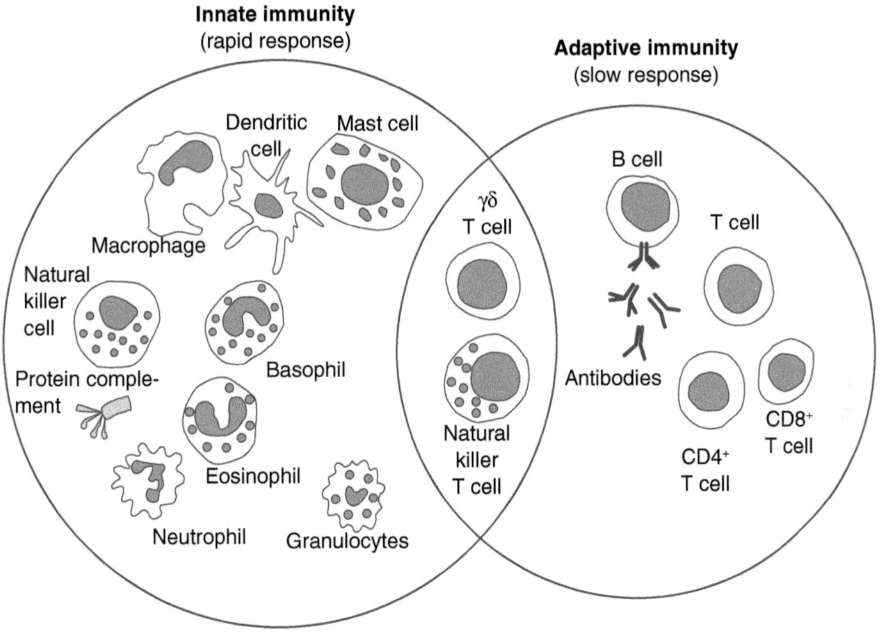

INNATE IMMUNE SYSTEM

The first line of defence that the immune system has are **natural barriers**. These barriers are designed to prevent entry of microbes into our bodies. The main barriers are:

- skin;
- respiratory tract;
- gastrointestinal tract.

They all contain a layer of epithelial cells that prevent microbes from entering, and also secrete **antibiotics** (substances that kill or inhibit bacteria). Moreover, the upper respiratory tract has goblet cells, which secrete a sticky mucous that can trap microbes. The trapped microbes are then removed by hair-like cilia through the mouth and nose, via coughing and sneezing. Within the stomach the acid pH of roughly 1.7 also kills microbes. Furthermore our normally harmless gut bacteria can also inhibit the growth of pathogenic bacteria.

A second part of the innate immune system are rapidly responding, **soluble factors**. These include the **type I interferons** (IFN), **complement** and **C-reactive protein**. The type I interferons are a family of antiviral agents that were originally recognized because they interfere with viral replication, making cells resistant to viral infection. The type I interferons IFN-α and IFN-β have the highest antiviral potency and are produced by a variety of cells, with dendritic cells a major source. These cells recognize double-stranded (ds)RNA, which is found in some viruses, through a receptor known as Toll-like receptor 3. DsRNA is a good marker of viral infection as it is not normally present in mammalian cells. Type I interferons then inhibit viral replication by degrading the viral genome and inhibiting viral mRNA transcription.

There are a large number of circulating and membrane-associated components in the **complement pathway** that can be activated through three separate **activation pathways**:

- lectin pathway;
- classical pathway;
- alternative pathway.

We will not cover the above activation pathways of the complement pathway in detail but hope that you will remember the names. There are also three main functions of the complement system, mediated by **complement components** which are proteins that are abbreviated as 'C':

- C3b acts as an **opsonin**. Opsonins coat microbes and promote the binding of **phagocytes** (see below).
- C5a and C3a act as **chemoattractants**. These act to attract phagocytes and recruit immune cells to the area of infection.
- C5–C9 form a so-called **membrane attack complex (MAC)**. This complex creates a pore in the plasma membrane of the targeted cell which destroys it.

A related molecule and opsonin is **C-reactive protein** (CRP). It is produced and secreted from the liver in response to infection, known as the **acute phase response**. Similarly to complement fragment C3b, CRP coats microbes, attracting phagocytes. This process is discussed here.

Put simply, **phagocytosis** is a process in which invading microbes are ingested by immune cells called **phagocytes** and are internally destroyed (Figure 12.2). Not only microbes, but nutrients, insoluble particles, damaged or dead host cells and activated clotting factors can be phagocytosed. There are two types of **circulating phagocytes**:

- monocytes (which can mature into macrophages) and
- neutrophils.

Neutrophils, also known as polymorphonuclear leukocytes, are the most plentiful leukocytes in the blood and are the fastest responders to the majority of infections. They ingest and kill microbes in the circulation and also migrate into inflamed tissues to ingest microbes. Neutrophils die after a few hours to form the **pus** that is common at sites of infection. There around 10-fold fewer monocytes in the blood than there are neutrophils, and subsequently monocytes have a lesser role in the circulation. Although **monocytes** do ingest microbes in the circulation, their main role occurs when they migrate into the tissues, both healthy and inflamed, and mature into **macrophages**. In contrast to the short-lived neutrophils, macrophages reside in tissues for months or even years, providing long-term phagocytic defence from invading microbes.

Figure 12.2 Phagocytosis and subsequent intracellular killing by phagocytic cells. (1) Pathogen recognized by one of the phagocyte receptors. (2) Pathogen engulfed by phagocyte to form the phagosome. (3) Phagosome fuses with lysosome to form a phagolysosome. (4) Pathogen destroyed by proteolytic enzymes, reactive oxygen species and nitric oxide.

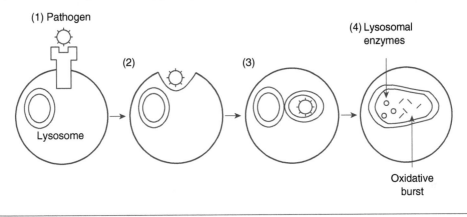

Macrophages residing within tissues remain in the resting state until they are stimulated by one of a variety of danger signals. Resident macrophages, when stimulated, release signalling molecules known as **cytokines** to recruit further monocytes and neutrophils to the infected site. For example, macrophages will secrete **interleukin-1 (IL-1)** and **tumour necrosis factor alpha (TNFα)**. These in turn will stimulate endothelial cells lining the blood vessels to increase the expression of adhesion molecules. These adhesion molecules bind circulating monoctyes and neutrophils to the endothelium, allowing them to pass through the vessel walls into the tissues. To help with this process, macrophages also secrete **interleukin 8 (IL-8)**, which makes the neutrophils more adherent to the endothelium, ensuring successful passage into the tissue.

How do phagocytes (i.e. monocytes and neutrophils) identify the molecules and pathogens that need to be destroyed? The answer is that phagocytes use cell membrane receptors to recognize molecules on pathogens before phagocytosis. The most important classes of phagocyte receptors and the molecules that are sensed by them are the following:

- **Toll-like receptors** (TLR) on phagocytes recognize pathogen-associated molecular patterns (PAMPs; i.e. molecules in pathogens but not host cells) on pathogens.
- **C-lectin receptors** on phagocytes recognize sugar moieties on glycolipids and glycoproteins on invading pathogens and dying host cells.
- **Complement receptors** and **immunoglobin receptors** on phagocytes recognize complement components (e.g. C3b; see above) or immunoglobulins, respectively. Thus the immune system marks pathogens first with complement components or immunoglobulins which are then detected by pathogens via complement and immunoglobin receptors.

This process of distinguishing foreign cells from host cells is an important principle of the immune system. Clearly, if healthy host cells are targeted then the immune system would destroy the body of the host. This is what happens in a number of disease states.

In autoimmune diseases, cells of the immune system destroy healthy cells of the host. In type I, or insulin-dependent, diabetes mellitus, for example, T cells of the immune system enter the islets of the pancreas and selectively destroy the β-cells, which are responsible for the production of insulin. Because pancreatic β-cells cannot regenerate, patients with type I diabetes mellitus have to begin lifelong insulin treatment. This condition is common and can reduce life expectancy. Although a genetic component has been identified, the precise cause of the disease is unknown.

And finally, how do phagocytes kill pathogens? There are three mechanisms:

- phagocytosis;
- oxidative burst;
- release of proteolytic enzymes (i.e. enzymes that digest proteins).

After membrane receptors bind the microbe, the phagocyte extends around and engulfs the microbe (see Figure 12.2). The membrane then closes and the microbe is internalized in a membrane-bound vesicle known as a **phagosome**, which fuses with **lysosomes** to form **phagolysosomes**. Lysosomes contain a variety of enzymes that are involved in breaking down microbes. Formed phagolysosomes are stimulated by signals from the phagocyte receptor to stimulate the enzymes phagocyte oxidase and nitric oxide synthase, which catalyse the production of reactive oxygen species and nitric oxide, respectively, both of which are toxic to the ingested microbes. Proteolytic enzymes within macrophages can be resynthesized and are stored in lysosomes, while in neutrophils they are stored as granules. These granules cannot be resynthesized and when the stores run out the cell dies.

In addition to neutrophils and monocytes (macrophages), we have another cell type within the innate immune system that contributes to defence through the killing of virally infected host cells. **Natural killer cells** recognize cells that are infected through markers of viral infection. They will only destroy cells with low levels of **major histocompatability complex I (MHC I)**, an important molecule in acquired immunity that will be covered in subsequent sections. This allows natural killer cells to determine healthy cells that are high in MHC I from virally infected cells that are low in MHC I. Natural killer cells kill infected cells through **perforin**, **granzyme** and **Fas ligand** expression, all of which induce programmed cell death, known as **apoptosis**.

Perforin moves from granules to the cell surface, polymerizes to form a pore that, like the complement membrane attack complex, can be inserted into the infected cell membrane. This allows water and solutes to flow freely into the cell. This pore also gives access to granzyme, which activates the caspase system of proteases to cause apoptosis. Fas ligand recruits the dauntingly named **fas-associated death domain protein** and **caspase-8**, which form the death-inducing signal complex that also promotes apoptosis. We can see that natural killer cells are therefore exceedingly effective killers but, along with macrophages, they can also secrete cytokines to stimulate the acquired immune system.

ACQUIRED IMMUNE SYSTEM

The **acquired immune system** is slower than the **innate immune system**: it can take days or even weeks before it is functions fully. Acquired immune systems are relatively new in an evolutionary sense and have developed to recognize and remember foreign antigens, but not resident cells. The main cells involved in the acquired immune system are the bone marrow **B** and thymus **T lymphocytes** and also **antigen-presenting cells**. All lymphocytes are derived from haematopoietic stem cells in the bone marrow. These cells have the potential to become many cells and, depending on the stimulating factor, can become either myeloid or lymphoid progenitor cells. Myeloid cells can differentiate into red blood cells, granulocytes and macrophages. Lymphoid cells can differentiate into natural killer cells and immature small lymphocytes that either mature in the bone marrow (B cells) or thymus (T cells).

For many systems within the immune system, the function boils down to two questions: How are pathogens recognized? How are pathogens killed? As for many cells in the innate immune system we now ask these questions for lymphocytes and antigen-presenting cells.

Both B and T cells express unique antigen receptors (**B** and **T cell receptors**) that are relatively specific for an antigen. The effectiveness of this system can be highlighted by the fact that these antigen receptors can recognize $\approx 100\,000\,000\,000$ (10^{11}) different antigens. This number is achieved via a cells with unique receptor specificities. Therefore, when the immune system is challenged there are an insufficient number of specific lymphocytes to mount a response. Any lymphocyte that has recognized the foreign antigen will subsequently divide to produce clones (i.e. daughter cells) of the original cell.

Another molecule that involved in antigen recognition/presentation is known as major histocompatibility complex (MHC; note MHC or MyHC is also used as an abbreviation for myosin heavy chain in other chapters). There are two classes of MHC, class I and class II. Class I MHC is found on all cells and class II mainly found in B cells, macrophages and dendritic cells. At this point some of you may find it strange that we are mentioning macrophages when discussing the acquired immune system, when they have already had a role in the innate system. This is just one example of the interactions between these two systems and highlights that they are not completely independent of each other. In fact this is one example of where the innate system, via macrophages, can activate the acquired system to help deal with any antigens that may have evaded its defence mechanisms.

Both classes of MHC have grooves to bind antigen. It is said that the antigens fit the grooves like hot dog in a bun! The main function of MHC is to present antigen to T cells. This is because the T cell receptor can only recognize antigen presented by the MHC and will not recognize free antigen. T cell receptors are not only specific for antigen but also MHC. During maturation in the thymus, T cells are exposed to host cells expressing MHC, so at a later time these cells do not activate the T cells. Any T cell that does not learn 'self' MHC will be eliminated at that point. During antigen processing the class of MHC employed will be dependent upon the derivation of antigen. In general, extracellular, or vesicle-derived, antigens are presented to T cells by MHC II and intracellular, or cytosol-derived, antigens by MHC class I. The main cell types involved in antigen processing, often named antigen-presenting cells, are macrophages, dendritic cells and also B cells.

As opposed to the T cell receptors, B cell receptors can detect free antigen without the need for the involvement of MHCs. When activated, B cells transform to become plasma cells that secrete **immunoglobins (Ig),** better known as **antibodies**. These antibodies all have the same basic structure but recognize different antigens. In humans, there are five main classes of immunoglobin, due to the five classes of heavy domain identified: IgM, IgD, IgG, IgE and IgA. Within these classes numerous antigens can be detected, as there are variable regions on both light and heavy chain regions. This means that IgA molecules, for example, all differ. Once produced and secreted by the plasma cells, the main function of antibodies is to bind antigens. This leads to several subsequent effects, in which the antibodies:

- act as opsonins, 'tagging' antigens for phagocytosis by neutrophils or macrophages (discussed as part of the innate immune system);
- cause bacteria to clump together, preventing them from entering host cells; and
- can activate the complement to initiate the killing of the antigen.

After describing how T and B cells recognize antigens, we now aim to answer the question: 'How do lymphocytes destroy these pathogens?' T cells are constantly monitoring the blood and secondary lymphoid organs for antigens on the surface of antigen present-ing cells. When an antigen is sensed on the surface of an antigen-presenting cell then **T helper (Th) cells** release a variety of signalling proteins known as cytokines. Dependent upon the **cytokines** released there will be either a Th1 or Th2 immune response some-times referred to as a cell-mediated (Th1) or humoral immune response (Th2). Recently, further Th responses, including a Th17 response, have been identified but we will not touch on these here. The reason for these different immune responses is to give the most appropriate response depending upon the pathogen identified. Extracellular antigens pre-sented on MHC class II will be recognized by Th cells and activate either Th1 or Th2 cells. If Th1 cells are activated this leads via the cytokine IFN-γ to the activation of mac-rophages, which then destroy antigens by phagocytosis. A further response is to cause B cells to increase their production of the antibody IgG, which acts as an opsonin, increasing phagocytic activity. On top of that a Th1 response will also stimulate another subset of T cells known as T cytotoxic cells. T cytotoxic cells kill virally infected cells in a mechanism similar to natural killer cells. These cells are activated by antigen presented on the cell surface of virally infected cells by MHC class I. During a Th2 response, directed mainly by the cytokines IL-4 and IL-5, B cells are stimulated to produce the antibody IgE and eosinophils are activated. IgE, in turn, activates mast cells, which alongside eosinophils are experts at the expulsion of parasites (e.g. helminths). Therefore infection with viruses or intracellular bacteria tends to result in the production of Th1 cytokines and that parasitic infections favour the production of Th2 cytokines.

IMMUNE SYSTEM SUMMARY

Confused? No wonder given the number of different cells and mechanisms within the immune system. To sum it up, the innate system will rapidly try to defend against the majority of pathogens, recognizing conserved structures such as PAMPs. This system will remove these pathogens mainly by phagocytosis or the cytotoxic activity of natural killer cells. Any pathogens that manage to evade the clutches of the innate system will, hopefully, be recognized by the acquired system, which recognizes and remembers specific bacteria and virus. Once the acquired immune system is activated there will be clonal expansion of T and B cells and cytokine signals will stimulate the humoral (i.e. antibody production) or cellular (T cytotoxic cells) defence mechanisms, depending on the pathogen presented.

EXERCISE AND THE IMMUNE SYSTEM

Early research investigating the effects of exercise on the immune system stemmed from anecdotal reports from athletes and researchers who noted that during periods of heavy training or after a marathon race they frequently fell ill with sore throats, coughs and colds. These illnesses are referred to as upper respiratory tract infections (URTI) and are a major cause of visits to physicians, days of sick and importantly missed training sessions. Over the last 30 years, numerous research studies have resulted in the hypothetical relationship between the intensity or volume of exercise and the risk of developing URTIs illustrated in Figure 12.3. This curve suggests that performing regular moderate intensity exercise reduces infection risk, compared to sedentary individuals. On the other hand when performing excessive amounts of high-intensity exercise there is an increase in the risk of URTI. While this model is based on scientific evidence, it is not possible to accept it fully at present because of the lack of appropriate studies in this area. One of the main problems is the identification of URTI. This is often carried out via self-assessment and rarely is this

Figure 12.3 Proposed model of the relationship between the intensity/volume of exercise and risk of developing upper respiratory tract infections (URTI). This suggests that the risk for URTI is lowest at moderate doses of exercise and is highest at heavy doses of exercise.

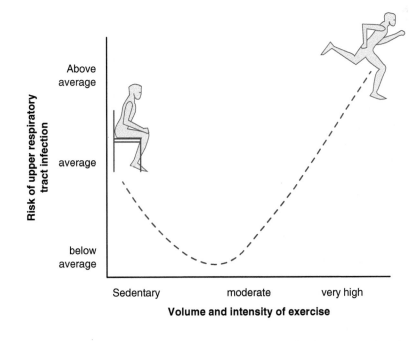

verified by a physician or appropriate biochemical analysis. Similarly, physical activity levels are often obtained via questionnaires and are rarely quantitatively measured. Other limitations are that participants are often asked to recall after a long period of time ($\approx$six months) the number of URTI episodes suffered and their severity.

We will not attempt to review all of this evidence here but will present a couple of investigations that have studied this model. For a recent review of the evidence supporting the J-shaped curve see Moreira et al. (2009). In a study by Nieman et al. (1990) it was shown that while women who performed 15 weeks moderate exercise training had similar number of URTI episodes to a control group, the duration of each episode was reduced. This suggests that regular exercise appears to aid in the recovery from colds. However, further work by Weidner et al. (1998) found that when directly injected with human rhinovirus, which causes the common cold, performing exercise every second day did not reduce the severity and duration of symptoms. One can immediately see that these two studies are contradictory and this is indicative of this area. A large-scale epidemiological study is needed in order to measure the relationship between the quantity of exercise and URTI risk.

Compared to the proposed beneficial effects of moderate intensity exercise in reducing the risk of URTI, more research has looked into the potentially deleterious effects of periods of heavy exercise. One example of research supporting the J-shaped curve was carried out following the 1982 Two Oceans Marathon by Peters and Bateman (1983). This research demonstrated that in the two-week period after the marathon a third of runners reported that they suffered from a URTI, with only 15% of controls reporting URTI. Interestingly this study also found a relationship between finishing time and URTI incidence, with those finishing in faster times having a greater URTI incidence. Similar findings were reported by Heath et al. (1991) who found that those running 486–865 miles over a 12-month period had a 2-fold increase in URTI risk, compared to controls. Those running more than 1388 miles a year had more than a 3-fold increase in risk of URTI.

Overall, the evidence is quite convincing that heavy exercise does increase the risk of UTRI and may suppress the immune system. However, in general, if we look at resting immune function in athletes compared to controls there is probably little difference, particularly within the acquired immune system. Looking at the innate system there is some evidence that the activity of natural killer cells is enhanced (Nieman et al., 1995), while neutrophil phagocytic activity and oxidative burst is reduced (Smith and Pyne, 1997). However, looking at the period immediately after exercise some definite changes in immune function can be observed. This has led to the proposal of 'the open window theory', which says that in the recovery period after exercise the immune system is suppressed, giving pathogens a chance to gain a foothold. We will now look at how exercise affects both innate and acquired systems and briefly introduce some nutritional strategies that have been proposed to counter this immune suppression.

ACUTE HIGH-INTENSITY EXERCISE AND IMMUNE FUNCTION

A major change that occurs within the immune system in response to high-intensity exercise is a change in the number of **leukocytes** (a crude measure as it encompasses many different cell types such as lymphocytes which again encompass B, T and natural killer cells) themselves (Figure 12.4). It can be seen from this figure that with

Figure 12.4 Changes in blood leukocyte numbers, which are most immune cells, after a bout of moderate and high-intensity exercise. This figure demonstrates a single rise and fall in leukocyte numbers at moderate intensity, with a biphasic response at high intensity. Redrawn after Robson et al. (1999).

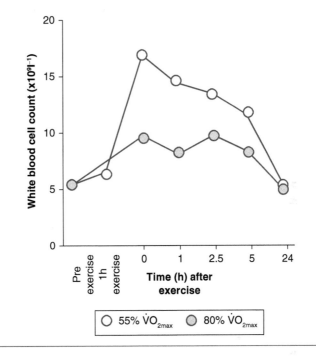

high-intensity exercise (80% $\dot{V}O_{2max}$ for 37 min) there is a biphasic response with an initial increase followed by a brief dip and subsequent rise in leukocyte numbers. At more moderate exercise intensities (i.e. ≈50% $\dot{V}O_{2max}$ for 164 min) a single rise and fall in leukocytes is seen. It can take over 24 h for leukocyte numbers to return to baseline levels. Now, as we know, the cells of the immune system all perform a variety of tasks and so we need to delve a little deeper and look at the response of individual cell types.

While there are small increases in monocyte numbers and little change in eosinophil numbers, there are marked changes in neutrophil and lymphocyte numbers (Figure 12.5). Immediately after exercise there is an elevation in neutrophil count that continues well into the recovery period. The initial rise in neutrophil count is due to demargination, meaning that neutrophils previously residing in areas of the lung, spleen, muscle and liver, adhering to the vascular endothelium, enter the circulation. This is likely the result of exercise-induced increases in cardiac output, shear stress and catecholamines (Boxer et al., 1980). Adrenaline epinephrine acts to increase neutrophil numbers through a few mechanisms: increased cardiac output, increased lung and muscle blood flow and by reducing the adherence of neutrophils by downregulation of cell surface adhesion molecules that act to 'stick' neutrophils to the endothelium. In the recovery period after exercise, as cardiac output and adrenaline begin to return to resting levels, the stress hormone cortisol begins to get involved. Cortisol, secreted from the adrenal cortex of the adrenal gland, acts on the bone marrow to increase the release of neutrophils.

Figure 12.5 Changes in neutrophil and lymphocyte numbers after bouts of exercise at 55% and 80% of the $\dot{V}O_{2max}$, respectively. At both intensities of exercise there is a rise in neutrophil numbers, which return to baseline levels after 24 h, while lymphocytes increase immediately post exercise and then fall to below resting levels in the recovery period after exercise. Redrawn after Robson et al. (1999).

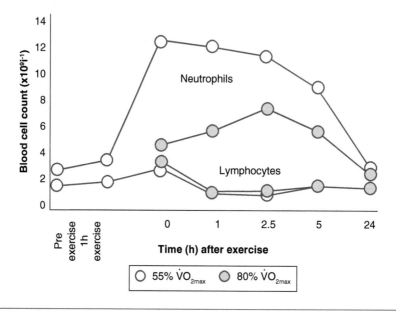

As we can see from Figure 12.5, immediately after exercise, similar to neutrophils and via the same mechanisms, there is an increase in lymphocytes (i.e. B, T and natural killer cells). However, 1 h into the recovery period lymphocyte numbers actually decrease to below resting levels. This pattern of response is seen in all lymphocyte subsets, with the largest responses seen in the natural killer and T cytotoxic cells (Simpson et al., 2006). There are two potential mechanisms responsible for the reduction in lymphocyte numbers:

- apoptosis (programmed cell death); or
- mobilization of lymphocytes to secondary lymphoid organs.

While early research suggested that apoptosis may be the mechanism, recent research has found that annexin V, a marker of apoptosis, expression does not change in response to either intensive, moderate or downhill running (Simpson et al., 2007). Further work by these authors has shown that lymphocytes after exercise have high levels of cellular adhesion and activation molecules that would facilitate the extravasation, or movement, of these cells out of the circulation into secondary lymphoid organs (Simpson et al., 2006). This research is by no means conclusive, with a more detailed investigation of lymphocyte trafficking in response to exercise needed, although this is not a simple task.

Sticking with lymphocytes, a decrease in cell numbers does not necessarily mean that there will be a decrease in lymphocyte function and so this needs to be investigated. Remember that, once activated by antigen, lymphocytes proliferate to produce daughter clone cells and secrete specific cytokines leading to either a Th1 or Th2 effector response. In general, the literature agrees that there is a decrease in lymphocyte proliferation in response to exercise (Nieman et al., 1994), although changes in cell numbers after exercise make these results tricky to interpret. Looking at lymphocyte cytokine production there appears to be a suppression of the Th1 (cell-mediated) response with no affect on Th2 (humoral) responses. These assertions were reached through several studies that have shown a decreased IL-2 and IFN-γ production with no change in IL-4 production (Baum et al., 1997; Lancaster et al., 2005). Further indication that lymphocyte function is reduced after exercise was shown by a recent and elegant study by Bishop et al. (2009). In this study the authors looked at the ability of T cells to migrate to epithelial cells infected with human rhinovirus and found that exercise resulted in a reduction in lymphocyte migration. In terms of function, this would suggest that high-intensity exercise suppresses the ability of the immune system to mount an effective response to any invading pathogen, particularly the ability of T cells to reach infected cells and activate a maximal cell-mediated immune response.

Moving on to the innate immune system, there are two main effects of exercise:

- a decrease in the oxidative burst of neutrophils after completing a marathon (Chinda et al., 2003), an effect that is not seen after more moderate duration exercise; and
- an initial rise in natural killer cell cytotoxic activity, with a subsequent decrease around 2 h into the recovery period (Kappel et al., 1991).

Both the oxidative burst and cytotoxic activity will destroy invading pathogens. Thus both changes, which other researchers do debate, would reduce the effectiveness of the innate immune system to clear any pathogens and such changes may contribute to the increased risk of infection.

In summary, high-intensity exercise results in a suppression of certain components of the innate and acquired immune system that may allow pathogens to result in subsequent URTIs. This research is currently somewhat superficial: measured cells are not always well characterized using molecular markers and the underlying mechanisms are incompletely understood. Also, researchers have found it difficult to show a link between athletes with the greatest immunosuppression and URTI.

NUTRITION AND IMMUNE FUNCTION

What interventions improve immune function and reduce the frequency of URTIs? Numerous nutritional strategies have been employed with the aim of improving immune function with a mixture of success. In this section we will review some of these interventions.

Carbohydrate

Glucose is an important fuel for immune cells and research in the 1980s demonstrated that the magnitude of the cortisol response to exercise is linked to the concentration of blood glucose (Tabata et al., 1984). Similar research has also shown that reductions in blood glucose also result in higher adrenaline responses to exercise. Cortisol and adrenaline have been associated with many of the changes in immune function associated with exercise. Thus it has been hypothesized that increasing carbohydrate consumption may reduce or improve exercise-induced immunosuppression. Studies directly testing this hypothesis have demonstrated that glucose ingestion does indeed retard the rise in cortisol and adrenaline during exercise, but this does not fully restore immune function.

An early study investigating this was carried out by Mitchell et al. (1998). These researchers compared the lymphocyte proliferative response and immune cell numbers after exercise in conditions of low glycogen, achieved through prior exercise and low-carbohydrate diet (0.5 g/kg) and high-glycogen diet (8.0 g/kg). They found that while there was no effect on lymphocyte proliferation there was a greater decrease in lymphocyte numbers when exercise was performed with low carbohydrate. Following these studies, people have investigated whether carbohydrate consumption during exercise (i.e. sports drinks) can be effective in alleviating the immunosuppression. The main effects associated with carbohydrate consumption are maintained immune cell counts, improved neutrophil degranulation and oxidative burst (Scharhag et al., 2002), reduced Th1 suppression (Lancaster et al., 2005) and greater lymphocyte migration toward human rhinovirus (Bishop et al., 2009). Another major effect of carbohydrate consumption is to reduce the magnitude of the cytokine response to exercise. This will be discussed in detail in the exercise and cytokines section below. To conclude, carbohydrate supplementation does improve a few aspects of post-exercise immunosuppression but, importantly, has been found to result in no improvement in the incidence of UTRI after a marathon race (Nieman et al., 2002). Also, when devising carbohydrate-based strategies it is important to keep in mind that glycogen is sensed by AMPK and will regulate adaptive responses.

Protein, glutamine

It was originally accepted that immune cells relied wholly on glucose for their energy requirements. This was until it was demonstrated that, even in the quiescent state, lymphocytes, macrophages and neutrophils also utilize the amino acid glutamine as an energy source (Newsholme et al., 1999). As leukocytes do not contain the enzymes responsible for glutamine synthesis these cells must rely on skeletal muscle to produce and supply glutamine. The 'glutamine hypothesis' was then established. This suggests that during exercise glutamine requirements are increased within skeletal muscle and other organs, leading to insufficient glutamine availability within the immune cells. In support of this hypothesis it has been shown that during intense exercise plasma glutamine levels do indeed fall (Parry-Billings et al., 1992). If this fall in glutamine is responsible for post-exercise immunosuppression then the obvious question would be: Does glutamine supplementation improve immune function after exercise? It appears the answer is no (Rohde et al., 1998); *in vitro* work has shown that lymphocytes function perfectly well when exposed to post-exercise glutamine concentrations.

It is generally now accepted that the glutamine hypothesis is not likely to be responsible for post-exercise immunosuppression. However, in general, research has demonstrated that insufficient dietary protein does lead to a reduction in immune function and an elevation in the number of infections gaining hold (Chandra, 1997). This is unlikely to be an issue in the majority of athletes who consume enough protein in their diet, but could present a problem in athletes undergoing caloric restriction or with a diet low in protein. Indeed protein deficiency has been demonstrated to have a variety of effects on the immune system. These alterations include a decrease in the proliferative response of T cells upon stimulation, reducing the number of daughter cells available to deal with infections. Within the innate immune system decreases in phagocytic or natural killer cell function have also been observed, reducing the ability of the innate system to destroy pathogen or virally infected cells. As mentioned, this is unlikely to be a major issue in athletes, except in those who require low body weight (e.g. gymnastics, marathon runners) or to make a certain weight category (e.g. boxing, weightlifting). Indeed, groups of athletes classified as having anorexia athletica, an eating disorder related to training and sports performance, do appear to have an elevated risk of infection (Beals and Manore, 1994).

Fat

Fat has a major role as a reservoir of energy, but it is also a major constituent of cell membranes, and derivatives of these cell membrane lipids are important in immune system communication. Eicosanoids (e.g. prostaglandins, leukotrienes, thromboxames) are derived from fatty acid precursors released from the phospholipids in cell membranes. The majority of cells have high amounts of the fatty acid arachidonic acid (*n*-6 polyunsaturated fatty acids: *n*-6 meaning the double bond is at the sixth carbon molecule from the end of the fatty acid chain) and it is the main precursor for eicosanoid synthesis. When arachidonic acid is the precursor fatty acid, the derivatives are known as 2 series prostaglandins (e.g. PGE-2) and thromboxanes and 4 series leukotrienes.

Probably the most important eicosanoid in exercise immunology is PGE-2, as the production of PGE-2 from monocytes increases following a bout of exercise (Pedersen et al., 1990). Of its many effects, the most relevant for us are the suppressive effects of PGE-2 on lymphocyte proliferation, natural killer cell activity and the production of Th1 cytokines (Calder et al., 2002). Referring to earlier sections, we remember that these are all effects that are observed after exercise. So it is possible the increased PGE-2 production may contribute to the exercise-induced immunosuppression. With this in mind, the group of Bente Pedersen, in Copenhagen, investigated natural killer cell activity after exercise when PGE-2 production was reduced by indomethacin, an anti-inflammatory drug that inhibits prostaglandin production. It was demonstrated that when indomethacin was given, natural killer cell activity was fully restored to pre-exercise levels, indicating that reduction of PGE-2 production may be of benefit to athletes.

This has led to the suggestion that athletes would benefit from an increase in the consumption of the *n*-3 polyunsaturated fatty acids (PUFAs) found in fish oil. The rationale for this is that the *n*-3 PUFAs eicosapentaenoic acid and docosahexaenoic acid found in fish oil displace arachidonic acid from the cell membranes and are used as precursors for eicosanoid synthesis. This results in a reduction in PGE-2 production that is replaced by the less immunosuppressive PGE-3. There is, however, very little research investigating fish oil supplementation in athletes, with that available demonstrating no change in cytokine production, immune cell numbers and salivary IgA production (Toft et al., 2000; Nieman et al., 2009). Recent research has demonstrated that fish oil supplementation increases natural killer cell cytotoxic activity and Th1 cytokine production in the recovery period after exercise (Gray et al., 2012). Further work is still needed to determine whether these changes translate into a reduction in URTI incidence,

However, what is clear is that if the fat content of the diet is increased, with no change in the type of fatty acids consumed, there are negative effects on immune function. When dietary fat in humans is reduced from 40–50% total energy to around 25–30% total energy, with no changes to exercise, studies have shown that lymphocyte proliferation and natural killer cell activity are enhanced. Similarly in a group of 10 young men who consumed 62% of their energy from fat and performed endurance training 3–4 times a week for seven weeks, there was a significant decrease in natural killer cell activity (Pedersen et al., 2000).

There has also been a wide range of research investigating the effects of vitamins, minerals and antioxidants on immune function, which we will not attempt to discuss here. We will, however, point in the direction of some new research investigating the effects of quercetin, an antioxidant found in several fruits, vegetables and leaves, on the incidence of URTI. So far, 12 weeks of supplementation with quercetin has been found to reduce the number of URTI sick days and severity of episodes in middle aged–elderly volunteers (Heinz et al., 2010). It will be interesting if these effects can be repeated in athletes and their mechanisms of action investigated.

Furthermore, there has also been increased interest in the use of prebiotics and probiotics as a tool to improve immune function. The definitions for these agents are:

- **Gut microbiota** are microorganisms (virus, bacteria etc) present in the gastrointestinal tract.
- **Probiotics** are live microorganisms, which when administered in adequate amounts, confer a health benefit to the host.
- **Prebiotics** are a selectively fermented ingredient that allows specific changes, both in the composition and/or activity in the gut microbiota that confers benefits upon host well-being and health.

The use of prebiotics and probiotics is based on the rationale that certain bacteria have a beneficial effect on our health, and there is clear evidence that the gut microbiota can help with the function of the immune system. One study so far has demonstrated that during a four-month study period probiotic supplementation with *Lactobacillus casei* Shirota resulted in a 50% reduction in the number of URTIs and improvements in the ability to train when a URTI was present (Gleeson et al., 2011). The mechanism behind this beneficial effect was not clear, but may have been the result of maintenance of salivary IgA levels in the probiotic group. We are sure further work will be aimed at uncovering these mechanisms and investigating the potential for other probiotics and prebiotics to improve post-exercise immune function and thus reduce URTI incidence.

MOLECULAR EXERCISE IMMUNOLOGY

In the first part of the chapter we introduced exercise immunology as a subfield of exercise physiology. There are clearly several areas where molecular exercise physiologists can advance this field. The challenge is to develop tighter molecular hypotheses, to identify the genetic variations that affect the relationship between exercise and immunity and to apply the resultant knowledge in order to improve exercise-induced immunosuppression or to use exercise and other interventions to improve immune function in diseases where the immune system malfunctions. In this section we will introduce several molecular exercise immunology research areas.

EXERCISE, CYTOKINES AND MYOKINES

We mentioned earlier that cytokines are the signalling molecules of the immune system, playing a major role in determining the level and form of immune response. In 1991 Northoff and Berg (1991) first reported that circulating levels of these cytokine are increased in response to exercise (Figure 12.6). Subsequent work has shown that the earliest and largest response to exercise is seen with IL-6. Within the immune system

Figure 12.6 Cytokine concentrations before and after exercise. Average plasma cytokine concentrations for 10 male subjects measured before, every half-hour during 2.5 h of treadmill running and every hour in the 6 h resting period after running. Substantial rises in IL-6 and IL-1ra are seen, with little effect on TNFα and IL-1. Redrawn after Ostrowski et al. (1998).

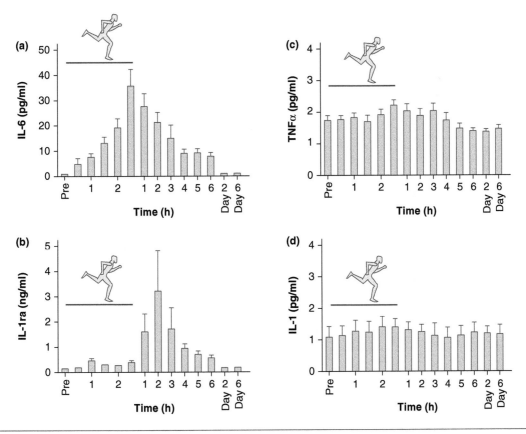

IL-6 is produced from many cell types and has several roles, including stimulating the growth of antibody-producing B cells. This led to a whole series of investigations, many of them by the group of Bente Pedersen, to determine the source and role of IL-6 and other exercise-induced cytokines during exercise.

Knowing that plasma levels of IL-6 increase during exercise, the first question that must be answered is: What cells produce and secrete IL-6 into the plasma? The initial thinking was that monocytes, the main source of IL-6 in the immune system, were stimulated due to exercise-induced muscle damage and were the source of this elevation IL-6. This was not the case, however, and research by Starkie et al. (2001) used flow cytometry to show that monocyte IL-6 production did not increase after exercise. Using molecular techniques, including reverse transcriptase polymerase chain reaction (RT-PCR), Western blots and immunohistochemistry, it was subsequently demonstrated that the active muscle was the major source of IL-6 during exercise. In one of these early studies it was shown, using RT-PCR, that intramuscular IL-6 increased almost 100-fold from resting levels after 3 h of two-legged knee extensor exercise (Figure 12.7). Note the lack of increase in TNFα also observed.

Figure 12.7 Myokines before and after exercise. Muscle interleukin 6 (IL-6) and tumour necrosis factor (TNF)-α mRNA before (Pre), during (30 and 90 min), and immediately after (180 min) 3 h of two-legged knee-extensor exercise. An approximate 100-fold increase in muscle IL-6 mRNA is seen post exercise, with little change in muscle TNFα mRNA. Redrawn after Steensberg et al. (2002).

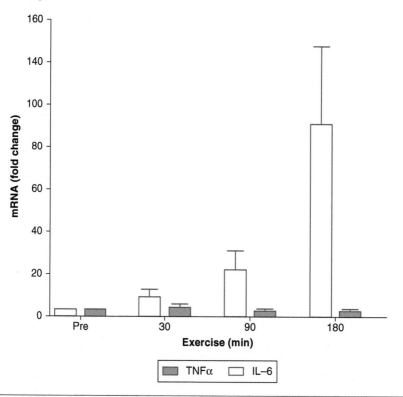

In order to demonstrate that this intramuscular-derived IL-6 was actually released into the circulation and could account for the increase in plasma levels, further work was carried out (Adam et al., 2000). In this study blood was sampled from both the femoral vein and artery, blood flow recorded, IL-6 measured and skeletal muscle IL-6 release calculated via the Fick principle, which all exercise physiologists are all too familiar with. The study did indeed confirm that skeletal muscle was the source of IL-6 during exercise and resulted in the naming of IL-6 as a '**myokine**'. The definition of a myokine given by Pedersen and colleagues is 'cytokines and other peptides that are produced, expressed and released by muscle fibres and exert either paracrine or endocrine effects.' Further work has also identified IL-8 and IL-15 as myokines.

We have now successfully answered the first question: What is the source of IL-6? The next question requiring an answer is, What is the physiological role of this exercise-induced IL-6? Again, originally researchers believed that the major role of IL-6 would be in the modulation of the immune system, for which there is evidence. Such immune roles include: stimulating the acute phase response, neutrophil production, B cell differentiation and in mediating the transition from innate to acquired immunity (to name but a few). However, roles for myokines have now been shown in a variety of exercise-related functions such as:

- metabolism;
- angiogenesis; and
- skeletal muscle size regulation.

These roles will now be discussed. For many years, exercise physiologists have searched for a contraction-induced factor that mediates the changes we see in other cells and organ systems during exercise. One of the major events that occur during exercise is that there is an increase in glucose transport into skeletal muscles (see Chapter 8). Also, glucose is synthesized by the liver during exercise, which is termed endogenous glucose production. Mainly insulin but also glucagon, cortisol and catecholamines are hormones that increase endogenous glucose production. Is there a role for myokines, as they would be a good hormone link between muscle contraction and responses in other organs? Because IL-6 production increases when skeletal muscles are glycogen depleted, it has been hypothesized that secreted IL-6 may regulate carbohydrate metabolism. This led Mark Febbraio and his colleagues (2004) to carry out an elegantly designed study testing the hypothesis that IL-6 may be the stimulus to increase endogenous glucose production during exercise. This study found that, as hypothesized, IL-6 infusion during exercise increases endogenous glucose production, although it could not account for 100% of the response. They also indicated that IL-6 may stimulate glucose uptake into skeletal muscle. Further work by this group confirmed this finding (Carey et al., 2006), and also demonstrated that the likely mechanism through which IL-6 worked was by increasing AMPK activity, although the activating mechanism was not characterized. At this point it is worth pointing out that the precise role for IL-6 in the control of glucose uptake is a contentious area within the literature and its role is by no means clear cut. Indeed several studies have shown that IL-6 can actually inhibit glucose uptake in resting tissues (e.g. Rotter et al., 2003).

Perhaps the most startling metabolic role for IL-6 is seen in mice in which the gene for IL-6 has been knocked out (Wallenius et al., 2002). At a young age there are no

Figure 12.8 IL-6 knockout results in mature-onset obesity in mice (Wallenius et al., 2002). IL-6 knockout mice at the age of 10 months are larger, have greater fat mass and develop insulin resistance, suggesting that IL-6 is important in maintenance of body mass and metabolic function.

Wildtype   IL-6 knock out
                                            (mature onset obesity)

obvious differences in these mice but as they mature, the IL-6 knockout mice begin to become obese compared to control animals. This is clearly demonstrated in Figure 12.8, where you can see that IL-6 knockout mice (IL-6$^{-/-}$) are far larger. These mice also develop insulin resistance, an early warning sign for the development of type II diabetes.

These mice also have a reduced endurance exercise capacity, both at a young and mature age. From this research it is therefore clear that sufficient levels of the myokine IL-6 are important for the control of body weight, endurance capacity and metabolic regulation during exercise.

As mentioned previously, although IL-6 is the focus of the majority of research, it is not the only exercise-induced myokine. IL-15, for example, is also increased during exercise, and although the precise physiological role for IL-15 is not as well characterized, we do have some initial suggestions. The most promising work comes from Quinn et al. (2008) who have shown that overexpressing IL-15 induces skeletal muscle hypertrophy. Perhaps more interestingly, oversecretion of IL-15 from skeletal muscle also reduces body fat, indicating that IL-15 may be an important molecule involved in muscle-to-fat cross-talk. Further work in humans is required to confirm this finding.

NF-κB AND THE CONTROL OF MUSCLE MASS

Several cytokines produced within the immune system have been noted for their ability to induce skeletal muscle atrophy, with **tumour necrosis factor alpha (TNFα)** probably the most widely studied. TNFα was originally named cachectin, in relation to cachexia, which is the loss of muscle mass associated with diseases such as cancer, AIDS and rheumatoid arthritis. In a cell culture model when muscle cells are exposed to increasing concentrations of TNFα there is a concurrent decrease in the amount of myosin heavy chain (major myofibrillar protein), highlighting its role in controlling muscle mass (Li et al., 1998). One of the main effects of TNFα is to activate NF-κB, and it has been demonstrated that if NF-κB activity is blocked, then the atrophying effect of TNFα on muscle is lost (Li et al., 1998).

NF-κB was originally identified as a transcription factor (for a description of a transcription factor see Chapter 1) with importance in activating the kappa light chain genes in B cells, hence the name. It has now been recognized as a universally expressed factor with roles in a host of cellular processes, including immune function, cell proliferation and cell differentiation. This family of dimeric (i.e. composed of two subunits) proteins is encoded by five gene members (RelA/p65, RelB, c-Rel, p50 and p52) that is activated in response to many stimuli via cell surface receptors. These include:

- cytokines,
- growth factors,
- bacteria,
- mitogens and
- oxidative stress.

When cells are not exposed to any of these stressors NF-κB is kept inactive through association with the inhibitor of NF-κB (IκB) proteins. When a cell is stimulated via one of the above stressors, IκB is phosphorylated by a complex of proteins known as IκB kinase (IKK). This phosphorylation of IκB results in its degradation, meaning that NF-κB is free to accumulate in the nucleus, and bind to and activate its target genes. This pathway is briefly summarized in Figure 12.9.

What is the relation of NF-κB signalling to exercise physiology? First, NF-κB activity is affected by exercise. Second, the NF-κB pathway has been shown to affect muscle mass, which may link the immune changes during exercise but also those associated with diseases such as AIDS, cancer and sepsis to muscle mass. While the precise aetiology of these conditions is unknown, they have all been associated with elevated levels of

Figure 12.9 Schematic overview of NF-κB signalling. (1) Activation of NF-κB pathway by, for example, bacteria, cytokines or oxidative stress. (2) IκB is phosphorylated by IKK 'freeing' NF-κB. (3) NF-κB moves into the nucleus and results in transcription of genes such as IL-1 and IL-6.

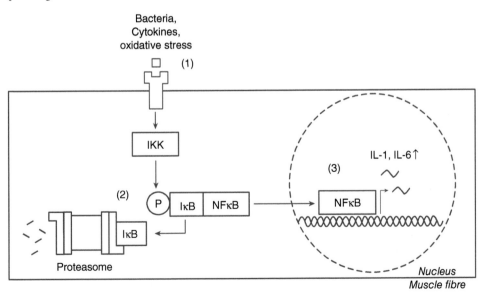

the cytokines IL-6, TNFα and IL-1, all which increase NF-κB activity. Further support for the involvement of NF-κB in ageing comes from the findings that it is activated by periods of disuse and that its levels are around 4-fold higher in skeletal muscles from elderly men compared to those from young men (Cuthbertson et al., 2005).

To test the effect of high NF-κB pathway activity, Cai et al. (2004) developed mice that had muscle-specific activation of IKK, which were termed MIKK mice. Looking at Figure 12.10, what effect would this have on NF-κB activity and subsequently on the size of skeletal muscles? Well, as activation of IKK results in the phosphorylation/degradation of IκB and transport of NF-κB to the nucleus, these mice will have increased activity of NF-κB.

This shows that IKK knockout (which presumably activates NF-κB) causes skeletal muscle atrophy. These effects are due to an enhancement of protein breakdown, with a protein breakdown assay showing a 2- to 2.5-fold increase in protein breakdown in IKK knockout mice. Surprisingly, protein synthesis was actually slightly increased but the key point is that protein breakdown was still higher than protein synthesis, which

Figure 12.10 The effects of IKK activation (MIKK) on the gross appearance of hindlimbs. This suggests that high IKK activity causes atrophy as MIKK mice have smaller muscles (a). Conversely, knock out of IKK increases twitch force (b), tetanic force (c) and time to exhaustion (d). Redrawn after Cai et al. (2004).

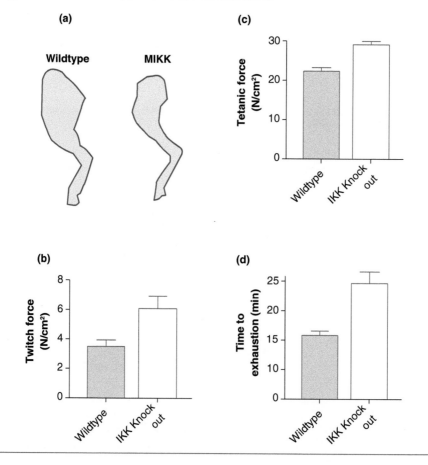

causes a loss of protein. The next stage of study in these mice was then to determine if pharmacological inhibition, using high-dose salicylates (the active ingredient in aspirin), of IKK and/or NF-κB would reverse these effects. The simple answer to this question is yes, muscle mass is maintained in these mice when IKK/NF-κB was inhibited. NF-κB inhibition also improves survival rates and maintains muscle mass in mice with pharmacologically induced tumours.

Taking this further it would be of clear interest to exercise physiologists to determine the effects of inhibition of NF-κB in normal healthy mice. In another experiment mice were developed where IKK was deleted and not present (Mourkioti et al., 2006). Again we can see that this deletion would mean that IκB would not be degraded and NF-κB would not be able to accumulate in the nucleus and activate any genes. So how did this affect these mice? First, soleus muscle cross-sectional area was greater after IKK deletion (Ikk2^{mko}). Not surprisingly, they were also able to produce a greater amount of force and during a running test to exhaustion were able to perform longer (Figure 12.10).

Summing up, we can clearly see that overactivation of the cytokine-activated NF-κB pathway promotes skeletal muscle atrophy, while NF-κB increases muscle size and function.

THE ROLE OF T CELLS IN SATELLITE CELL FUNCTION

Before reading this section it would be good to refresh yourself with the function of satellite cells in skeletal muscle. One of the main factors determining the function of satellite cells within injured skeletal muscle is infiltrating immune cells. There are very few immune cells in developing muscle, but when muscle is regenerating there can be up to 100 000 immune cells per mm^3 of muscle. In response to injury, the most rapidly responding immune cell that infiltrates the muscle is the neutrophil, the numbers of which peak within a day and then rapidly decline thereafter. The second day after injury is characterized by an influx of monocytes, which mature into macrophages, with numbers normally declining at around 48–72 h after damage. About 3 days after injury we begin to see the infiltration of T cells that can remain for up to 10 days.

There is some debate about whether neutrophils or macrophages, which you remember are phagocytes, contribute to muscle damage or actually help with the regeneration of muscle through the clearance of cellular debris. We will not cover this topic here but interested readers are directed to an excellent review on this topic by Tidball and Villalta (2010). We are going to focus on the emerging role of the infiltrating T cells in skeletal muscle repair.

While the majority of research has focused on the role of neutrophils and macrophages, T cells are likely to play a role as the time course of their infiltration overlaps the time course of skeletal muscle repair. Satellite cell migration and proliferation begin at around 24 h after damage and so early signals attracting these cells to the site of injury are likely derived in the neutrophils and macrophages. When T cells are activated around day 3 post damage, the cytokines produced lead to an increase in both migration and proliferation of satellite cells isolated from young rats. Interestingly, when the same experiments were carried out in satellite cells isolated from older rats, there was no such response (Dumke and Lees, 2011). These authors also investigated whether T cells could modulate myogenesis, which is the development of skeletal muscle. In young muscles, T cells had no influence on myogenesis, examined in this case by myosin heavy chain expression. However, in older muscle there was a reduction in myosin heavy chain expression when exposed to T cells. Overall, this research indicates that in young muscles T cells contribute to the migration and proliferation of satellite cells during muscle regeneration. However this response is absent in older muscles and may contribute to the reduction in adaptations that we see in this population. Furthermore, although T cells have no effect on myogenesis in young muscle, they have a deleterious effect in old muscle.

While this research has indicated that immune cells have a role in the regeneration of skeletal muscle, there remains a great deal of information that needs to be uncovered. The precise role of immune cells in the adaptation to exercise is high on the list of areas requiring further research as the responses after exercise may be different from the responses we see to experimentally induced muscle damage in laboratory animals.

Genetic variation and future research

There have been very few studies investigating whether genetic variation within the immune system may have an effect on its responses and subsequent URTIs after exercise. This is surprising as there have been several recent high-impact studies showing that such genetic variation is of the utmost importance in determining immune function.

As a reminder, TLRs are key recognition receptors of the innate immune system and as shown by Beutler and Hoffmann are key in mounting an effective immune response to pathogens. It has been shown that there are several single-nucleotide polymorphisms (SNPs) found in TLR genes and there are now several studies which have investigated how these variants may alter immune function and disease risk. One such study investigated genetic variation in the Toll-like receptor 4 (TLR4) and susceptibility to the tropical infection called melioidosis. In this study it was shown that two SNPs on the TLR4 gene were associated with an increased incidence of melioidosis (West et al., 2011). This highlights that genetic variation within the immune system can be important in its effective operation.

In relation to exercise it is known that monocyte TLR4 expression is transiently reduced, with the authors of this study speculating that this may contribute to exercise-induced immunosuppression (Oliveira and Gleeson, 2010). It has also been demonstrated that TLR4 mediates skeletal muscle catabolism in response to lipopolysaccharide, which, as described earlier, is an endotoxin secreted by bacteria and thought to be partly responsible for the muscle loss in many inflammatory conditions such as cancer, type 2 diabetes and sepsis (Doyle et al., 2010). It is therefore possible that individuals with TLR4 SNPs may be more susceptible to exercise-induced immunosuppression, subsequent URTIs and skeletal muscle mass loss during the aforementioned disease states.

This is just one example of genetic variation within the immune system and a hypothetical link to potential exercise-induced adaptations in healthy and clinical populations. Clearly, molecular exercise immunology projects are required to investigate the potential for genetic variation within the immune system to alter immune function, infection risk and adaptations to exercise. Another example of the importance of genetic control in exercise immunology comes from twin studies (e.g. de Craen et al., 2005).

In this study it was shown that for all cytokines, over 50% of the variance in circulating levels could be accounted for by genetics, with IL-1β having the highest level with an estimate of 86% of the variance being genetically determined. As we have seen that exercise-induced cytokines may have important metabolic and performance effects, then it is likely that genetics may play a role in the magnitude of the cytokine, and subsequent metabolic/performance, responses to exercise.

SUMMARY

The immune system is a system that normally protects us from the many bacteria and viruses that exist in the world. In this chapter we have seen that while regular moderate exercise can enhance immune function, periods of heavy exercise can actually lead to a suppression of the immune system (the J-shaped curve). Many nutritional strategies have been used in an attempt to restore post-exercise immune function; such as carbohydrates, glutamine, fatty acids and the relatively new and promising area of prebiotics and probiotics. Delving into the world of molecular exercise immunology we have seen that skeletal muscle can also be a source of cytokines, such as IL-6, during exercise, and that these cytokines play important roles in endurance capacity, obesity and glucose metabolisms. Further roles for the immune system in exercise physiology include the transcription factor NF-κB, which has a role in controlling muscle mass, force production and endurance performance. Finally, when skeletal muscles are damaged (i.e. through eccentric exercise) we have shown that the T cells of the immune system can move into the muscle and help with repairing the damage, although this response is absent in elderly muscles.

REVIEW QUESTIONS

1. What are the main differences between the innate and acquired immune systems?
2. Name two cell types that perform phagocytosis.
3. Describe the relationship between the volume and/or intensity of exercise and the risk of upper respiratory tract infection?
4. What happens to the number of neutrophils and lymphocytes in the blood after exercise?
5. What is meant by the term 'myokine'?
6. Describe and explain the phenotype observed in mice with the IL-6 gene knocked out.
7. If NF-κB activity is suppressed what effect would that have on muscle size?
8. Describe the differences between the role of T cells in young and old damaged muscle.

REFERENCES

Adam S, van Hall G, Osada T, Sacchetti M, Saltin B and Pedersen BK (2000). Production of interleukin-6 in contracting human skeletal muscles can account for the exercise-induced increase in plasma interleukin-6. *J Physiol* 529, 237–242.

Baum M, Muller-Steinhardt M, Liesen H and Kirchner H (1997). Moderate and exhaustive endurance exercise influences the interferon-gamma levels in whole-blood culture supernatants. *Eur J Appl Physiol Occup Physiol* 76, 165–169.

Beals KA and Manore MM (1994). The prevalence and consequences of subclinical eating disorders in female athletes. *Int J Sport Nutr* 4, 175–195.

Bishop NC, Walker GJ, Gleeson M, Wallace FA and Hewitt CR (2009). Human T lymphocyte migration towards the supernatants of human rhinovirus infected airway epithelial cells: influence of exercise and carbohydrate intake. *Exerc Immunol Rev* 15, 127–144.

Boxer LA, Allen JM and Baehner RL (1980). Diminished polymorphonuclear leukocyte adherence. Function dependent on release of cyclic AMP by endothelial cells after stimulation of beta-receptors by epinephrine. *J Clin Invest* 66, 268–274.

Cai D, Frantz JD, Tawa NE, Jr, Melendez PA, Oh BC, Lidov HG, et al. (2004). IKKbeta/NF-kappaB activation causes severe muscle wasting in mice. *Cell* 119, 285–298.

Calder PC, Yaqoob P, Thies F, Wallace FA and Miles EA (2002). Fatty acids and lymphocyte functions. *Br J Nutr* 87 Suppl 1, S31–S48.

Carey AL, Steinberg GR, Macaulay SL, Thomas WG, Holmes AG, Ramm G, et al. (2006). Interleukin-6 increases insulin-stimulated glucose disposal in humans and glucose uptake and fatty acid oxidation in vitro via AMP-activated protein kinase. *Diabetes* 55, 2688–2697.

Chandra RK (1997). Nutrition and the immune system: an introduction. *Am J Clin Nutr* 66, 460S–463S.

Chinda D, Nakaji S, Umeda T, Shimoyama T, Kurakake S, Okamura N, et al. (2003). A competitive marathon race decreases neutrophil functions in athletes. *Luminescence* 18, 324–329.

Cuthbertson D, Smith K, Babraj J, Leese G, Waddell T, Atherton P, et al. (2005). Anabolic signaling deficits underlie amino acid resistance of wasting, aging muscle. *FASEB J* 19, 422–424.

de Craen AJ, Posthuma D, Remarque EJ, van den Biggelaar AH, Westendorp RG and Boomsma DI (2005). Heritability estimates of innate immunity: an extended twin study. *Genes Immun* 6, 167–170.

Doyle A, Zhang G, Abdel Fattah EA, Tony Eissa N and Li YP (2011). Toll-like receptor 4 mediates lipopolysaccharide-induced muscle catabolism via coordinate activation of ubiquitin-proteasome and autophagy-lysosome pathways. *FASEB J* 25, 99–110.

Dumke BR and Lees SJ (2011). Age-related impairment of T cell-induced skeletal muscle precursor cell function. *Am J Physiol Cell Physiol* 300, C1226–C1233.

Febbraio MA, Hiscock N, Sacchetti M, Fischer CP and Pedersen BK (2004). Interleukin-6 is a novel factor mediating glucose homeostasis during skeletal muscle contraction. *Diabetes* 53, 1643–1648.

Gleeson M, Bishop NC, Oliveira M and Tauler P (2011). Daily probiotic's (Lactobacillus casei Shirota) reduction of infection incidence in athletes. *Int J Sport Nutr Exerc Metab* 21, 55–64.

Gray P, Gabriel B, Thies F and Gray SR (2012). Fish oil supplementation augments post-exercise immune function in young males. *Brain Behav Immun* 26, 1265–1272.

Heath GW, Ford ES, Craven TE, Macera CA, Jackson KL and Pate RR (1991). Exercise and the incidence of upper respiratory tract infections. *Med Sci Sports Exerc* 23, 152–157.

Heinz SA, Henson DA, Austin MD, Jin F and Nieman DC (2010). Quercetin supplementation and upper respiratory tract infection: A randomized community clinical trial. *Pharmacol Res* 62, 237–242.

Kappel M, Tvede N, Galbo H, Haahr PM, Kjaer M, Linstow M, et al. (1991). Evidence that the effect of physical exercise on NK cell activity is mediated by epinephrine. *J Appl Physiol* 70, 2530–2534.

Lancaster GI, Khan Q, Drysdale PT, Wallace F, Jeukendrup AE, Drayson MT, et al. (2005). Effect of prolonged exercise and carbohydrate ingestion on type 1 and type 2 T lymphocyte distribution and intracellular cytokine production in humans. *J Appl Physiol* 98, 565–571.

Li YP, Schwartz RJ, Waddell ID, Holloway BR and Reid MB (1998). Skeletal muscle myocytes undergo protein loss and reactive oxygen-mediated NF-κB activation in response to tumor necrosis factor α. *FASEB J* 12, 871–880.

Mitchell JB, Pizza FX, Paquet A, Davis BJ, Forrest MB and Braun WA (1998). Influence of carbohydrate status on immune responses before and after endurance exercise. *J Appl Physiol* 84, 1917–1925.

Moreira A, Delgado LS, Moreira P and Haahtela T (2009). Does exercise increase the risk of upper respiratory tract infections? *Br Med Bull* 90, 111–131.

Mourkioti F, Kratsios P, Luedde T, Song YH, Delafontaine P, Adami R, et al. (2006). Targeted ablation of IKK2 improves skeletal muscle strength, maintains mass, and promotes regeneration. *J Clin Invest* 116, 2945–2954.

Newsholme P, Curi R, Pithon Curi TC, Murphy CJ, Garcia C and Pires de Melo M (1999). Glutamine metabolism by lymphocytes, macrophages, and neutrophils: its importance in health and disease. *J Nutr Biochem* 10, 316–324.

Nieman DC, Nehlsen-Cannarella SL, Markoff PA, Balk-Lamberton AJ, Yang H, Chritton DB, et al. (1990). The effects of moderate exercise training on natural killer cells and acute upper respiratory tract infections. *Int J Sports Med* 11, 467–473.

Nieman DC, Miller AR, Henson DA, Warren BJ, Gusewitch G, Johnson RL, et al. (1994). Effect of high- versus moderate-intensity exercise on lymphocyte subpopulations and proliferative response. *Int J Sports Med* 15, 199–206.

Nieman DC, Buckley KS, Henson DA, Warren BJ, Suttles J, Ahle JC, et al. (1995). Immune function in marathon runners versus sedentary controls. *Med Sci Sports Exerc* 27, 986–992.

Nieman DC, Henson DA, Fagoaga OR, Utter AC, Vinci DM, Davis JM, et al. (2002). Change in salivary IgA following a competitive marathon race. *Int J Sports Med* 23, 69–75.

Nieman DC, Henson DA, McAnulty SR, Jin F and Maxwell KR (2009). n-3 polyunsaturated fatty acids do not alter immune and inflammation measures in endurance athletes. *Int J Sport Nutr Exerc Metab* 19, 536–546.

Northoff H and Berg A (1991). Immunologic mediators as parameters of the reaction to strenuous exercise. *Int J Sports Med* 12 Suppl 1, S9–15.

Oliveira M and Gleeson M (2010). The influence of prolonged cycling on monocyte Toll-like receptor 2 and 4 expression in healthy men. *Eur J Appl Physiol* 109, 251–257.

Ostrowski K, Hermann C, Bangash A, Schjerling P, Nielsen JN and Pedersen BK (1998). A trauma-like elevation of plasma cytokines in humans in response to treadmill running. *J Physiol* 513, 889–894.

Parry-Billings M, Budgett R, Koutedakis Y, Blomstrand E, Brooks S, Williams C, et al. (1992). Plasma amino acid concentrations in the overtraining syndrome: possible effects on the immune system. *Med Sci Sports Exerc* 24, 1353–1358.

Pedersen BK, Tvede N, Klarlund K, Christensen LD, Hansen FR, Galbo H, et al. (1990). Indomethacin in vitro and in vivo abolishes post-exercise suppression of natural killer cell activity in peripheral blood. *Int J Sports Med* 11, 127–131.

Pedersen BK, Helge JW, Richter EA, Rohde T and Kiens B (2000). Training and natural immunity: effects of diets rich in fat or carbohydrate. *Eur J Appl Physiol* 82, 98–102.

Peters EM and Bateman ED (1983). Ultramarathon running and upper respiratory tract infections. An epidemiological survey. *S Afr Med J* 64, 582–584.

Quinn LS, Anderson BG, Strait-Bodey L, Stroud AM and Argiles JM (2009). Oversecretion of interleukin-15 from skeletal muscle reduces adiposity. *Am J Physiol* 296, E191–202.

Robson PJ, Blannin AK, Walsh NP, Castell LM and Gleeson M (1999). Effects of exercise intensity, duration and recovery on in vitro neutrophil function in male athletes. *Int J Sports Med* 20, 128–135.

Rohde T, Asp S, MacLean DA and Pedersen BK (1998). Competitive sustained exercise in humans, lymphokine activated killer cell activity, and glutamine: an intervention study. *Eur J Appl Physiol Occup Physiol* 78, 448–453.

Rotter V, Nagaev I and Smith U (2003). Interleukin-6 (IL-6) induces insulin resistance in 3T3-L1 adipocytes and is, like IL-8 and tumor necrosis factor-alpha, overexpressed in human fat cells from insulin-resistant subjects. *J Biol Chem* 278, 45777–45784.

Scharhag J, Meyer T, Gabriel HH, Auracher M and Kindermann W (2002). Mobilization and oxidative burst of neutrophils are influenced by carbohydrate supplementation during prolonged cycling in humans. *Eur J Appl Physiol* 87, 584–587.

Simpson R, Florida-James G, Whyte G and Guy K (2006). The effects of intensive, moderate and downhill treadmill running on human blood lymphocytes expressing the adhesion/activation molecules CD54 (ICAM-1), CD18 (β2 integrin) and CD53. *Eur J Appl Physiol* 97, 109–121.

Simpson RJ, Florida-James GD, Whyte GP, Black JR, Ross JA, et al. (2007). Apoptosis does not contribute to the blood lymphocytopenia observed after intensive and downhill treadmill running in humans. *Res Sports Med* 15, 157–174.

Smith JA and Pyne DB (1997). Exercise, training, and neutrophil function. *Exerc Immunol Rev* 3, 96–116.

Starkie RL, Rolland J, Angus DJ, Anderson MJ and Febbraio MA (2001). Circulating monocytes are not the source of elevations in plasma IL-6 and TNF-alpha levels after prolonged running. *Am J Physiol* 280, C769–C774.

Steensberg A, Keller C, Starkie RL, Osada T, Febbraio MA and Pedersen BK (2002). IL-6 and TNF-alpha expression in, and release from, contracting human skeletal muscle. *Am J Physiol* 283, E1272–E1278.

Tabata I, Atomi Y and Miyashita M (1984). Blood glucose concentration dependent ACTH and cortisol responses to prolonged exercise. *Clin Physiol* 4, 299–307.

Tidball JG and Villalta SA (2010). Regulatory interactions between muscle and the immune system during muscle regeneration. *Am J Physiol* 298, R1173–R1187.

Toft AD, Thorn M, Ostrowski K, Asp S, Moller K, Iversen S, et al. (2000). N-3 polyunsaturated fatty acids do not affect cytokine response to strenuous exercise. *J Appl Physiol* 89, 2401–2406.

Wallenius V, Wallenius K, Ahren B, Rudling M, Carlsten H, Dickson SL, et al. (2002). Interleukin-6-deficient mice develop mature-onset obesity. *Nat Med* 8, 75–79.

Weidner TG, Cranston T, Schurr T and Kaminsky LA (1998). The effect of exercise training on the severity and duration of a viral upper respiratory illness. *Med Sci Sports Exerc* 30, 1578–1583.

West TE, Chierakul W, Chantratita N, Limmathurotsakul D, Wuthiekanun V, Emond MJ, et al. (2012). Toll-like receptor 4 region genetic variants are associated with susceptibility to melioidosis. *Genes Immun* 13, 38–46

Glossary

4E-BP1 Eukaryotic translation initiation factor 4E-binding protein 1; involved in the regulation of protein synthesis

ACE I/D Polymorphism in the angiotensin-converting enzyme with either an insertion or deletion of a DNA sequence

Acetylation Addition of an acetyl (CH_3CO) group to a protein, usually on a lysine. Histone aceytlation is a key event in gene regulation

Acquired (adaptive) immune system The acquired (adaptive) immune system fights against microbes that may have evaded the innate system

Acrylamide gel Used for gel electrophoresis of proteins especially during Western blots. Acrylamide is toxic and carcinogenic

Action potential A rapid rise and fall of the membrane potential; occurs in neurons and in muscle fibres and leads to neurotransmitter release and muscle contraction

ACTN3 R577X Polymorphism in the *ACTN3* gene; R stands for arginine and X for a stop codon. If a stop codon is encoded then no functional ACTN3 is produced

Adipokines Cell-to-cell signalling proteins that are secreted by adipose tissue (fat)

Agarose gel Used for gel electrophoresis of DNA

Allele One form or variant of a gene

AMPK AMP-activated kinase; a serine/threonine kinase involved in the regulation of the adaptation to endurance exercise

Anabolic resistance A decreased response especially of skeletal muscle to anabolic stimuli such as resistance exercise, hormones or nutrients; contributes to sarcopenia

Angiogenesis Formation of new blood vessels from existing blood vessels; an adaptation to endurance exercise

Antibody A protein also known as immunoglobulin produced by B cells. Antibodies bind specifically binds to target proteins; also used as a method to visualize specific proteins for example during immunohistochemistry and Western blotting

Apoptosis Programmed cell death

Association study A study where subjects are genotyped for one genotype and this is then compared to the trait investigated

ATPase stain A histochemical method to stain different types of muscle fibres; usually involves a alkaline or acid preincubation step to only stain one type of muscle fibre

Autophagy Breakdown of cells to ensure cellular survival during starvation

Autosome A chromosome that is not a sex chromosome (i.e. not X or Y). There are 22 pairs of autosomes in the human genome

Axon A projection that usually conveys action potential away from the neuron cell body

B cell B cells are immune cells produced by the bone marrow

β-cell dysfunction   A state where β-cells of the pancreas release less insulin at a given blood glucose concentration

Bergström needle   A needle used especially for skeletal muscle biopsies

Bioinformatics   Retrieval and analysis of biological data using mathematics and computer science

C2C12   A mouse muscle myoblast cell line that can be differentiated into myotubes

C/EBPβ   A transcription factor believed to be involved in the genesis of the athlete's heart

Calcineurin   A protein phosphatase that dephosphorylates the transcription factor NFAT

Caloric restriction   A reduction of nutrient intake without causing malnutrition; increases lifespan in several species

CaMK Ca^{2+}/calmodulin-dependent protein kinase; a serine/threonine kinase; involved in the adaptation to endurance exercise

Case–control study   A study where genotype frequencies in subjects with and without a trait are compared

Catalase   An antioxidant enzyme that catalyses the break up of hydrogen peroxide to water and oxygen

cDNA   Complementary DNA; in a reverse transcription reaction, a mRNA template is transcribed into cDNA

Cell line   Cell culture originating from a single cell

Central pattern generator   A neural network that generates rhythmic patterns for movements such as walking or running

Chromatin remodelling   The process of opening up or closing the packaged chromatin, for example by histone modification; a key step in the regulation of gene transcription

Chromosome   Chromosomes are made of DNA tightly coiled DNA

Cyclosporin A   Calcineurin inhibitor

Cytokine   Proteins secreted by leukocytes and other immune cells

Dendrite   Projections of neurons that receive inputs from synapses of other neurons and convey this information to the cell body of the neuron

Dendritic spines   Small protrusions from dendrites that receive input from a single synapse. They are believed to change during learning processes

Diabetes mellitus (types 1 and 2)   A metabolic disease characterized by high blood glucose (hyperglycaemia). In type 1 diabetes, insulin is not produced due to β-cell loss; type 2 diabetes results from insulin resistance and β-cell dysfunction

Diploid   Two copies of each chromosome per cell

DNA   Deoxyribonucleic acid made from guanine, adenine, thymine and cytosine (bases) with a backbone of a deoxyribose sugar connected by phosphate groups

Encode project   Large-scale project to systematically find functional elements in the human genome

Endorphin   Endogenous peptides (opoids) believed to be involved in the 'runner's high'

Enhancer   In contrast to (proximal) promoters, enhancers are distal regulatory stretches of DNA to which transcription factors bind. They can be far away from the regulated gene and interact by looping

Epigenetics   Heritable changes in gene expression that are not due to changes in DNA sequence. A loser definition is that epigenetics is genetic control without the need that this is heritable

Epitope   Part of an antigen (antibody-detected protein/molecule) that is recognized by antibodies and other immune system cells

EpoR   Erythropoietin receptor; an activating mutation in this receptor increases the haematocrit in heterozygous carriers

Ethidium bromide   DNA-binding agent that will fluoresce in ultraviolet light. It is a mutagen, carcinogen and teratogen

Excitatory postsynaptic potentials (EPSP) and inhibitory postsynaptic potential (IPSP) An excitatory or inhibitory postsynaptic membrane potential change

Exon DNA sequence in a gene that is present in the final mRNA after introns have been removed

Gene doping Genetic modification with the aim of enhancing performance

Genetic fingerprinting Also known as DNA profiling. A technique used to identify individuals via their DNA profiles

Genome–lifestyle divergence hypothesis A hypothesis postulating a difference between the lifestyle we live and the genome humans have been selected for

Genome-wide association studies (GWAS) Usually a case–control study where individuals are genotyped with SNP arrays to analyse millions of SNPs (one-base DNA sequence variations)

Gluconeogenesis Generation of glucose from substrates such as pyruvate, lactate, glycerol, amino acids or fatty acids

Glucokinase Phosphorylates glucose to glucose-6-phosphate

GLUT4 Glucose transporter type 4; responsive to insulin and exercise. Other glucose transporters exist

Glutathione peroxidase (GPx) Antioxidant enzymes that reduce lipid hydroperoxides

Glycogenolysis Breakdown of glycogen to glucose-1-phosphate

Granulocyte White blood cells marked by granules in their cytoplasm

Haematoxylin & Eosin A stain which stains nuclei dark purple and protein pink

Hayflick limit Describes the number of times a human cell can divide until cell division stops; a mechanism linked to ageing

Henneman size principle States that motor units are recruited from smallest (S) via intermediate (FR) to fastest (FF). Also known as ramp-like recruitment

Heritability Proportion of the variation of a trait that is due to DNA sequence variations (genetic factors)

Heritage Family Study Most important study into the genetics of endurance training, led by Prof. Claude Bouchard

Heterozygous Difference in two alleles (forms or variants of a gene)

Hif-1 Hypoxia-inducible factor 1; transcription factor that is stabilized in response to hypoxia. Regulates the expression of erythropoietin (EPO) and angiogenesis

Histone Proteins found in cell nuclei that package DNA into nucleosomes; histones are modified to package or unpackage DNA

Hominins Members of the human tribe; hominin evolution is believed to have started 7 million years ago to the east of the East African rift

Homozygous Identical alleles (forms or variants of a gene)

Hyperplasia An increase of cell number

Hypertrophy An increase of cell size

Hypertrophic cardiomyopathy A pathological form of cardiac hypertrophy

Hypothalamus Part of the nervous system that connects to the endocrine system

IGF-1 Insulin-like growth factor 1; increases protein synthesis and causes hypertrophy via the mTOR pathway

IKK IκB kinase is upstream of NF-κB proteins and sequesters them in an inactive state in the cytoplasm; it causes atrophy

Innate immune system First defence system; it defends from infection in a non-specific manner

Insulator Insulators in the DNA are elements that block the interaction between enhancers and promoters

Insulin resistance Cells fail to exhibit the normal response to insulin which is glucose uptake and glycogen synthesis

Interleukins Cytokines that are produced by leukocytes and other cells

Intron   DNA sequence of a gene which is removed by splicing

Kinase   An enzyme that transfers phosphate groups, especially from ATP to other molecules. Protein kinases are key signal transduction molecules

Knockout mouse   A genetically engineered mouse where a gene has been inactivated (knocked out)

Leptin   A hormone that regulates energy intake and expenditure

Leukocyte   White blood cells that defend the body against infectious disease

Linkage analysis   A genetic method that allows researchers to identify the location of a disease-causing gene relative to genetic markers

Lipodystrophic   Conditions where the adipose tissue (fat) is absent or reduced

Liquid nitrogen (N_2)   Nitrogen in the liquid state; nitrogen boils at $-196°C$

LKB1   Liver kinase B1; phosphorylates AMPK at Thr172

Long-term potentiation (LTP)   Long-lasting signal transmission between two neurons in response to a stimulus

Lysosome   Organelles that contain acid hydrolase enzymes that break down waste materials

Macrophage   Immune cells that engulf and digest debris and pathogens

MAPK   Mitogen-activated protein kinase; these serine-threonine kinases are involved in signal transduction

Mechanotransduction   Signal transduction events triggered by mechanical stimuli; resistance exercise is one example where mechanotransduction occurs

Metformin   A drug used to treat type 2 diabetes mellitus; activates AMPK

Methylation   Addition of a methyl group to DNA or protein

DNA microarray   Printed DNA spots on a solid surface. It is used to measure the expression of thousands of genes or to detect thousands of SNPs

miRNA   microRNA; small non-coding RNA; regulates gene expression

Mitochondrial biogenesis   Process by which new mitochondria are formed; an adaptation to endurance exercise

Mitochondrial haplogroup   DNA sequence variations in human mitochondrial DNA; can be used to trace ancestry

Mitochondrion   Organelle which is the site of oxidative phosphorylation

Mitosis   Separation of two sets of chromosomes during cell division

Molecular exercise physiology   Molecular exercise physiology is the study of exercise physiology using molecular biology methods

Monoclonal antibody   Antibodies made by identical immune cells

Monogenic trait   Such traits depend mainly on one DNA sequence variation

Motor unit (S, FR, FF)   A motor neuron and the muscle fibres innervated by it. Motor units can e distinguished into slow (S), fatigue-resistant (FR) and fast-fatiguing (FF)

mRNA   Messenger RNA, produced as a result of transcription or gene expression

MtDNA   Mitochondrial DNA; mitochondria have their own DNA which is roughly 16 000 bases long in humans.

mTOR   Mammalian complex of rapamycin; this kinase is a regulator of protein synthesis. It forms two complexes named mTORC1 and mTORC2

Muscle fibre (I, IIa, IIx, IIb)   Multinucleated, differentiated muscle cell. The classification as I, IIa etc. is based on the predominant myosin heavy chain isoform that is expressed

Muscle fibre grouping   A phenomenon typically observed in old muscle. It is due to a loss of one kind of muscle fibre

Myoblast   A mononucleated muscle cell derived, for example, from satellite cells. Myoblasts differentiate to form myotubes and fully differentiated muscle fibres

Myogenic regulatory factors (Myf5, MyoD, myogenin, Mrf4)   Transcription factors that regulate the determination (a cell entering the muscle lineage) and differentiation which is the process where myoblasts form myotubes and muscle fibres

Myokine   Cytokine secreted by muscle cells

Myosin heavy chain   Main part of skeletal muscle myosin proteins. It consumes ATP to produce work and heat. Type I, IIa, IIx and IIb isoforms exist (type IIb is not expressed in many human muscles)

Myostatin   Secreted, muscle-produced protein that inhibits muscle growth

Myotube   Myotubes are formed from myoblasts during differentiation. Myotubes are immature muscle fibres

Natural killer T cell   A set of immune cells; their inactivation can lead to autoimmunity

Neurogenesis   The event where neurons are generated from neural stem and progenitor cells

Neuron   Key cell of the nervous system; electrically excitable cell that transmits information electrically and through chemical signals (neurotransmitter)

Neurotransmitter   Molecules that transmit signals from neurons to target cells which can be other neurons or cells such as muscle fibres

Neutrophil (granulocyte)   Key white blood cell which is part of the innate immune system

Next-generation sequencing (NGS)   High capacity DNA-sequencing methods

NFAT   Nuclear factor of activated T cells; transcription factor activated by dephosphorylation by calcineurin

NF-κB   Nuclear factor-kappa B; involved in immune system regulation and muscle atrophy

Notch pathway   A pathway where cell–cell signals lead to the cleavage of Notch. The intracellular part then migrates to the nucleus and co-activates a transcription factor and regulates gene expression

Nuclear localisation sequence (NLS)   An amino acid sequence that directs a protein to the nucleus

p53   Protein 53; a protein that detects DNA-damaging signals and then regulates DNA repair, cell cycle arrest, or apotosus. Also termed the 'guardian of the genome' as it is a tumour suppressor

p70 S6k   Protein 70 S6 kinase; key part of the mTOR pathway; it regulates protein synthesis

Pax7   A marker and identity regulator of satellite cells; Pax7 knockout mice have few satellite cells

PCR   Polymerase chain reaction; a method to amplify or make more of a small stretch of DNA

PEPCK   Phosphoenolpyruvate carboxykinase; an enzyme involved in gluconeogenesis. *PEPCK* overexpression in mouse muscle leads to an increase in endurance capacity

Personalized medicine   Tailoring medicine to the individual for example based on genetic testing

PGC-1α   Peroxisome proliferator-activated receptor gamma coactivator 1-alpha; a gene that is expressed in response to endurance exercise and that drives mitochondrial biogenesis

Phagocytosis   Engulfment of a particle by a cell; macrophages undergo phagocytosis

Phosphatase   An enzyme that removes a phosphate group

Phosphorylation   The process where a serine, threonine or tyrosine on a protein is phosphorylated (PO_4 group). A key regulatory event

Physiological cardiac hypertrophy (athlete's) heart   Exercise-induced eccentric hypertrophy, especially of the left ventricle

PI3K   Phosphoinositide 3-kinase; a kinase which regulates glucose uptake, glycogen synthesis and protein synthesis

PKB (Akt)   Protein kinase B (also known as Akt); a serine/threonine kinase that regulates glucose uptake and protein synthesis via mTOR

Polyclonal antibody   Antibodies produced by B cells against an antigen; the antibodies vary and are frequently produced by inoculation of a suitable mammal

Polygenic trait   Such traits depend on multiple DNA sequence variation

Polymorphism   A common DNA sequence variation that occurs >1% of the population

Proteomics   The high-throughput analysis of proteins

Primary antibody   Antibody that targets the primary target (usually protein of interest) in a Western blot or immunohistochemistry reaction

Primary cell culture   Culture of cells directly obtained from an organ; such cells generally have a limited lifespan in contrast to cell lines

Primer   A short stretch of DNA that binds to the start or end of a target sequence; a key ingredient for the polymerase chain reaction

Promoter   A region of DNA that initiates the transcription of a gene. Transcription factors bind to the promoter

Quantitative trait locus (QTL)   Stretches of DNA where DNA sequence variation can cause the variation of a trait

Rapamycin   An inhibitor of mTOR

Rare mutation   A DNA sequence variation that is rare (<1% of the population is a carrier)

Restriction enzyme   Enzymes that locate and cleave specific, short DNA sequences

Resveratrol   A compound that activates sirtuins and may extend lifespan and increase mitochondrial biogenesis

Reverse transcription   The transcription of RNA into DNA; retroviruses are capable of this

Ribosome   The organelle that synthesizes peptides and proteins from mRNA. Ribosomes are mainly rRNA but also include some ribosomal proteins

RNA   Ribonucleic acid; single stranded, similar to DNA. Made from guanine, adenine, uracil and cytosine

RNA polymerase   An enzyme that reads DNA and produces a RNA template

RONS   Reactive oxygen and nitrogen species

rRNA   Ribosomal RNA; a building block of ribosomes

RT-PCR   Reverse transcription polymerase chain reaction; a method to detect the expression of a gene

RT-qPCR   Reverse transcription quantitative polymerase chain reaction; a method to quantify the expression of a gene

Sanger DNA sequencing   Classical DNA sequencing method

Sarcomere   The basic contraction unit of a muscle; comprises mainly of myosin and actin

Sarcopenia   The loss of muscle mass and function during normal ageing

Satellite cell   The resident, adult muscle stem cell; capable of extensive self renewal and differentiation

Secondary antibody   An antibody to detect a primary antibody; usually conjugated to a visualisation reaction

Senescence   The phenomenon that cells cease to divide after a certain number of divisions

Single nucleotide polymorphism (SNP)   One base DNA sequence variation

Sirtuin   Ribosyltransferases or deacetylases that are involved in the regulation of adaptation to exercise and ageing

Smad   The names comes from the *C. elegans* gene name *SMA* and the *Drosophila* gene name mothers against decapentaplegic. In muscle, Smads are regulated among other by myostatin

Sport and exercise genetics (kinesiogenomics)   The study of heredity in a sport and exercise context

Stable isotope   Chemical isotopes that are not radioactive

Stem cell   A cell that is capable of extensive self-renewal and differentiation. Stem cells are frequently key for the repair of organs

Supercompensation   A time course that describes the decrease of glycogen during exercise and the recovery and overshoot of glycogen during regeneration and after a meal. Such time course does neither describe nor explain common adaptations to exercise

Superoxide dismutase   An enzyme that catalyses the break down of superoxide (O_2^-) into oxygen and hydrogen peroxide

Synapse   A structure usually at the end of an axon/neuron that triggers the release of neurotransmitter in response to electrical signals

Synergist ablation   Surgical removal of a synergist muscle. The result is frequently the hypertrophy of the remaining muscle

Taq polymerase   *Thermus aquaticus* DNA polymerase. This enzyme can synthesize a DNA double strand on the basis of a single-strand template and is not degraded by heat as the organism lives in hot wells. Key enzyme for the polymerase chain reaction

T cell   Immune cell that plays a key role in cell-mediated immunity. T cells mature in the thymus, hence the 'T' in the name

Telomerase   A ribonucleoprotein enzyme that elongates telomeres

Telomere   6–8 base pair repeats (TTAGGG in vertebrates) at the end of chromosomes; telomeres shorten during each cell division

Tfam   Mitochondrial transcription factor A; a transcription factor that regulates the transcription of mtDNA

Trainability   The magnitude of adaptation to a given exercise training regime. Individuals with high trainability adapt a lot

Trait   A trait is a variable phenotypic characteristic. Examples are eye colour, maximal oxygen uptake or muscle mass

Transcription   Process by which DNA is copied into RNA

Transgenic mouse   A mouse whose genome has been altered by genetic engineering

Translation   Refers to the process where a peptide or protein is synthesized on the basis of an mRNA template on the basis of the ribosome

TNFα   Tumour necrosis factor α; regulates protein breakdown

Twin studies   A study type used to estimate the heritability of a trait

Ubiquitination   The transfer of a ubiquitin group to a protein. Ubiquitination regulates protein breakdown by the proteasome but can also be a signalling event

VEGF   Vascular endothelial growth factor; a key regulator of angiogenesis for example as an adaptation to exercise

Western blot   A method combining gel electrophoresis and antibody-based detection of specific proteins

X-ray crystallography   A method to determine the structure of a molecule

Index

Note: Page numbers in *italics* refer to figures. Page numbers in **bold** refer to tables.